Surface Contamination

Genesis, Detection, and Control

Volume 2

Surface Contamination

Genesis, Detection, and Control

Volume 2

Edited by

K. L. Mittal

IBM Corporation
East Fishkill Facility
Hopewell Junction, New York

PLENUM PRESS · NEW YORK AND LONDON

Library of Congress Cataloging in Publication Data

Main entry under title:

Surface contamination.

Proceedings of a symposium on Surface Contamination: Its Genesis, Detection, and Control held at the fourth International Symposium on Contamination Control in Washington, D.C., on September 10—14, 1978, sponsored by the Institute of Environmental Sciences and the International Committee of Contamination Control Societies
   Includes index.
   1. Surface contamination—Congresses I. Mittal K. L., 1945-   II. International Symposium on Contamination Control, 4th, Washington, D.C., 1978. III. Institute of Environmental Sciences. IV. International Committee of Contamination Control Societies.
TA418.7.S8   620.1'129   79-15433
ISBN 0-306-40177-0 (v. 2)

Proceedings of a Symposium on Surface Contamination: Its Genesis, Detection, and Control, held at the Fourth International Symposium on Contamination Control in Washington, D.C., on September 10—14, 1978, published in two volumes, of which this is volume two.

©1979 Plenum Press, New York
A Division of Plenum Publishing Corporation
227 West 17th Street, New York, N.Y. 10011

Printed in the United States of America

Preface

The present volume and its companion Volume 1 document the
proceedings of the Symposium on Surface Contamination: Its
Genesis, Detection and Control held in Washington, D.C., September
10-13, 1978. This Symposium was a part of the 4th International
Symposium on Contamination Control held under the auspices of the
International Committee of Contamination Control Societies, and
the Institute of Environmental Sciences (U.S.A.) was the offical
host.

The ubiquitous nature of surface contamination causes concern
to everyone dealing with surfaces, and the world of surfaces is
wide and open-ended. The technological areas where surface cleaning is of cardinal importance are too many and very diversified.
To people working in areas such as adhesion, composites, adsorption, friction, lubrication, soldering, device fabrication,
printed circuit boards, etc., surface contamination has always
been a bete noire. In short, people dealing with surfaces are
afflicted with molysmophobia†, and rightfully so.

In the past, the subject of surface contamination had been
discussed in various meetings, but this symposium was hailed as
the most comprehensive symposium ever held on this important topic,
as the technical program comprised 70 papers of more than 100
authors from 10 countries. The symposium was truly international
in scope and spirits and was very well attended. The attendees
represented a broad spectrum of backgrounds, interests, and professional affiliations, but all had a common interest and concern
about surface contamination and cleaning.

This Symposium was organized with the following objectives in
mind: to bring together scientists, technologists, and engineers

---

†Molysmophobia means fear of dirt or contamination, from Mrs.
Byrne's Dictionary of Unusual, Obscure, and Preposterous Words,
University Books, Secaucus, N.J. 1974.

interested in all aspects of surface contamination, to review and
assess the current state of the knowledge, to provide a forum for
exchange and cross-fertilization of ideas, and to define problem
areas which need intensified efforts; the comments from the audience confirm that these objectives were definitely fulfilled. For
example, there were brisk and enlightening - not exothermic - discussions and information exchanges between the presenters and the
audience. It should be added that the purpose of a symposium is
to present the state of the knowledge of the topic under consideration, and this is best accomplished by inviting the leading workers
to present overview papers covering topics of their special interest; these should be complemented and augmented by original research
papers. This is exactly what was done for this symposium as there
were 20 invited overviews covering a wide range of subtopics within
the purview of surface contamination. May it be added that there
are three components of the subject of surface contamination:
(i) cleaning of surfaces, (ii) characterization of the degree of
surface cleanliness, and (iii) storage of clean surfaces or the
kinetics of recontamination. All these components were covered in
this symposium.

The present proceedings volumes contain a total of 64 papers
(some papers from the original program are not included for a
variety of reasons, and a few papers which were not presented but
are included) which were reviewed by at least two qualified reviewers, and most of these were modified and revised in light of
the reviewers' comments before inclusion in these volumes. Only
about half of these papers (unreviewed) in their abbreviated form
(long abstract) were included in the document distributed to the
registrants at the time of the meeting. So the present volumes
contain a great deal more material. The papers included are in
their full form and are divided into four sections as follows:
General Papers; Cleaning of Surfaces; Surface Contamination Detection, Identification, Characterization, and Control; and Implications of Surface Contamination. Sections I & II are embodied in
Volume 1, and Volume 2 contains Sections III & IV. Broadly speaking, all kinds of surfaces (metal, oxide, ceramic, glass, alloy,
polymer,and liquid) and contaminants (organic film, inorganic,
particulate, microbial,and radioactive) are covered in these volumes as the intent of the Symposium was to cover the topic of surface contamination in general and broad terms rather than to con-
centrate on a particular surface/contamination combination.

The topics covered include: sources, forms, and mechanism(s)
of surface contamination; various techniques (including solvent,
plasma, UV/ozone, chemical, mechanical, ion milling, and surface
chemical) for cleaning surfaces; radioactive and microbial contamination of surfaces; atomically-clean surfaces; preparation of
clean mineral surfaces; particle adhesion; preparation of clean

water surfaces; cleaning of polymeric surfaces; various techniques
[including contact angle or wettability, evaporative rate analysis,
indium adhesion test, surface potential difference, spectroscopy
(Auger, ESCA, ISS, SIMS), ellipsometry, plasma chromatography, ion
chromatography, and microfluorescence] for characterizing the degree
of surface cleanliness; ionic contamination detection and characterization; and implications of surface contamination.

It should be added that the availability of sophisticated
surface analytical tools has been a boon in the area of surface
contamination and a series of papers deal with the utility of such
techniques for monitoring and characterizing microamounts of surface contaminants. The papers dealing with implications of surface
contamination cover topics ranging from microelectronics to the
public health sector.

In essence, these proceedings volumes present a very comprehensive coverage of the latest state of the knowledge relative to
the important subject of surface contamination. A special feature
of these volumes is the inclusion of 20 invited overviews which
should provide a veritable gold mine of valuable information, and
these overviews coupled with the 44 contributed research papers should
should serve as a vade mecum for anyone interested in surface contamination and cleaning. It should be added that originally it was
intended to include a Discussion at the end of each section, but in
spite of the continuous exhortation, the number of questions received in written form (although there was a brisk verbal discussion at the end of each paper) did not warrant inclusion of written
Discussions in these volumes.

Acknowledgements: First of all, I am thankful to the management of IBM Corporation for permitting me to organize this symposium and to edit these volumes. Particularly, I would like to
acknowledge the understanding and patience of my manager, Dr. H. R.
Anderson, Jr., during this activity. Special thanks are due to
Mr. R. W. Martin for his help and cooperation during the various
stages of putting together the technical program, to Mrs. Betty
Peterson of the Institute of Environmental Sciences for her ready
and willing help whenever it was needed. The reviewers should be
thanked for their sacrifice of time and many valuable comments.
The cooperation and enthusiasm of the authors is a must for any
proceedings volume and I would like to thank them for their efforts
and cooperation in submitting the manuscripts. Thanks are due to
all the secretaries who helped with the correspondence typing. I
am thankful to my wife, Usha, for letting me work late hours during
the tenure of editing these volumes and also for helping with the
subject index; to my daughter, Anita, and son, Rajesh, for letting
me spend those hours which rightfully belonged to them.

Last, but not least, I would like to express my thanks to Mrs. Edith Oakley (Forbes Services) for meeting, without complaint, various deadlines for typing of the manuscripts.

K. L. Mittal
Symposium Organizer and Chairman

IBM Corporation
East Fishkill Facility
Hopewell Junction, N.Y.   12533

Contents of Volume 2

Contents of Volume 1

Part III
Surface Contamination Detection, Identification, Characterization, and Control

CONTAMINANT DETECTION, CHARACTERIZATION, AND REMOVAL BASED ON

SOLUBILITY PARAMETERS

Lloyd C. Jackson

The Bendix Corporation, Kansas City Division*

P. O. Box 1159, Kansas City, Missouri 64141

Contamination removal is critical in the electronics industry because foreign material on surfaces
has the potential of causing electrical surface
leakage or failures in encapsulation, coating, or
bonding applications. In the past, solvents for
contamination removal generally were selected on
the basis of adhesion tests and production evaluations. Using the relatively new technologies of
evaporative rate analysis and solubility parameter,
a new approach to surface cleanliness has been
achieved.

CONTAMINANTS

An organic residue contaminant is a material which usually
occurs on electrical components as a very thin film. These types
of contaminants are usually unknown materials. Bacteria and dust
particles are also contaminants, but they are not included as
part of this study. Identification of unknown contamination is
quite often speculative or not possible.
Cleaning solutions, if used over and over, will contain
increasing amounts of contaminant that will be redeposited on a
surface as it is removed from the solvent. Surface cleanliness
requires a technical approach to remove the various contaminants

*Operated by the U. S. Department of Energy by The Bendix
Corporation, Kansas City Division, under Contract No. EY-76-C-
04-613. This article was supported by the Division of Military
Application.

that may be present.   The trial-and-error method of evaluation is
no longer acceptable.

SOLUBILITY PARAMETER TECHNOLOGY

Solubility parameter technology involves the study of how
well materials are compatible with each other.   Similar molecules
will dissolve more easily in each other than will dissimilar
molecules.   When the solubility parameter of a material to be
dissolved is known, it becomes the "target area," as shown in
Figure 1.   The selection of a solvent with a solubility parameter
value reasonably close to that of the target solubility parameter
will provide for increased solubility.   However, if a solvent is
chosen that has a solubility parameter too far away from the
target area, the molecules will not match or fit together, and
there will be little diffusion, molecular attraction, or solubility.   The solubility parameter of a material depends on the
polar forces, the nonpolar forces, and the hydrogen bonding
potential of each molecule.

EVAPORATIVE RATE ANALYSIS

Evaporative rate analysis is a relatively new technique for
the examination of surfaces.   It provides a measure of the
presence of contaminants.   It involves measuring the rate of
evaporation of a carbon-14 tagged radioactive chemical from a

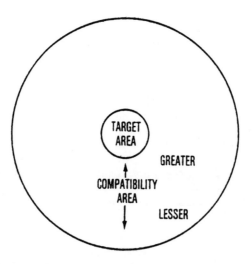

Figure 1.   Solubility Parameter Solubility-Incompatibility
            Concept.

surface by a Geiger counter. The degree of retention of the radioactive chemical is a measure of that surface or contamination. The Meseran Surface Analyzer uses this principle and was used for this investigation.

CONTAMINANT REMOVAL BY SOLVENT SELECTION

It is possible to characterize contaminants so that the exact solvent or solvents can be selected to do an effective job of cleaning. A series of solvents was chosen which represented a major portion of the solubility parameter range of organic material, that is, 7 to 12. Selection was also based on obtaining solvents which represented the weak, moderate, and strong categories of hydrogen bonding.

After each contaminant was rinsed by the respective solvent, the level of radioactivity counts was determined by the Meseran Surface Analyzer. Bar chart examples are shown for two contaminants in Figures 2 and 3. The most effective contaminant removal is indicated by the lowest Meseran values obtained.

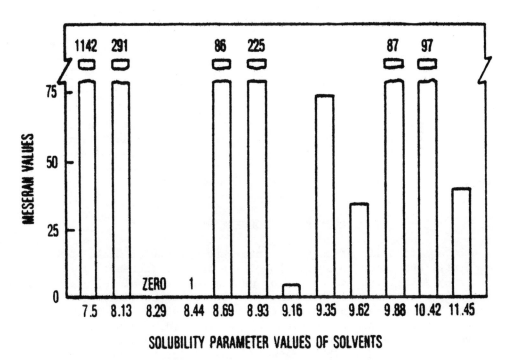

Figure 2. Effectiveness of Solder Flux Removal by Series of Solvents Characteristic of Solubility Parameter Range.

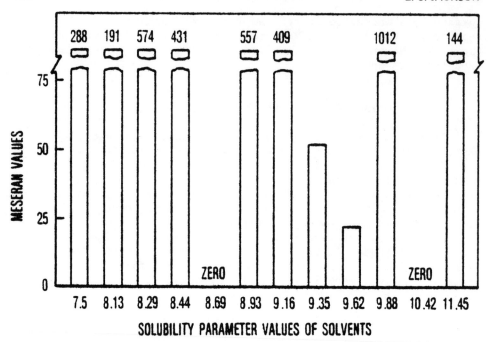

Figure 3.   Effectiveness of Photo Resist Removal by Series of
            Solvents Characteristic of Solubility Parameter
            Range.

     In some cases, complete removal of a contaminant by a sol-
vent is not quite achieved because of insufficient time. Very
low Meseran values are obtained by the more effective solvents.
According to this analysis, selection of the best solvents can
be made for the most effective contaminant removal.

SOLVENTS AND SOLVENT BLENDS

     Classification of solvents into three groups of weak,
moderate, and strong hydrogen bonding potential increases the
accuracy in solvent selection for contaminant removal. It tends
to group solvents into their electron charge capability. The
strong hydrogen bonding molecules are the alcohols. These have
high solubility parameter values. The weak hydrogen bonding
solvents are the chlorinated solvents and solvents such as
heptane and toluene. These have low solubility parameter
values. In between are the moderate hydrogen bonding molecules
such as esters and ketones.

Table I. Effects of Solvents on Contaminants*, Relative Ease of Removal of Contaminants, by Solvent Series

Contaminants	Weak hydrogen bonding class							Moderate hydrogen bonding class				Strong hydrogen bonding class
	Heptane	2-Ethyl-hexyl chloride	Methyl chloroform	Toluene	1,1,2 Trichloroethylene	Vinyl trichloride	2-Ethyl-hexyl acetate	n-Butyl acetate	Cellosolve acetate	Acetone	Cyclohexanone	Isopropyl alcohol
Solder flux types												
Abietic acid	C	C	A	C	A	B	A	B	B	B	B	B
Stearic acid	A	A	A	A	A	C	A	B	B	B	C	A
Photo resist	C	C	C	C	C	C	C	A	B	B	A	C
Spray Coating	C	A	A	A	A	B	C	B	B	A	B	C
Fingerprint	B	A	A	B	B	B	A	A	B	A	B	A
Hand Lotion	B	A	B	A	A	C	A	A	C	C	C	C
Detergent	C	C	C	C	C	B	B	C	B	C	C	A
Mold Release Wax	C	C	C	C	C	C	C	C	C	C	C	C
Grease	B	C	C	C	C	B	B	C	C	B	C	C
Silicone grease	C	C	C	C	C	C	C	C	C	C	C	C

*Levels of contaminant solubility based on Meseran values after solvent rinsing: A = contaminant effectively removed (16 or less Meseran units), B = partly soluble; longer rinse times should improve contaminant removal (17 to 99 Meseran units), and C = limited contaminant solubility in solvent (greater than 100 Meseran units).

Cleaning efficiency of solvents for contamination removal is based on how close the solubility parameter value within a hydrogen bonding class matches that of the contaminant. The relative ease of removal of contaminants by solvents is shown in Table 1. Solvent blends can be formulated from individual solvents to provide polarity and solubility parameter composites. Selection of solvents or solvent blends can be made to remove specific contaminants (organic residues) without damaging plastics. Solvent blending, based on solubility parameter principles, has been demonstrated to be more effective in removing specific groups of contaminants than individual solvents. A specific solvent blend (C), for example, is best suited for various contaminants that might be encountered in electrical component manufacture.

Solvent Blend C	Volume Percent
1,1,1,-Trichloroethane (Methylchloroform)	17
n-Butyl Acetate	33
1,1,2,-Trichloroethylene	33
Isopropyl Alcohol	17

SWELL INDEX OF PLASTICS

Some plastics are not damaged by solvents; other plastics are sensitive to solvents in different ways and in varying degrees. If a polymer is crosslinked, as are plastics derived from thermosetting resins, or is highly crystalline, solvency will be less likely to occur. An amorphous or loosely structured plastic would probably be soluble in several solvents.

Where plastics are used as substrates, solvents selected to remove contaminants must not dissolve the plastic or cause crazing of the surface. Swell index is a measure of these effects of solvents on plastics.

Swell index data on a variety of thermoplastic and thermosetting plastics, including a glass-filled printed circuit board laminate (Table 2) when used with contaminant solvency (Table 1), provide for solvent selection best suited for a particular application. For example, if the plastic being used is sensitive to moderate hydrogen bonding solvents, then it may be necessary to use a weak hydrogen bonding solvent to remove contaminants.

Table II. Effects of Solvents on Plastics,* Relative Solvent, by Solvent Series

Plastics	Weak hydrogen bonding class							Moderate hydrogen bonding class				Strong hydrogen bonding class
	Heptane	2-Ethylhexyl chloride	Methyl chloroform	Toluene	1,1,2 Trichloroethylene	Vinyl trichloride	2-Ethylhexyl acetate	n-Butyl acetate	Cellosolve acetate	Acetone none	Cyclohexanone none	Isopropyl alcohol
Polyethylene	1	1	1	2	2	2	1	1	1	1	1	1
Polyvinylchloride	1	1	1	2	3	4	2	4	2	4	5	1
Nylon	1	1	1	1	1	1	1	1	1	1	1	1
Acrylic	1	1	1	1	5	5	1	1	5	5	1	1
Polystyrene	1	5	5	5	5	5	5	5	5	5	5	1
Cellulose acetate	1	1	2	2	3	5	1	3	5	5	5	2
Polyphenylenesulfide	1	1	1	1	1	1	1	1	1	1	1	1
Polycarbonate	1	1	3	3	5	5	1	3	3	3	5	1
Diallylphthalate	1	1	1	1	1	1	1	1	1	1	1	1
Phenolic	1	1	1	1	1	1	1	1	1	1	1	1
Polyphenylene oxide	1	2	5	5	5	5	2	3	3	3	4	1

*Levels of plastic solubility are based on percent swell in 24 hours: 1 = little or no solvent absorption (swell = less than 1%), 2 = slight absorption of solvent (swell = 1 to 10%), 3 = moderate absorption of solvent (swell = 11 to 49%), 4 = excessive solvent absorption (swell = greater than 50%), and 5 = plastic dissolved partially or completely in solvent.

CHARACTERIZATION OF SURFACES

Solubility parameter technology and the evaporative rate analysis technique have been combined to characterize surface contaminants. This technology can be expanded to include direct characterization of surfaces in terms of their balance of forces based on their polarity and nonpolarity attributes.

Solubility parameter technology is the study of how well materials are compatible with each other. Similar molecules will dissolve more easily in each other than will dissimilar molecules. Applying the same principle to surface charac- terization, it follows that polar solvents emphasize polar characteristics of surfaces and nonpolar solvents emphasize nonpolar characteristics. A four-quadrant chart analysis technique has been established that can be applied to surface characterization of plastics as well as contaminants.

The relative level of polarity of molecules can be estimated from the electronegativity differences in the functional groups. As the electronegativity differences in the elements of a chemical bond increase, the bond becomes more polar.

Three radioactive chemicals were selected for use with the Meseran Surface Analyzer according to their level of polarity. The radiochemical diethylsuccinate with its C-0 and C=0 bonds, will have a moderately large electronegativity difference of 1.0. Tetrabromoethane, on the other hand, with its C-Br bonds, will have a small electronegativity difference of 0.3. In these two cases, the electronegativities of the terminal atoms (oxygen and bromine) are more negative than the carbon atom and are considered as being of negative polarity. Diethylsuccinate has a greater charge potential than tetrabromoethane and therefore is more polar. Tridecane, however, with its C-H bonds can be considered as being nonpolar, because the terminal hydrogen atom is less electronegative than the carbon atom.

Because the bulk vapor pressure of the radioactive chemicals is substantially equivalent, the observed amount of measured retained radioactivity results primarily from the ability of a substrate to interact with the functional group of the radio- active chemical. In this regard, the three radioactive chemicals (tridecane, tetrabromoethane, and diethylsuccinate) have similar vapor pressure properties.

The three radioactive chemicals can be used as surface energy probes to determine which one reflects the polarity or nonpolarity condition of a surface. In this manner, the level of polarity or balance of forces that exist on a surface can be ascertained. The largest Meseran number obtained by one of the three radioactive chemicals will show the strongest charac- teristics of the surface. Thus, if the tetrabromoethane number is largest, the predominant characteristic of the surface will be of weak polarity. If diethylsuccinate provides the largest number, the surface has a moderate to strong polarity. If the surface is nonpolar, the tridecane number will be the largest. Therefore, if A stands for tetrabromoethane, C stands for diethylsuccinate, and B stands for tridecane, then:

A > C > B equals a weak polar surface,

C > A > B equals a moderate to strong polar surface, and

B > A or C equals a nonpolar surface.

Investigators have used 2- or 3-dimensional graphs to identify compatibility of solvents, resins, and polymers based on units of solubility parameter, polarity, and hydrogen bonding. This technique characterizes materials according to their regions of solubility. In this work, however, the degree of affinity for the surface by each radioactive chemical is expressed by radioactive counts and is the extension of solubility parameter compatibility principles. The radioactivity counts are a measure of the surface and since there is a degree of nonpolarity in every molecule, the nonpolar tridecane data can be subtracted from the Meseran numbers obtained with the polar radioactive chemicals to reveal the relative polarity level present on a surface. Thus, A minus B (Formula 1) equals weak polar data and C minus B (Formula 2) equals moderate to strong polar data.

This principle was applied to known pure organic chemical films on aluminum foil surfaces, to plastic surfaces, and to various contaminants on aluminum foil surfaces. The maximum benefit was obtained by locating the specific data according to Formulas 1 and 2, on a four-quadrant chart (Figure 4). In this manner, surfaces of unknown contaminants can be determined for their relative level of polarity and nonpolarity.

SURFACE ELECTRICAL CHARACTERISTICS

Since organic contaminants can have varying degrees of polarity or nonpolarity, it can be expected that the electrical properties of these materials would respond to voltage in terms of surface leakage or voltage breakdown (arc over). Using a printed circuit board test pattern, electrical tests were made on a known series of chemicals and on contaminants.

The series of chemicals represented the solubility parameter range which should cover most types of contaminants. Surface leakage tests were made over a 10 to 500 applied voltage range. A definite pattern of electrical activity was found based on the nonpolarity and polarity/hydrogen bonding characteristics in terms of solubility parameter values. A summary of the data is given in Table 3, which characterizes chemical type with surface leakage properties.

A large variety of contaminants were similarly measured

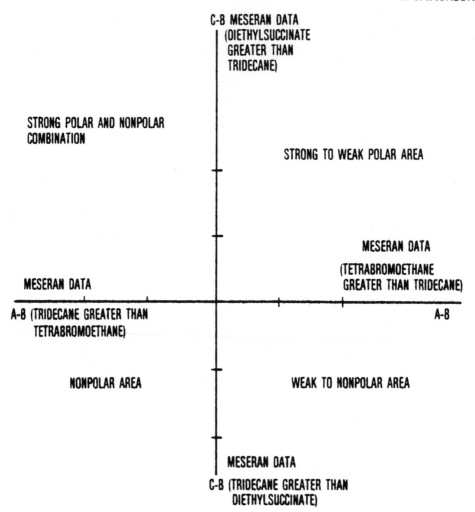

Figure 4.   Polarity Indications According to Four-Quadrant
            Chart Analysis.

for their surface leakage characteristics. The principles
developed with the known chemicals are also seen with the
contaminants according to their level of polaritynonpolarity.
     Major arc over problems occur with materials which are
primary amines, alcohols (hydroxyls), and ketones. The greater
the hydrogen bonding characteristic is of the chemical, the
lower the breakdown voltage.
     The various levels of polarity of known chemicals are
related to surface electrical leakage in a predictable manner.
Arc over limits are less sensitive to polarity differences. In

both cases contaminants and known chemicals having high solubility parameter values yield greater electrical surface leakage
and least arc over (voltage) resistance.

Table III. Classification of Organic
Residue Chemical Type With
Surface Leakage

Types of Organic Residue	Leakage at 250 to 500 V (Amperes)
Aliphatics, Aromatics	1 µA to 1 pA
Chlorides, Secondary Amines	10 nA
Esters, Ethers, Ketones	0.1 mA to 1 µA
Alcohols	0.1 A to 0.1 µA
Primary Amines	>0.1 A

SUMMARY

Surface cleanliness, in the past, has generally been evaluated indirectly by adhesion or electrical tests. As cleanliness
requirements for electrical parts increase, contamination
detection and removal become more critical. Using the relatively
new technologies of evaporative rate analysis and solubility
parameter, a new approach to surface cleanliness has been
achieved. Based on the principles of solubility parameters,
compatibility of contaminants with solvents is described.
Greater accuracy in solvent selection can be achieved when
solvents are classed according to their hydrogen bonding
potential. Blending solvents according to solubility parameter
principles permits specific solubility characteristics to be
obtained for specific contaminant removal. The swell index
of plastics also follows solubility parameter principles of
solvent compatibility. Surface measurements by the evaporative
rate analysis technique permit contaminant detection as low as
0.1 microgram for each 1.0 square centimeter of surface area.
Using surface energy probes of various polarity levels, surface
characterization can be achieved. Application of four-quadrant
chart analyses serves as a means of surface characterization.

The surface electrical properties of contaminants (surface leakage and arc over) can be shown to relate to the various levels of polarity-nonpolarity of materials. This work now permits actual measurement of surface cleanliness.

ACKNOWLEDGEMENT

Portions of this paper have been reproduced by special permission of Adhesives Age magazine from several issues as follows:

L. C. Jackson, "How to Select a Substrate Cleaning Solvent - Contaminant Removal Using Solubility Parameter Technology," Adhesives Age, Vol. 17, pp. 22-31, 1974.

L. C. Jackson, "A Scientific Technique for Adhesive Development," Adhesives Age, Vol. 18, pp. 30-31, 1975.

L. C. Jackson, "Solvent Cleaning Process Efficiency - Contaminant Removal in the Electronics Industry," Adhesives Age, Vol. 19, pp. 31-34, 1976.

L. C. Jackson, "Surface Characterization Based on Solubility Parameters," Adhesives Age, Vol. 19, pp. 17-19, 1976.

L. C. Jackson, "Removal of Silicone Grease and Oil Contaminant," Adhesives Age, Vol. 28, pp. 29-32, 1977.

BIBLIOGRAPHY

H. Burrell, "Solubility of Polymers," Encyclopedia of Polymer Science and Technology, Vol. 12, pp. 618-626, 1970.

J. D. Crowley, and others, "A Three Dimensional Approach to Solubility," Journal of Paint Technology, 1966.

J. L. Gardon, "Cohesive-Energy Density," Encyclopedia of Polymer Science and Technology, Vol. 3, pp. 833-862.

J. L. Gardon, "The Influence of Polarity Upon the Solubility Parameter Concept," Journal of Paint Technology, 1966.

C. M. Hansen, and A. Beerbower, "Solubility Parameters," Encyclopedia of Chemical Technology, John Wiley and Sons, 1971.

C. M. Hansen, "The Three Dimensional Solubility Parameter - Key to Paint Component Affinities," Journal of Paint Technology, 1967.

J. H. Hildebrand, and others, Regular and Related Solutions, Van Nostrand Reinhold.

J. H. Hildebrand, and R. L. Scott, <u>Regular Solutions</u>, Prentice-Hall, 1962.

K. L. Hoy, "New Values of the Solubility Parameters From Vapor Pressure Data," Journal of Paint Technology, 1970.

D. L. Kaelble, <u>Physical Chemistry of Adhesion</u>, Wiley, 1971

SURFACE-CONTAMINATION DETECTION THROUGH WETTABILITY MEASUREMENTS

Malcolm E. Schrader

David W. Taylor Naval Ship Research
and Development Center
Annapolis, Maryland   21402

Wettability measurement is described as the most
convenient method available for detecting surface
contamination. It is, in addition, extremely surface
specific (exceeding the electron spectroscopies in this
respect), and sensitive to a fraction of a monolayer.
Forces governing the contact angle (a measure of
wettability) of a drop in mechanical equilibrium on a
solid surface are discussed. The behavior of clean
surfaces of hard inorganic materials such as metals,
glasses, and ceramics, which spread (0° contact angle)
all liquids except mercury, is contrasted with that of
organic materials. The handling of wettability data
for various liquids on low energy (organic) surfaces
by means of Zisman's critical surface tension concept
is described. Applicability of the Girifalco-Good
theory of interfacial forces and of Fowkes' modification of the theory, to wettability of low energy
surfaces is discussed. Contamination of a high energy
inorganic surface with low energy organic material will
raise the contact angle, whereas contamination of a low
energy organic surface with high energy impurities will
lower the contact angle. The most dramatic changes in
wettability resulting from contamination, however, occur
upon organic contamination of an inorganic surface, where
the contact angle changes from zero to a positive value.
Theories of the effect of oxide contamination of a
metallic surface are discussed, as are recent ultrahigh
vacuum experiments measuring the wettability of oxide-
free metal surfaces. The results of recent Lifshitz
theory calculations showing the effect of minimal organic
contamination on metal surfaces are presented.

INTRODUCTION

Wettability measurements are a convenient and extremely rapid
method for probing the surface constitution of macroscopic solids.
The results they yield are reflective of the composition of the
outermost atomic layer, and are in addition, sensitive to impurities within this layer. The most widely used quantitative measure
of wettability is the contact angle. For a sessile drop on a smooth
solid surface viewed parallel to the plane of the surface, Figure 1,
it is defined as the angle Θ between the surface and the tangent to
the drop profile at the point of contact with the surface. If the
drop is in stable mechanical equilibrium, the interfacial force
relationship can be given by the Young Equation[1]

$$\gamma_{SL} + \gamma_{LV} \cos \Theta = \gamma_{SV}$$

where γ_{LV} is the surface tension of the liquid-vapor interface,
γ_{SL} of the liquid-solid interface, γ_{SV} that of the solid-vapor
interface, and Θ the contact angle. It states that the sum of the
surface tension of the liquid-solid interface and the horizontal
component of the surface tension of the liquid equals the surface
tension of the solid-vapor interface. Transposing,

$$\gamma_{LV} \cos \Theta = \gamma_{SV} - \gamma_{SL}$$

so that the right side of the equation is an adhesion parameter,
describing the tendency of the liquid to cling to the solid, while
the left side consists of experimentally measurable quantities,
i.e., the surface tension of the liquid and the contact angle.

Contact angle values can be divided into the following three
categories:

$$\Theta = 0°, \quad 0° < \Theta < 90°, \quad \Theta > 90°$$

It can be seen from the Young Equation that at $\Theta = 0°$, the drop in
surface tension of the solid as the liquid advances over it, is
equal to or greater than the surface tension of the liquid.
Consequently, enough free energy becomes available to spread the
liquid and form a film. When Θ is between $0°$ and $90°$, $\gamma_{SV} - \gamma_{SL}$
is smaller than the surface tension of the liquid, so that a film
cannot be formed. However, $\gamma_{SV} - \gamma_{SL}$ is still positive, i.e., there
is still a drop in free energy on transposing the solid-vapor interface to a solid-liquid interface, so that asperities in the solid
will generally be filled by the advancing liquid. At $\Theta > 90°$,
$\gamma_{SV} - \gamma_{SL}$ is negative, indicating an increase in free energy on
transforming the solid-vapor to a solid-liquid interface, so that
asperities in the solid are not generally filled by the advancing
liquid. While all these categories are important in terms of wettability phenomena, the classification of Θ into $0°$, and $>0°$, i.e.,
spreading and nonspreading, is the most important in terms of
utilization of contact angles to classify solid surfaces, as will
be seen in the next section.

CRITICAL SURFACE TENSION

It has been commonly supposed that the most important factor affecting the wettability of solids is compatibility, or lack thereof, in the nature of the liquid and solid. According to this concept, polar liquids would interact with, and therefore tend to spread, on polar solids, while nonpolar liquids would spread only on nonpolar solids. In fact, however, while it is true for example, that water, a polar liquid, will generally spread more easily on polar than on nonpolar solids, it is also true that nonpolar oils will generally spread more easily on polar than on nonpolar solids. The major consideration is the solid surface energy, γ_S, that is available to yield a large decrease in free energy upon interaction with the liquid (right hand side of the Young Equation). Accordingly, solid surfaces have been classified into the categories of high and low energy with respect to characteristics affecting their wettability.[2,3] The high energy surfaces include metals, metal oxides, and siliceous glasses, while those of low energy consist mainly of organic materials. As a rule, compounds which are liquid at room temperature spread on high energy surfaces, since their surface tensions are considerably less than the surface energies of these solid substrates. On the other hand, they may or may not spread on low energy surfaces.

The first systematic quantitative treatment of the wettability of organic or low energy surfaces, resulted from the investigations of W. A. Zisman and coworkers.[4] They found a linear relationship between the cosine of the contact angle and the surface tension of the liquid on measuring contact angles of various hydrocarbon liquids of a homologous family upon a single organic surface. On extrapolating the plot of cos θ vs γ_{LV} to cos $\theta = 1$ (zero contact angle) they obtained the value of γ_{LV} at or below which all liquids in that family would spread on the given surface. Extrapolation based on different families of liquids generally yielded approximately the same intercept on a given organic surface. This value of γ_{LV} was therefore a characteristic constant for the surface of a given low-energy material and was named γ_C, the critical surface tension. γ_C is not a thermodynamic quantity, but rather an empirical constant which is equal to or less than the surface free energy of the solid. The critical surface tension concept, therefore, achieves two major advances in understanding and utilizing the wettability data. First, it enables prediction, by interpolation or extrapolation, of the contact angle of any hydrocarbon liquid of known surface tension on a low energy surface if the contact angles of other liquids in its homologous series have been measured. Second, after measurement of the contact angles of only one homologous series on a given low energy surface, it allows one to predict, from the surface tension of any liquid of any series, whether the liquid will or will not spread (zero contact angle) on

the given surface. While the system works with maximum accuracy
for hydrocarbons, it also applies to other liquids which do not
have hydrogen-bonding capability.

GOOD-GIRIFALCO THEORY

The first modern theoretical treatment of wettability as a
function of constitution was provided by Girifalco and Good in
their treatment of general interfacial forces.[5] They referred to
interfaces in systems in which the "dominant cohesive and adhesive
forces are of the same type" as "regular interfaces." For these
interfaces, it was proposed that the interaction energy was equal
to $2(\gamma_a\gamma_b)^{\frac{1}{2}}$, i.e., twice the geometric mean of the respective surface tensions of the two components at the interface. For the
case of "nonregular systems," this interaction term was multiplied
by Φ, an "empirical property of the interface." The equation
describing the interfacial tension is then

$$\gamma_{12} = \gamma_1 + \gamma_2 - 2\Phi \ (\gamma_1 \ \gamma_2)^{\frac{1}{2}}$$

where γ_{12} is the interfacial tension between the two components
(two liquids or a liquid and solid), γ_1 and γ_2 the respective
surface tensions of each component, and Φ a constant characteristic
of the interface. For regular interfaces, of course, $\Phi = 1$.
Girifalco and Good then carried out theoretical calculations of Φ
by summing the energy of interaction of pairs of molecules across
the interface. Good and Elbing obtained Φ in terms of Hamaker
constants, which, for nonpolar molecules, were calculated by the
method of London from the polarizability and ionization energy of
the molecules of each component.[6] These calculations provided the
approach with predictive value for nonregular as well as regular
interfaces.

FOWKES' γ^d

Fowkes[7,8] subsequently proposed a modification of the
Good-Girifalco theory, which involved separating the surface
tension of each phase into additive components,

$$\gamma = \gamma^d + \gamma^h + \gamma^m$$

so that, for example, the surface tension of water, γ, would be
equal to $\gamma^d + \gamma^h$ where γ^d is a component of the surface tension
of water resulting from dispersion forces and γ^h the component
resulting from hydrogen bonding forces. Likewise, the surface
tension of mercury was described as equal to $\gamma^d + \gamma^m$ where γ^d is
the dispersion component and γ^m the so-called metallic force

component. Assuming that the dispersion component, γ^d, of the surface tension of one substance would interact with only the dispersion component of the surface tension of other substances, Fowkes used the geometric mean of the dispersion components as an interaction term. The total interfacial tension of two phases interacting through London dispersion forces only was then

$$\gamma_{12} = \gamma_1 + \gamma_2 - 2(\gamma_1^d \gamma_2^d)^{\frac{1}{2}}.$$

Thus, the use of Φ for nonregular interfaces was avoided by using fractional surface tensions in the interaction term.

By assuming $\gamma = \gamma^d$ for hydrocarbons and that water interacts with hydrocarbons via the γ^d component only, a γ^d for water of 21.8 erg/cm^2 was calculated, by utilizing its known interfacial tension against one or more hydrocarbons. Substituting 21.8 erg/cm^2 for γ_L^d in

$$\gamma_L = (\gamma_L^d \gamma_S^d)^{\frac{1}{2}}$$

it can be seen that any solid surface interacting with water by means of London forces only must have $\gamma^d \geq 236$ erg/cm^2 to spread the water on its surface. Since organic surfaces in general have values that are substantially less than 100 erg/cm^2, and since furthermore, the low energy surfaces interact primarily by means of London forces, it is clear on theoretical grounds what had previously been known experimentally, namely, that water will almost never spread on the surface of an organic material.

Fowkes also assumed that mercury interacts with hydrocarbons via the γ^d component only. As with water, he utilized the known interfacial tension of mercury against various hydrocarbons, to calculate γ^d for mercury. Utilizing these values of γ^d for water and mercury, Fowkes then calculated an interfacial tension for the mercury-water interface on the basis of the speculative assumption that water and mercury interact with each other in the same manner that each interacts with hydrocarbons. The interfacial tension thus calculated was quite close to the experimentally measured interfacial tension of water and mercury, thus yielding the rather startling conclusion that water and mercury interact by means of dispersion forces only.

The apparent discovery that interaction at the interface of water and mercury involves only dispersion forces led Fowkes to extrapolate this principle to metal surfaces in general. Previous experience with metal surfaces had shown that in the absence of a contaminating organic layer all metals are hydrophilic, i.e., water will spontaneously spread on their surfaces with a zero contact angle.[3] However, the "real" metal surfaces heretofore investigated contain combined oxygen (ranging in nature from a monolayer of

chemisorbed oxygen to a thick layer of surface oxide) as a result
of their exposure to the atmosphere. These surface metal oxides
are capable of strong hydrogen bonding interaction with water.
Fowkes hypothesized that an atomically clean oxygen-free metal
surface without this hydrogen bonding capability would interact by
means of dispersion forces only, which would be inadequate to yield
a zero contact angle.

The surface of gold seemed to provide a convenient test for
this hypothesis, since it is uniquely inert to oxygen and does not
form a stable oxide phase. The other criterion it would have to
meet, of course, is that its γ^d be less than the critical value of
236 erg/cm^2 (above which it could spread water by dispersion forces
alone). Fowkes, utilizing a pair summation method, calculated γ^d
for gold from results in the literature. These results consisted
of Hamaker coefficient calculations reported by Reerink and Overbeek
in 1954,[9] based on data reported by Westgren in 1918,[10] and Tourila
in 1926 and 1928.[11] The Hamaker coefficient values were 5 x 10^{-14}
erg, 1 x 10^{-13} erg, and 6 x 10^{-13} erg. The highest γ^d value
calculated by Fowkes from these Hamaker coefficients was 120
erg/cm^2, much too low, according to his theory, to allow gold to
spread water by dispersion forces alone. While some experimental
support appeared initially for this theory of hydrophobic gold,[12,13]
subsequent investigation by Bernett and Zisman[14] continued support
for the traditional concept of hydrophilic gold.

WETTABILITY OF ULTRACLEAN METAL SURFACES:
ULTRAHIGH VACUUM TECHNIQUES

In the late 1960's, a technique was developed for in situ
measurement of contact angles after surface preparation in ultrahigh vacuum.[15] It involved the admission of suitably purified vapor
to the system, condensation of the vapor in the sample chamber by
means of a cold finger, and depositing a drop on the sample surface
through magnetic manipulation. In 1970 results were reported on
the use of this technique to determine the contact angle of water
on gold.[16] Gold surfaces were prepared in the ultrahigh vacuum
apparatus by heating polished gold disks to successively higher
temperatures in the presence of oxygen followed by vacuum, or by
evaporating gold films on smooth substrates in situ. The gold disk
method was repeated in a conventional vacuum system for comparison.
A striking feature of the results was the hysteresis of the water
contact angle which was observed during the various stages of surface activation of the gold disk. The hysteresis effect was
observed in a number of different ways: (a) Freezing out the water
vapor in the vacuum system after the drop was on the gold surface.
As the drop evaporated, a receding angle was observed. Manipula-
tions of the vapor pressure in vacuum eliminated the possibility

that this was due to the effect of vapor pressure on the contact
angle;   (b) Observing the contact angle change with time.   The
drop may continue to spread for many hours; and (c) Agitation of
the drop.   This caused immediate partial spreading.

Upon raising the temperature and time of heating of the gold
disk in air followed by evacuation, the receding angle decreased
to zero, while the advancing angle decreased to about 20°-30°.
With more activation (increased heating) the advancing angle
decreased further, while the receding remained zero.   In the conventional vacuum system a limit was reached in the decrease of the
advancing angle.   Further activation either raised the angle or
ceased to lower it.   In the ultrahigh-vacuum system on the other
hand, increased activation decreased the advancing angle to zero
degrees.

For the case of gold films evaporated in situ, the contact
angle of water on a gold film[16] evaporated onto the surface of a
polished fused silica disk was zero.   The system was evacuated
again and another layer of gold deposited on top of the original.
The contact angle was zero once more.   A few minutes was sometimes
required for the drop to reach the equilibrium zero value.

In another experiment the gold was deposited on a polished
graphite surface.   A zero contact angle was again observed a few
minutes after deposition of the water drop.   The contact angle of
water on the graphite disk surface which was shielded from the gold
vapor flux was approximately 22°.

The existence of contact angle hysteresis as a result of
contamination or heterogeneity on a smooth surface has been discussed by a number of authors.[16]   The hysteresis observed in the
work on gold can be attributed to the presence on these surfaces
of both hydrophilic and hydrophobic areas, or site clusters.
Increased surface activation resulted in an increased ratio of
hydrophilic to hydrophobic areas, with an accompanying decrease in
both advancing and receding angles, until a zero receding angle
was obtained.   At this point the average surface free energy less
that of the liquid-solid interface was sufficient to overcome that
of the water and keep it spread.   The advancing angle was nevertheless still relatively high due to the inability of the drop
periphery to advance across hydrophobic regions.   The situation
may be pictured on the basis of hydrophilic "islands" in a hydrophobic "sea."   As the surface was activated further, the hydrophobic
area continued to diminish with an accompanying decrease in the
advancing angle until the hysteresis disappeared as the advancing
angle reached zero.

It is clear from a comparison of the conventional high-vacuum and ultrahigh-vacuum experiments that the gradual increase in ratio of hydrophilic to hydrophobic area which occurs upon heating can be interpreted in terms of a gradual removal of hydrophobic organic contamination from the real gold surface. For the case of the measurements in conventional high vacuum, the activation procedure of high temperature oxidation and evacuation gradually cleans the surface, until there are sufficient uncontaminated hydrophilic areas to yield a low or zero receding angle, and an advancing angle of 20°-30°. A point is reached at which additional activation is ineffective and may even result in additional contamination. This is due to the fact that the conventional high vacuum is not an ultraclean system. Once the system is evacuated, the residual organic vapors have ready access to the gold surface, unimpeded by the presence of air. When the surface is partially cleaned, a steady state is established at which the rate of contamination equals the rate of vacuum cleanup. However, in the ultrahigh-vacuum system which consists solely of metal and glass, there are no organic vapors to recontaminate the gold. The high temperature oxidation-evacuation procedure consequently continues to clean the surface until the hysteresis disappears and the advancing as well as receding contact angle is zero.

The inability of the conventional high-vacuum system to clean the gold surface is not observed for the case of a polished fused-silica surface. In the latter case preliminary heating to 200° is adequate to yield a zero contact angle with water at room temperature. Using water contact angles as a measure of cleanliness, it is apparent that it is far more difficult to remove organic contamination from a gold than from a fused-silica surface. It is clear, then, that a set of conditions which is sufficient to decontaminate one particular surface will not necessarily succeed for a different type.

There is a possible objection to interpretation of these results in terms of removal of organic contamination. Assuming that the surface of real gold is hydrophobic rather than hydrophilic, the ultrahigh vacuum results for the solid gold sample could be explained in terms of hydrophilic impurities diffusing to the surface during the heat cleaning process, causing hysteresis at first, followed by a zero advancing angle as the surface becomes completely contaminated with the hydrophilic impurities. It would then have to be assumed that in the conventional high vacuum (where a zero degree contact angle could not be attained) the hydrophilic impurities which make their way to the surface subsequently become contaminated by the residual organic vapors. However, the method used of evaporation and condensation of a gold film in situ completely bypasses this possibility of surface segregation, since the deposited film is very thin and furthermore is not heated.[17]

For the case of any metal other than gold, an uncontaminated real surface would contain chemically combined oxygen, either as a built-up surface oxide or as a chemisorbed monolayer. The attainment of a surface which is clean (oxygen free) as well as uncontaminated with organic material, would then entail a procedure such as ion bombardment to remove any chemically combined oxygen already present, followed by maintenance of a suitable ultrahigh vacuum to prevent oxygen from recombining with the surface before the measurement is completed. Of course, for the case of a film evaporated in situ, the ion bombardment is not necessary. In the present work, where water vapor is introduced for the measurement, extensive degassing of the water source is not necessarily sufficient to avoid introduction of a sufficient number of oxygen molecules to chemisorb to a few square centimeters of an active metal surface. For the case of metals in general, therefore, a special gettering technique would have to be devised to ensure measurement on an oxygen-free surface.

The surface of gold, however, is unique among all metals in its relative lack of affinity for oxygen. It is the only metal which does not form a bulk oxide which is thermodynamically stable at room temperature. While this does not preclude the existence of a chemisorbed monolayer on the gold surface, available literature[18,19,20] indicates that such a monolayer will not form at room temperature, even at high oxygen pressure and much longer periods of time than the duration of these wettability experiments on evaporated gold films.

Investigation of the contact angle of water on clean metal surfaces was subsequently extended to copper and silver, active metals which form surface oxides, or chemisorbed oxygen monolayers, when exposed to oxygen.[21] In particular, the possibility was considered that clean metals do indeed interact physically with water due to dispersion forces only, and that gold spreads water due to an especially high γ^d resulting from its high molecular weight. The main experimental problem in measuring the wettability of clean copper and silver consisted of admitting water vapor without any accompanying trace of oxygen. The extreme precautions taken to accomplish this consisted of the following. (1) Degassing the liquid water into the sorption pump by momentarily opening a valve and then closing it as the top water layer started to freeze from evaporation. This operation was performed 60 times. (2) The water was exposed to a chamber evacuated to the ultrahigh vacuum region by means of the ion pump. This operation was performed seven times. The water reservoir remaining after treatment according to this and the previous step served as the original source of water vapor throughout all the experiments with copper and silver. For some of the experiments with copper, the following step was added. (3) Prior to admittance to the sample chamber for contact

angle measurement, water vapor from the degassed liquid was adsorbed
onto clean (oxygen-free) germanium powder in an intermediate chamber
and allowed to equilibrate for at least 1 hour. The germanium had
been previously cleaned by heating in vacuo at 700°,[22] and was
regenerated after each run. Clean germanium rapidly chemisorbs a
monolayer of oxygen. The efficiency of the germanium powder was
monitored in a separate experiment by deliberate adsorption of
oxygen (after one of the 700° cleanings) which was measured by
pressure difference with a thermistor pressure sensor. The amount
of oxygen chemisorbed was approximately equal to that which would
have been dissolved in the entire water reservoir if it were in
equilibrium with the atmosphere. Since the reservoir was actually
thoroughly degassed and since, furthermore, only a small portion of
it vaporizes into the vacuum chamber for contact angle measurement,
the capacity of the germanium powder far exceeded that necessary to
completely free the water vapor of any possible oxygen residue.

The contact angle of water, purified in this manner, on copper
film deposited in ultrahigh vacuum was found to be zero degrees.
The water contact angle for silver deposited by evaporation in
ultrahigh vacuum was also found to be zero degrees. In the silver
experiments, however, the germanium purification step was omitted
after analysis with a quadrupole residual gas analyzer did not
detect any molecular oxygen in the degassed water. The residual
gas analysis also indicated that hydrogen and carbon monoxide were
released from the evaporated silver and copper surfaces after
admission of water vapor.

As for the case of gold, it is seen that measurements of the
contact angle of water on evaporated films of copper and silver in
ultrahigh vacuum fail to yield any evidence that clean, oxygen-free,
metallic surfaces are hydrophobic. In fact, it is not completely
certain that water can come into contact with these active surfaces
without decomposing, since the exposure of clean copper or silver
to even small amounts of moisture results in evolution of hydrogen.
However, while this may be due to reduction of the water molecule
by the surface, it is most probable that the phenomenon results
from displacement of weakly chemisorbed hydrogen from the surface
by the water molecules.

RECENT HAMAKER COEFFICIENTS

In 1969 Derjaguin[23] published a revised calculation of experimental data in which a Hamaker coefficient for gold of 4.1×10^{-12}
erg was reported. Recent developments in the theory of long-range
van der Waals forces have enabled the calculation of Hamaker
coefficients on a theoretical basis, utilizing spectral data and a
macroscopic continuum model. Parsegian et al[24] have calculated a

Hamaker coefficient of 2.4×10^{-12} erg for gold in water in this
manner. Both these values yield γ^d s for gold which are greater
than the minimum of 236 erg/cm^2 required to spread water by
dispersion forces alone. Matsunaga and Tamai,[25] also using macroscopic theory, calculate Hamaker coefficients for aluminum, copper,
silver, and gold, which they interpret as consistent with the
hypothesis that the clean surfaces of these metals can spread water
by means of dispersion forces only. There is consequently no
experimental or theoretical support at present, for the supposition
that gold, or any other clean solid metal surface thus far
investigated, is hydrophobic.

MACROSCOPIC CONTINUUM THEORY

On the basis of present state-of-the-art knowledge, therefore,
water will spread on glasses and ceramics, oxides, and on metal
surfaces with or without an oxide layer. Since water does not
spread on organic surfaces, this provides, in principle, a simple
way of differentiating these surfaces from organic surfaces.
Unlike the actual measurement of finite contact angles, the
difference between spreading and nonspreading (i.e., between zero
and nonzero contact angles) can easily be determined with the naked
eye. Now, since these inorganic surfaces are high energy, and
organic surfaces low energy, it is expected that, given the
opportunity, these inorganic surfaces will have the tendency to
collect films of organic contamination. Due to the ubiquitous
presence of organic particles in the air, there is ample opportunity for this to occur. In fact, experience tells us that it is
usually impossible to expose a smooth high energy surface to air
for even one minute without formation of an invisible layer of
organic contamination which results in a nonzero water contact
angle. The theory governing the efficacy of the water contact
angle in detecting the presence of an organic film is conveniently
approached through the macroscopic continuum model for long range
van der Waals forces.

The key component of van der Waals forces is the London,[26] or
"dispersion" force, which is necessary to explain the attraction
between completely nonpolar molecules or atoms which occurs, for
example, on formation of liquid argon. The London theory describes
this attraction between the individual gas-phase molecules which
arises from second order perturbation theory applied to the electrostatic interaction between two dipoles. The physical picture of its
origin involves the concept of instantaneous dipoles formed during
the motion of an electron around the nucleous. Alternatively, it
may be regarded as oscillation of the charge cloud which describes
the average position of the electron. The instantaneous dipole
interacts with an instantaneous dipole in the other molecule formed
naturally and by induction. These correlated fluctuations result
in the mutual attraction of nonpolar molecules.

The macroscopic-continuum approach deals with the attraction between two macroscopic liquid or solid bodies at a distance much greater than the size of individual molecules. Building on early contributions from Lebedev[27] and then Casimir and Polder,[28] Lifshitz[29] published a comprehensive theory, in 1956, of these long-range van der Waals forces. Electronic fluctuations leading to the existence of London forces between the two bodies are regarded as arising from each body treated as a continuum. Physical characteristics of bulk materials such as absorption spectra and refractive index are related, by the rigorous methods of continuum physics, to the long-range electrodynamic forces acting between the two bodies. Due to the complexity of the resulting equations, for some years use of the Lifshitz theory was limited to specific systems which could be approached by approximation.[30] Recent advances in computer techniques, however, have facilitated innova- tions which enable accurate calculations of long-range van der Waals forces by the macroscopic method.[31,32,33] One of the geometries treated was that of two semi=infinite slabs of material separated by a medium. If one slab is a solid (or liquid) substrate, the medium a liquid, and the other slab the vapor of the liquid medium,[30] the theory gives the van der Waals force, or free energy arising from that force, between the vapor and substrate. Attrac- tion between vapor and substrate results in thinning of the liquid, so that it is unstable, while repulsion of vapor and substrate results in thickening of the liquid, so that it is stable. This stability of course, refers only to the effect of these long-range forces. Wettability of a liquid on the clean surface of a solid substrate normally depends primarily on short-range interfacial forces operating over distances between atoms. For the case of a layer of organic contamination on a solid inorganic substrate, however, this organic layer will separate a supernatant liquid such as water, for example, from the surface to a distance beyond the reach of these short-range forces. The remaining question then, is, can the long-range force of the inorganic substrate "shine through" the contamination to cause spreading, or stable film formation, of a liquid which would not spread on an infinitely thick layer of the contamination itself.

Parsegian et al considered this question for the case of water on contaminated gold.[24] The difference between G_{ghw}, the free energy due to long-range forces in a system consisting of gold sub- strate, tetradecane (a typical hydrocarbon) contaminant of finite thickness, and an infinitely thick supernatant water film in equi- librium with water vapor, and G_{ghv}, the free energy due to long- range forces in a system consisting of gold substrate, tetradecane contaminant of finite thickness, and tetradecane vapor, is calculated, from spectra of gold, tetradecane, and water, for various thicknesses of tetradecane. The result is plotted as the absolute value of $G_{ghw} - G_{ghv}$ versus "a," the thickness of the tetradecane layer (Figure 2). The resulting curve is compared to the value 95.6 erg/cm^2, which is the short range contribution to

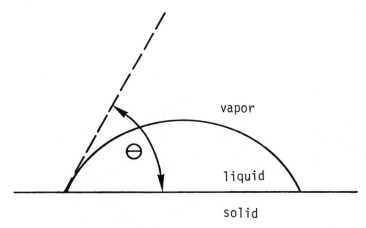

Figure 1. Contact angle of sessile drop on smooth solid surface.

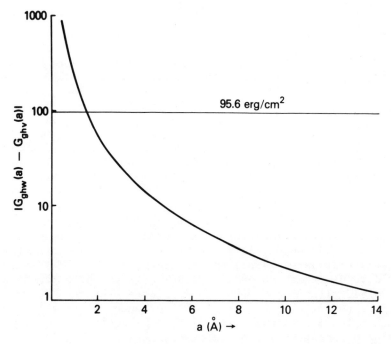

Figure 2. Comparison of long- and short-range contributions to the energy of water film formation on a tetradecane-coated gold surface. [Reprinted from Parsegian, V. A., G. H. Weiss, and M. E. Schrader, J. Colloid Interface Sci. 61, 356 (1977) by courtesy of Academic Press, Inc.]

water film formation on a hydrocarbon (tetradecane) – coated gold
surface. The actual values of $G_{ghw} - G_{ghv}$ are negative, indicating
water-film stability which increases with decreasing thickness of
the hydrocarbon, while the short-range value of 95.6 erg/cm^2 is
positive, indicating a constant contribution to instability at all
distances. It can be seen that the long range contribution becomes
comparable in magnitude to the short-range component only at
tetradecane thickness near 2Å, which is less than the expected
thickness of any physisorbed monolayer. In other words, the long-
range gold force does not "shine through" even one monolayer of
hydrocarbon contamination to any significant extent. The thinnest
possible uniform layer of hydrocarbon contamination, therefore,
screens the gold surface forces sufficiently to prevent the spreading (formation of a stable film) of water. This means that the
spreading or nonspreading of water, which are phenomena visible to
the naked eye, can not only be used to differentiate between
organic and inorganic surfaces, but will detect the thinnest
possible layer of organic contamination on an inorganic surface.

SUMMARY

      Contact angle measurements are a convenient and extremely
rapid method for probing the surface constitution of macroscopic
solids. Typical hard solids such as glasses, ceramics, metals,
and metal oxides yield high-energy surfaces. Organic compounds,
on the other hand, which are softer materials, generally yield low
energy surfaces. As a rule, compounds which are liquid at room-
temperature (except mercury) spread on high-energy surfaces, but
may or may not spread on low-energy surfaces. Water, however,
almost never spreads on low-energy surfaces. The magnitude of
contact angles on these low-energy surfaces can be described and
often predicted by utilizing the Zisman critical surface tension
concept, the Good-Girifalco theory of interfacial forces, or the
Fowkes dispersion component of surface tension. Metallic high
energy solids containing oxide or chemisorbed oxygen on the surface,
but free of other contaminants, all spread water. Water has also
been found to spread on all oxygen-free clean metallic surfaces
thus far investigated (i.e., evaporated films of copper, silver,
and gold in ultrahigh vacuum). When a high-energy surface is
contaminated with even a monolayer of low-energy organic material,
the substrate forces are essentially screened, so that the water
contact angle is determined largely by the contaminant. The
spreading or nonspreading of water can thus not only be used to
differentiate between organic low-energy and inorganic high-energy
surfaces, but will detect the thinnest possible layer of organic
contamination on a high-energy surface.

REFERENCES

1. T. Young, Phil. Trans. Roy. Soc. (London) 95, 65 (1805).
2. W. D. Harkins and A. Feldman, J. Amer. Chem. Soc. 44, 2665 (1922).
3. H. W. Fox and W. A. Zisman, J. Colloid Sci. 5, 514 (1950).
4. W. A. Zisman, Advan. Chem. Ser. No. 43, Am. Chem. Soc., Washington, D. C., 1964, p. 1, and references therein.
5. L. A. Girifalco and R. J. Good, J. Phys. Chem 61, 904 (1957).
6. R. J. Good and E. Elbing, Ind. Eng. Chem. 62 (3), 54 (1970) and references therein.
7. F. M. Fowkes, J. Phys. Chem 66, 382 (1962)
8. F. M. Fowkes, Ind. Eng. Chem. 56 (12), 40 (1964) and references therein.
9. H. Reerink and J. Th. Overbeek, Discuss. Faraday Soc. 18, 74 (1954).
10. Westgren, Ark. Kemi, Min. Geol. 7 No. 6 (1918)
11. Tuorila, Kolloidchem. Beih. 22, 191 (1926); 27, 44 (1928).
12. M. L. White, J. Phys. Chem. 68, 3083 (1964).
13. R. A. Erb, J. Phys. Chem. 69, 1306 (1965).
14. M. K. Bernett and W. A. Zisman, J. Phys. Chem. 74, 2309 (1970).
15. M. E. Schrader, J. Colloid Interface Sci 27, 743 (1968).
16. M. E. Schrader, J. Phys. Chem. 74, 2313 (1970).
17. P. W. Palmberg and T. N. Rhodin, Phys. Rev. 161, 586 (1967).
18. N. V. Kul'kova and L. L. Levchenko, Kinet. Katal. 6, 765, 688 (1965).
19. B. J. Hopkins, C. H. B. Mee and D. Parker, Brit. J. Appl. Phys. 15, 865 (1964).
20. W. M. H. Sachtler, G. J. H. Dorgelo and A. A. Holscher, Surface Sci. 5, 221 (1966).
21. M. E. Schrader, J. Phys. Chem. 78, 87 (1974).
22. A. J. Rosenberg, P. H. Robinson and H. C. Gatos, J. Appl. Phys. 29, 771 (1958).
23. B. V. Derjaguin, V. M. Muller and Ya. I. Rabinovich, Kolloid Zh. 31, 304 (1969).
24. V. A. Parsegian, G. H. Weiss and M. E. Schrader, J. Colloid Interface Sci. 61, 356 (1977).
25. T. Matsunaga and Y. Tamai, Surface Sci. 57, 431 (1976).
26. F. London, Z. Phys. 60, 491 (1930).
27. B. V. Derjaguin, I. I. Abrikosova and E. M. Lifshitz, Quart. Rev. (London) 10, 295 (1956).
28. H. G. B. Casimir and D. Polder, Phys. Rev. 73, 360 (1948).
29. E. M. Lifshitz, Sov. Phys. JETP 2, 73 (1956).
30. I. E. Dzyaloshinskii, E. M. Lifshitz and L. P. Pitaevskii, Adv. Phys. 10, 165 (1961).
31. V. A. Parsegian, Trans. Faraday Soc. 62, 848 (1966)
32. V. A. Parsegian, Science 156, 939 (1967).
33. V. A. Parsegian and B. W. Ninham, Nature 224, 1197 (1969).

MICROSCOPICAL IDENTIFICATION OF SURFACE CONTAMINANTS

Walter C. McCrone

McCrone Associates, Inc.

2820 South Michigan Avenue, Chicago, IL   60616

Surface contamination may result from deterioration, corrosion or attrition and from the introduction of contaminants from the air or contact with other objects. One must first see the contaminant, usually microscopically, and be able to analyze it in situ or remove it for analysis elsewhere. The microscope will also help suggest an appropriate analytical approach.

Identification may often be done by light microscopy using shape and optical properties as well as microchemical tests. If the analyst lacks that background, he may turn to x-ray diffraction, SEM/EDXRA, EMA, IMA or the new laser Raman microprobe.

INTRODUCTION

Microscopy, as defined by microscopists at McCrone Associates, includes all tools and techniques that characterize and identify microscopic objects. We include all microprobes, even ESCA and micro-XRD along with light and electron microscopes.

Ideally, all microscopists should use the most appropriate of these tools for each job they tackle. Unfortunately, this is not generally possible because of the complex nature of the different microscopes and microprobes. One person would have difficulty mastering all of these tools and the cost of a complete laboratory is very high — in excess of $2,000,000. Still, it is very difficult to recommend one or two of these tools over the others because each has special capabilities and limitations and because contamination problems are so varied.

Surface contaminants may be homogeneous thin films, heterogeneous thin films or they may be particulate. They may be crystalline or amorphous, monomolecular in thickness or several micrometers thick, organic or inorganic in composition etc. Often, several different microanalytical approaches are necessary. McCrone
Associates personnel are fortunate to have in-house at least one of
each of these specialized tools for contamination studies. Table I
lists these tools with their major capabilities and limitations.

POLARIZING LIGHT MICROSCOPE (PLM)

The light microscope has a number of direct applications to
the detection, characterization and identification of surface contaminants. It is also a helpful adjunct to studies made by other
tools. All samples should probably be examined by light microscopy
(stereobinocular or PLM) before more detailed study by the electron
or ion beam instruments or ESCA. Questions of purity often are
answered and the chemical and physical nature of the contaminant
can often be determined better at lower magnification and with visible light. Color, thickness, homogeneity and often composition of
films can be determined. Particle contaminants can often be identified by shape, color etc. At worst, the use of the light microscope will help to suggest the proper approach for further study.

The presence of a thin transparent film on a surface can often
be detected and its thickness measured by reflected light micros-

Table I.   Capabilities of Microscopes and
Microprobes for Contamination Analysis

	General particle identification	Minimum particle size	Film identification[†]	Minimum film thickness
PLM	excellent	1 μm	poor	25 nm
SEM (alone)	poor	50 nm	poor	25 nm
SEM (+ EDXRA)	good	50 nm	excellent	25 nm
TEM (alone)	poor	2 nm	poor	2 nm
TEM (+ SAED)	good	2 nm	good	20 nm
TEM (+ EDXRA)	good	2 nm	excellent	25 nm
EMA	excellent	100 nm	excellent	25 nm
IMA	good	20 nm	excellent	25 nm
ESCA	poor	*	excellent	1 nm

† assumes film is inorganic.
* ESCA requires a 10^{-9} g sample spread as 1-5 nm film.

Note: Acronyms explained later in text.

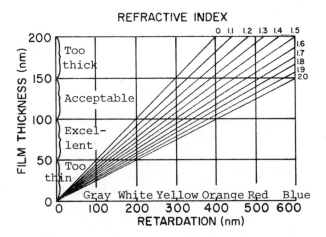

Figure 1. Relationship between thin film thickness, retardation, refractive index and its effectiveness as a TEM grid particle support.

copy.[1] Film presence is indicated by interference colors and thickness by the actual color observed (Figure 1). Strictly speaking, the thickness measured in this way is the optical thickness but knowledge of the refractive index permits easy calculation of the actual physical thickness:

$$\text{Thickness} = t = m\lambda/2n$$

where m is the number of wavelengths corresponding to the thickness, λ = wavelength and n the refractive index. This can be used to estimate the thickness of oxide films on metal, oil films on any surface etc. The same relationship is used when analyzing particle samples mounted on a TEM grid or microprobe plate. It is essential that films not be so thick the electron beam is hindered but still strong enough to ensure retention of the sample. Figure 1 has notations along the ordinate indicating suitability of different thickness films for their use with TEM samples. Films showing first-order red interference colors will be too thick for the usual 100-200 kV electron beam to penetrate, whereas low first-order gray films may well break in such beams.

Films have also been intentionally prepared by anodic oxidation of metals and alloys.[2,3] The resulting colored areas delineate individual grains and help to identify alloy phases and grain orientation.

Contaminant particles in decorative or protective coatings such as paint can also be observed by light microscopy. The offend-

ing particles can usually be easily excised with a needle for identification, often also by PLM.

A more unusual surface contaminant is a result of high humidity acting on the surface of, at least, slightly hygroscopic substances to form a sticky surface layer that causes agglomeration
problems and prevents free flowing of the substance.[4] Pharmaceuti-
cal compounds are often afflicted in this way and the surface film
is often identifiable by PLM as a liquid crystalline hydrate.

Extraction replicas are another excellent way to study contaminants. By replicating bearing races or electrical contacts,
deposits that cause failure are removed for analysis (Figure 2).
The varnish deposited on electrical contacts often results from
the oxidation and polymerization of organic vapors such as plasti-
cizers in the polymers used in relays etc. Relay and bearing, or
other similar, surfaces may be coated with collodion which, after
drying, can then be peeled off along with the surface contaminant.
This extraction replica is then examined by reflected and transmitted PLM. If necessary, the particles or films thus removed can
be further analyzed by XRD or one of the microprobes.

The light microscope often identifies surface contaminants
quickly and with certainty without the need for the more electronically sophisticated electron or ion beam instruments. Needless to
say, considerable time and money are saved when the PLM does the
job. Whether it does or does not identify such contaminants depends
almost entirely on the skill and background of the microscopist.

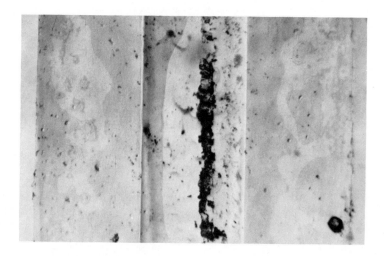

Figure 2. Collodion replica of ball race with varnish deposit and
particles removed with the replica.

SCANNING ELECTRON MICROSCOPY (SEM)

Although the PLM characterizes small particles and films much more completely than does the SEM (Figure 3), the latter, if fitted with EDXRA, has one major advantage, an elemental analysis capability. All elements above fluorine can be quickly detected with the energy dispersive x-ray analyzer (EDXRA) and all above beryllium with the wavelength dispersive system (WDXRA). The analysis is usually accurate enough to yield the stoichiometry of the compound involved although mixtures may cause a problem.

An interesting refinement of the elemental analysis by SEM is the use of electron beams of different energies in order to vary the penetrating power. This may yield different elemental analyses for the same surface area of a sample, thus indicating a surface layer or film having a composition different from the substrate. Surface films in the range of a few hundred to 4-5,000 Angstroms may be detected in this way.

Another way in which a surface film may be detected is by the relative absorption of the low energy x-rays of an element in the substrate as the electron beam energy is varied. The Mα x-rays of lead, for example, are much more readily absorbed in a thin layer coating lead than the higher energy L lines. This method is particularly useful for organic films or for films with other low atomic number elements not otherwise detected by EDXRA. Most SEMs are not fitted with WDXRA which would detect thin films containing elements with z>4.

Figure 3.   The Cambridge Stereoscan fitted with Ortec EDXRA.

Because organic films in general are not detected directly by
SEM, one sometimes removes them in order to detect inorganic substances in such films. The pigments in a paint or polymer coating,
if inorganic, can be exposed for analysis by a prior treatment with
plasma etching. This is most readily accomplished as an external
pre-treatment in a low temperature asher.

Contaminants on hair such as sprays and other treatments are
easily detected by SEM; this has some application in criminalistics
since hair is often found as trace evidence at a crime scene or
associated with a suspect or victim. Such "contaminants" may
assist in tagging that hair to a particular person.

The SEM is a very useful instrument, especially for both morphology and composition of surfaces.

TRANSMISSION ELECTRON MICROSCOPE (TEM)

Very fine particulate surface contamination is often best studied with the TEM. Particles even smaller than one nanometer can be
resolved and, if crystalline and slightly larger, usually identified
by selected area electron diffraction (SAED). This much neglected
technique yields crystal lattice parameters as does XRD but on
particles several hundreds of times smaller. The lattice parameters
measured on the diffraction pattern are checked against the known
data tabulated and published by the ASTM for about 25,000 substances.

Some TEMs have also been fitted with EDXRA for elemental analyses. Several firms also manufacture a combination TEM-electron microprobe analyzer (EMA). One such instrument produced by AEI, termed an electron microscope-microprobe analyzer (EMMA), has two WDXRA
detectors. When also fitted with EDXRA as is the McCrone Associates
instrument, the combination of high resolution (1 nm), SAED, EDXRA
and WDXRA will identify nearly any thin film or small particle. Detection limits are in the femtogram to attogram levels (10^{-15} to
10^{-18} g).

ELECTRON MICROPROBE ANALYZER (EMA)

The EMMA at McCrone Associates has been pretty completely dedi-
cated to the identification and counting of asbestos, hence most
surface contamination problems have been solved with one of the
other EMAs (Figure 4). EMMA could, however, have been used for any
of the following problems. Indeed, in some cases, the additional
SAED capability would have been very helpful. On the other hand,
EMA also functions as a scanning instrument whereas EMMA does not.
The resolving power of EMA for SEM-mode viewing (back-scattered
electron imaging) is comparable with PLM rather than SEM, but it
does yield a good electron micrograph of a surface or particle with
an indication of different average z number areas by the intensity

Figure 4. ARL electron microprobe analyzer, also fitted with the Ortec EDXRA as well as stage and bean controls and hardware to permit computer operation and data handling.

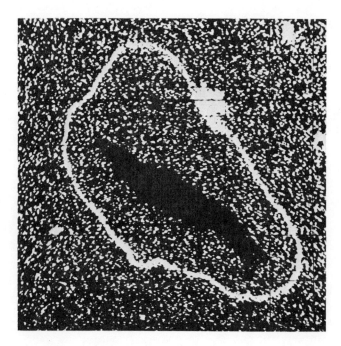

Figure 5. A back-scattered electron image of a nickel-cored, gold coated silver wire - 360X.

of the back-scattered electrons (Figure 5). Micrographs of the
same areas with the x-rays of any constituent elements above z = 4
can be observed on the CRT and recorded photographically. Complex
corrosion products, relay contact deposits or other contaminants
can be analyzed in this way. Associated elements will appear together throughout the micrograph. Successive color exposures on
the same film can show, for example, iron in red, molybdenum in blue
and carbon in green; iron carbide and molybdenum carbide, if present,
would then appear yellow and magenta, respectively. The precise
locations of chlorides on a relay contact or integrated circuit,
the thickness of a chromium plated part or the distribution of
carbon in a steel can all be shown as color maps of the surface.

Very thin uniform surface films are often detected as described
above for the SEM: use of low vs high energy electron beams to show
variation in elemental composition of the top layer; or variations
of intensity of low and high energy x-rays from the same element as
a result of absorption of the lower energy x-rays in the interven-
ing thin surface film. Another simple but direct indication of a
thin surface film is to first scratch the surface with a needle and
examine the unaltered surface and the scratch for composition differences.

ION MICROPROBE ANALYZER (IMA)

Certainly the most sensitive of the microanalytical tools of
those capable of detecting more than a few elements individually is
IMA (Figure 6). Erwin Mueller's atom-probe based on field-ion emission microscopy can, of course, detect and identify single atoms of

Figure 6.   The ARL ion microprobe analyzer.

any element but only one at a time. It is not, therefore, a generally useful microanalytical tool. IMA, on the other hand, will detect a few hundred to a few thousand atoms of any element in the
periodic table. Furthermore, it can detect ppm of nearly all elements in a matrix and ppb of some elements.

IMA, a sputter-source mass spectrometer, is the only one of
the tools covered thus far that can be said to be destructive. However, it would take many minutes to sputter away a single μm^3 during
which time many individual analyses would have been recorded. At
the same time, since the surface is slowly sputtered away, depth
profiles can be plotted with depth resolution of a few Angstroms.
This is accomplished when necessary by taking data as a function of
time under fixed conditions of beam current, accelerating voltage
and beam focus to ensure a constant rate of erosion and finally
measuring the depth of the resulting hole with the SEM. The extreme sensitivity and its excellent depth resolution makes IMA an
excellent way to look for trace elemental contaminants in a surface
or to study diffusion of one element into a second.

A fascinating problem illustrates the sensitivity of the method. We were asked if we could find a method for determining whether canned vegetables had been adequately sterilized by heat. We
decided to look for changes in the lacquer coating on the insides
of the cans during sterilization. IMA showed very clearly that
sodium from the canned food diffuses into the lacquer during processing and the sodium profile is then a measure of the sterilization time and temperature (Figure 7).

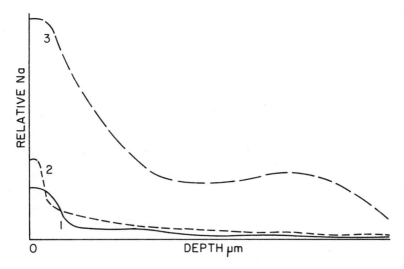

Figure 7. Depth profiles for sodium on lacquer coating on insides
of soup cans before heat sterilization (1), after filling but not
heating (2), and after heating (3).

ELECTRON SPECTROSCOPY FOR CHEMICAL ANALYSIS (ESCA)

ESCA, or photoelectron spectroscopy, is, in a sense, just the reverse of EMA. The latter generates x-rays by impacting a sample with electrons whereas ESCA generates electrons by irradiating a surface with x-rays. An advantage of ESCA is the fact that the emitted electrons have energies characteristic not only of a particular element, but of the binding energy for that element in the composition in which it exists. Nitrogen, for example, can be identified as such by ESCA, but nitrogen in the NO_3^- is also differentiated from nitrogen in NH_4^+ or NO_2^-. ESCA, then, yields not only elemental but molecular composition information.

On the other hand, an electron beam penetrates several micrometers into a surface and generates x-rays from sample volumes from about 10-100 μm^3. The x-ray beam in ESCA, however, penetrates only 1-5 nm, about 10^3 less penetration than the electron beam in the EMA. This means that ESCA analyzes a very thin surface layer and, as a consequence, it must cover a larger area of several mm^2 to build up a detectable signal. The sensitivity is in the ng-pg range but only as a very thin layer a few molecules thick and several million molecules in both lateral dimensions. The sample must then be very large in extent compared with SEM, EMA or IMA.

Still, surface contaminants often cover large areas and ESCA then usually solves such problems. It is not a rapid method and is often used in conjunction with other instruments. For example, a complete periodic table scan by ESCA takes 30 minutes or more depending on the instrument and the desired sensitivity. To use ESCA most effectively the sample would first be analyzed by SEM/EDXRA, EMA or IMA for elemental composition. ESCA would then more quickly yield valence states and other molecular configurations of the elements already known to be present.

ESCA is very helpful when contamination layers are very thin. It will detect surface oxidation of a fresh lead surface at room temperature in 4 seconds. It will detect thin layers of molybdenum sulfate on the sulfide so thin that an EMA cannot detect the oxygen and IMA would have difficulty.

ESCA is able to detect Auger electrons as well as the inner valence shell electrons, hence a laboratory with ESCA and a microprobe may be able to manage without Auger. ESCA is the only tool capable of analyzing 10-50 Angstrom films nondestructively but it is usually fitted with an ion gun to etch away successive monomolecular layers so that substrate material can be identified or even depth profiles determined. One must be careful in this instance to consider the possibility that the ion etching may modify the composition and yield spurious results.

REFERENCES

1. W. C. McCrone, and J. G. Delly, "The Particle Atlas," Vol. 1, 2nd ed., p. 251, Ann Arbor Science Publishers, Ann Arbor, Michigan, 1973. Much of the background for this paper is covered in the six volumes of "The Particle Atlas."
2. R. Hasson, Microscope 16, 329 (1968); 22, 317 (1974).
3. T. R. Almand and D. H. Houseman, Microscope 18, 11 (1970).
4. J. S. G. Cox, G. D. Woodard, and W. C. McCrone, J. Pharm. Sci., 60, 1458 (1971).

IDENTIFICATION OF CONTAMINANTS WITH ENERGETIC BEAM TECHNIQUES

David W. Dwight[*] and James P. Wightman[†]

Virginia Polytechnic Institute and State University

Blacksburg, Virginia   24061

A brief review of some modern techniques to obtain structure and bonding analysis in 1nm-1µm phases with ion, photon, or electron beams mentions advantages and limitations of TEM, SEM, ESCA, AES, and UPS. Highlights from recent applications in the fields of adhesion, corrosion, and wear demonstrate a broad range of new information on structure/property relationships in solid microphases now accessible. For example, quantitative analysis of weak boundary layers and segregation from bulk indicates the control of adhesion at the 0.2-0.5nm level. Also, microscopic heterogeneities at interfaces are basic in electrochemical mechanisms of corrosion. Finally, wear specimens show nearly ubiquitous atomic or molecular-level transfer upon touch, and surface chemical changes inevitably accompany changes in friction and wear behavior.

New results in our studies of both polymer and metal adherends illustrate the importance of depth-profile studies. Grazing angle ESCA of an ion-sputtered fluoropolymer demonstrate chemical structure effects confined to the top few atom layers. In a case where greater depth of analysis is important, titanium 6-4 adherends showed threefold changes in oxide layer thickness in the top 100nm probed by ion-milling/Auger Electron Spectroscopy.

[*]Department of Materials Engineering
[†]Department of Chemistry

INTRODUCTION

Energetic beam techniques promoted major advances in the study of the structure and properties unique to materials of dimensions 1nm–1μm. Transmission Electron Microscopy (TEM) and Scanning Electron Microscopy (SEM) led with morphology determination. Chemical analysis in these instruments usually analyzes x-ray emission and is limited to dimensions >100nm. However, the current decade has produced a burgeoning of electron, photon, and ion bombardment methods, and now two surface science journals are devoted to elucidation of "microscopic force laws that govern the properties of surfaces and interfaces...and the technological implications..."[1] Auger Electron Spectroscopy (AES) has sensitivity of a monolayer *and* can be focused laterally to identify features <0.5μm in diameter. Scanning the electron beam in coordination with a CRT image allows both physical and chemical heterogeneities of the sample surface to be recorded. Profiles of composition *vs* depth can be obtained by combining ion-beam milling with simultaneous AES. Unfortunately, ion and electron beams may damage polymer surfaces during the measurement. Photoelectron spectroscopy using x-ray (XPS or ESCA), or ultraviolet (UPS) radiation gives additional chemical bonding information, but cannot resolve lateral heterogeneities easily. Synergistic combinations of some of the methods in one apparatus offer thorough microstructure characterization.

This review first presents a brief primer on the most utilized energetic beam methods. Selected examples then are cited of applications of modern methodology in adhesion, corrosion, and wear. Finally, data on angle-resolved photoemission from an ion-sputtered fluoropolymer and Auger depth profiles on titanium 6-4 alloy after treatments designed to improve adhesion bonding, emphasize the importance of in-depth studies.

Techniques

Electron Microscopy. A simplified schematic (Figure 1) illustrates the two types of electron microscopes. Transmission Electron Microscopy has been the mainstay of microstructure analysis since the 1930's. The use of TEM in contamination research usually requires replication; an adhesive tape or solvent-carried polymer placed on the surface of interest produces a negative image of the surface with spacial resolution up to 6nm. Consecutive replication of the same contaminated surface with solvent-softened cellulose acetate tape first removes particles, then loosely held contaminants for subsequent analysis in the Scanning Electron Microscope (SEM). Finally, a replica of the underlying substrate can be made, "shadowed" with vacuum

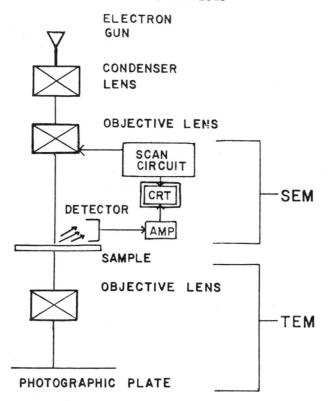

Figure 1. Simplified schematic diagram showing the basic components of Scanning (SEM) and Transmission (TEM) Electron Microscopes.

evaporated metal, and studied with high resolution.[2] The SEM has an advantage in depth-of-field, and relatively large sections of material can be fit in the analysis chamber. Practical samples can be analyzed without replication because SEM creates an image from surface emission of secondary or backscattered electrons; thus samples do not need to be transparent to the electron beam as is the case in TEM.

Electron Spectroscopy. In principle, measurement of the electron energy spectrum identifies elemental constituents and bonding states in a material. Figure 2 illustrates the excitation/emission process for three common spectroscopic techniques. In ESCA, x-rays excite core level photoemission while in UPS ultraviolet light removes valence band electrons. After core holes are created, relaxation processes produce both fluorescent x-rays and Auger electrons; probability favors Auger transitions for low atomic numbers; only hydrogen cannot be detected. Details of bonding are seen in shifts in the position of photoelectron peaks,

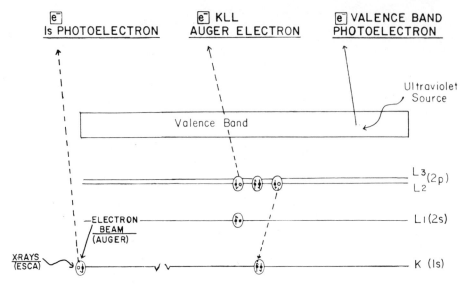

Figure 2. Simplified schematic diagram of the electronic energy
levels in any material. Superimposed are illustrations of energetic
beams (x-ray, electron, and ultraviolet) commonly used in three
types of electron spectroscopy (ESCA, Auger, and UPS).

and semiquantitative conclusions can be derived from peak intensities.

Electron excitation is used for Auger spectroscopy giving the
advantage of a beam that can be focused to very small lateral
dimensions to provide resolution of heterogeneities across the
sample surface. ESCA and UPS are limited to areas greater than
$\approx 5mm^2$, but have the advantage of providing less radiation damage
to substrates than Auger spectroscopy. During conventional Auger
spectroscopy, decomposition of adsorbed species, reduction of
surface oxides, diffusion to or from the irradiated area, and
electron induced desorption are possible complications.[3] These
effects are thought to arise from high electron flux or impurities
in the apparatus.

Illustrating the importance of multiple techniques, small
amounts of oxygen, fluorine, and chlorine have been found on the
sample after collection of a typical Auger spectrum.[4] These
contaminants were *not* found in the AES or the ESCA spectra, but
with the sensitivity of static SIMS. The build-up of these
contaminants was proportional to the decrease in the substrate's
signal obtained from *0-100 eV* by pulse counting of secondary
electrons with *low* total electron dose density.

Depth Profile. In contamination research it is important to

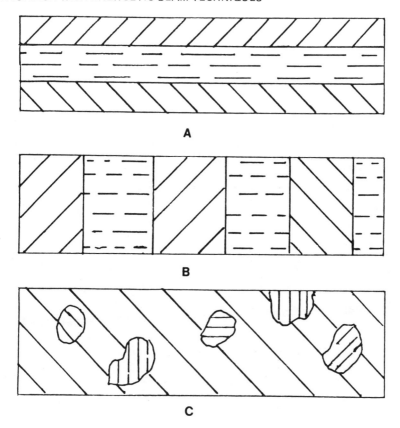

Figure 3.   Idealized cross-sections of three major types of samples
requiring analysis in contamination research.   A.   Laterally homogeneous but vertically inhomogeneous, B.   Laterally inhomogeneous
but vertically homogeneous, C.   Inhomogeneous in both dimensions.

determine the variation in composition and morphology with depth.
Figure 3 shows simplified cross-sections illustrating the classes
of samples encountered.   With ESCA one can determine trends with
depth (up to at most 10 or 20 atom layers) by changing the
photoelectron take-off angle relative to the analyzer, diagrammed
on the left in Figure 4.   Unfortunately, precise distinctions may
be difficult.   Angle-resolved photoemission of a class A sample
(carbon contaminating film on gold) determined the ratio of the
average contamination layer thickness to the mean free path of
photoelectrons, but found that the variable take-off geometry had
an error limit of at least ±50%.[5]   Ion bombardment can be used to
conduct deeper profile studies, although specimens may be distorted
from the true structure.[6]   As ion beams remove atom layers, the
sputtered products can be analyzed by mass spectrometry (SIMS).

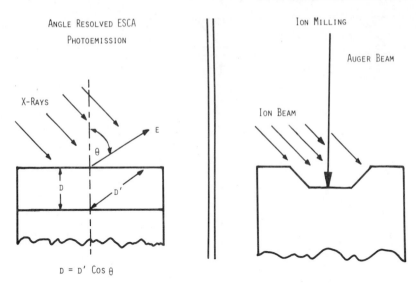

Figure 4.   Simplified schematic diagram illustrating two techniques
that provide in-depth study of heterogeneous samples.   Angle-
resolved photoemission (on the left) is non-destructive but limited
to the top few monolayers.   On the right, milling a crater in the
sample surface with an ion beam while analyzing the new surface
with Auger electron emission, can provide a chemical profile many
μm into the surface, but is destructive.

Details of the ion beam methods are covered elsewhere, but it
should be noted that ion scattering spectroscopy (ISS) has
sensitivity to a small fraction of the first atom layer and can
distinguish between isotopes.[7]

Other Surface-Sensitive Techniques.   Classical methods to
characterize surfaces, such as gas adsorption on finely divided
solids[8], heat of adsorption[9], contact angles[10], and zeta potential[11]
techniques are usually analyzed in terms of thermodynamic interactions between bulk homogeneous phases.   Fundamental intertreta-
tion of these measurements in terms of dispersion force and
acid-base interactions between phases at the interface[12] promises
to provide correlations with atomic properties obtained from
energetic beam techniques.[13]

Many new energetic beam techniques are developing for specific
purposes.   For example, surface - EXAFS (Extended X-ray Absorption
Fine Structures) promises to determine directly the bond lengths
and positions of atoms adsorbed on substrates.[14]   Also, Fourier
Transform technique and digital substraction promises to expand the
use of Infrared analysis of surfaces.[15]

Applications

Adhesion. Energetic beam techniques have been used in fractographic analysis of such diverse systems as Ta_2N-TiPdAu films on Al_2O_3 substrates[16] and epoxy adhesives on "Teflon" fluoro-polymer.[17] A unified theory of adhesion has been promoted which considers strength as a product of properties representing bulk mechanical behavior and interfacial bonding.[18] Qualitative analysis seems to indicate whether the latter is the weak link in adhesive failure. In the case of a structural epoxy adhesive, the initial locus of failure was found to be primarily cohesive through the adhesive, using SEM and electron probe microanalysis with a TiO_2 tracer in the adhesive. However, after water immersion, a complex locus of failure was found. The path of fracture occurred between the oxide surface layer and the epoxy adhesive, alternating onto the adhesive layer.[19] Carrying on further, Auger, ESCA, and SIMS were employed to ascertain the mechanism of durability conferred by silane-based primers. Fractures were carried out in the analysis chamber, and profiling more clearly delineated epoxy primer, oxide, or other surface layers. Environmental resistance correlated directly with $FeSiO^+$ radicals detected in SIMS.[20]

D. T. Clark and co-workers at Durham (U.K.) pioneered application of ESCA to structure and bonding studies in polymers.[21] A systematic investigation of homopolymers, combined with molecular orbital calculations confirmed the ability of ESCA to elucidate polymer-surface chemistry.[22] Critical to precise study of organic surface layers is the mean free path of photoelectrons, now carefully determined to be 1.3nm and 2.9nm at 969 eV and 1430 eV, respectively, for uniformly deposited films of poly-(p-xylyene).[23] This background was used to analyze a variety of surface effects in polymers, such as contamination from mold release agents, entrained emulsifiers, plasticisers, and catalysts. Further, chemical modification by oxidation and fluorination were defined in terms of kinetics and mechanisms as these reactions proceed into the top 3-5nm. Finally, these workers have explicated a variety of effects of inert and reactive gas plasmas upon polymers, including mechanisms of energy transfer, cross-linking, graft polymerization, and oxidative degradation.[24]

With more specific interest in adhesive bonding of polyethylene, studies showed a thick oxidized film as molded in air against aluminum foil. But if the film was peeled from the foil, essentially bulk failure occurred and the oxidized layer was confined to the top monolayer.[25] Composition data from ESCA on commercial polymers subjected to plasmas[26] or chemical etches[27,28] have provided new information to design surface treatments for adhesive bonding applications in particular.

The usefulness of modern energetic beam analysis to micro-

electronics processing was illustrated by Holloway in a review of
applications of ESCA, AES, ISS, and SIMS to substrate and substrate
processing, deposited films, patterning, interconnection, and
compatability.[29] Because over 50% of device failures are related
to interfacial phenomena, particular effort has been made to
understand cleaning techniques. Generally, it has been found that
plasmas, ultraviolet radiation, and ozone are more effective than
ion sputtering or chemical etches in removing carbon contaminants.[30,31] Quantitative AES studies indicated that about 0.5nm
was the maximum contaminant thickness allowable and still retain
practical bondability, although there was a marked dependence on
the type of bonding technique.[32]

     Troublesome inorganic contaminants, notably Ag and Na, are
well known to segregate to the surface from the bulk during
processing.[33] These effects can lead to device instability and
poor adhesion.[34] On the other hand, a beneficial effect was found
for the adhesion to Al_2O_3 when bulk impurities Ca and Si were
found to segregate to the surface.[16] In a similar system, the
unique properties of high energy ion backscattering were used to
monitor the migration of silicon through Au, Ag, and Al films at
temperatures less than half the eutectic temperature.[35] Bulk
properties such as ductility minima in Zircaloy seem to be related
to carbon segregation and metal carbide formation, as analyzed by
AES.[36]

     Surface cleaning studies on glass[37] and titanium alloy[38]
demonstrate that a redeposit of carbon contamination forms on these
surfaces in the ambient environment. Storage under ultraviolet
light will minimize the effect.

     Corrosion. Corrosion is closely related to adhesion; the
appearance of corrosion is indicative of loss of adhesion between
the paint and the underlying substrate. General corrosion mechanisms

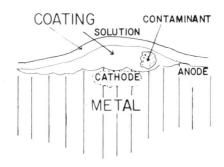

Figure 5. Schematic model of a general mechanism of corrosion.
Inorganic contaminants and moisture accelerate electrochemical
reactions that saponify the coating and corrode the substrate.

have been proposed to account for both anodic (oxidation) and
cathodic (reduction) reactions.[39] It has been shown that adhesion
failure under anodic conditions is due to displacement of the
primer by hydroxide ions electrochemically generated at the paint/
metal interface.[40] Resin composition markedly affected the reaction
rate. Recently, the use of photoelectron spectroscopy in corrosion
science has been reviewed, including applications to passive films,
solid state reactions, electro-chemistry, and aqueous corrosion,
concluding with examples of the concurrent use of ESCA and ion
beam etching.[41] Figure 5 presents a simplified model of the
general corrosion mechanism than can be inferred. Moisture
accelerates the electrochemical reactions, and inorganic
contaminants may draw moisture through the film as well as conduct
current and react with film or substrate. Recent ESCA studies on
the locus of failure in the loss of paint adhesion specifically
identified saponification of the polymer at the interface by
hydroxide ions generated in anodic corrosion.[42]

A thorough study of contamination and corrosion will require
not only multiple energetic beam techniques, but also *in situ*
reactions, to form identifiable derivatives. For example, Ion
profiling/Auger spectroscopy revealed the interactions between
sulfur dioxide and modified 440C stainless steel at different
temperatures.[43] SO_2 adsorbed dissociatively with equal quantities
of sulfur and oxygen on the surface, and the sulfur did not form a
metallic surface phase. Instead, the oxide layer thickness
increased the presence of SO_2, especially between 500° and 600°C,
where the oxide became thicker by a factor of seven.

The use of freshly cleaved *in situ* bulk samples as controls
for contamination and surface studies is important. This was
illustrated in the Auger analysis of E and S glass fibers.[44] In
S glass, magnesium and aluminum are concentrated on the fiber glass
surface, whereas in E glass, chlorine, silica, and aluminum show
surface segregation. In order to control progressive beam-induced
depletion of cations in soda lime-silica glass analysis, profiles
were obtained by making a linear series of analyses along the
surface of a sloping ramp etched into the glass in a separate ion
bombardment operation.[45]

Wear. The interactions between polymers and metals in touch-
and sliding-contact were elucidated by Field-Ion Microscopy (FIM)
and AES. Strong adhesion between the organics and all metals
(clean or oxidized) was observed.[46] Using field-ion microscopy
with increasing voltage, electron induced desorption in the
vicinity of 20keV indicated chemical bonding of polymer to the
metal surface. ESCA studies in friction, lubrication, and wear
are exemplified by a variety of experiments on wear conditions.[47]
A sulfide was formed at the expense of oxide under severe wear.
In mild wear scars, however, there was no evidence of sulfide or

mercaptide, but the oxide layer thickness doubled. Further, it
was determined that surface chemistry was a function of wear rate
rather than load.

We conclude this review with two illustrations using new data
from our continuing, fundamental studies on adhesion (including
synthesis, surface analysis, and mechanics). This work
characterized extensively the surfaces of both titanium 6-4
alloy[38,48,49] and "Teflon" fluoropolymers[17,27,50], using energetic
beam techniques (primarily SEM and ESCA). In all of the early
work, no information was collected relating to *sub*-surface
constituents. The following results emphasize the differences
that often exist between surface and sub-surface composition, and
thus the necessity of obtaining depth profile data.

EXPERIMENTAL

Materials

The fluoropolymer sample was supplied by NASA Lewis Research
Center. A piece of commercial "Teflon" FEP sheet stock was used as
a target for ion-beam-sputter modification studies.[51] We analyzed
both sides of the target: Side 1 represents a case of intensional
contamination, while Side 2 had been roughened by an argon beam
from an 8cm ion thruster. Also, we collected spectra from a thin
fluoropolymer film that was deposited on aluminum foil placed near
the target during sputtering.

The titanium 6-4 was obtained from NASA Langley Research
Center after grit blasting and degreasing. A variety of oxidative
surface treatments was   studied in detail[48]; a comparison between
two types: phosphate/fluoride and anodized illustrates the
importance of depth profiles.

Apparatus

A combined ESCA/Auger instrument (PHI Model 550, Physical
Electronics Industries, Inc., Eden Prairie, MN) with a double pass
Cylindrical Mirror Analyzer (CMA) described by Palmberg[52] was used.
Auger depth profiles were obtained with a coaxial electron gun and
simultaneous ion beam etching using 2 keV Ar^+ ions from a
differentially pumped PHI gun. A magnesium x-ray source (1254 eV)
operated at 10 kV and 60ma produced the ESCA results. Differentia-
tion of surface from sub-surface components was made with angle-
resolved spectra using Lapeyre's drum/aperature arrangement[53]
restricting the cone of acceptance of the CMA to normal or grazing
angles.

RESULTS AND DISCUSSION

Ion Sputtered Fluoropolymers

The grazing angle ESCA photoemission from both sides of the
"Teflon" FEP ion-sputtered target, shown in Figures 6 and 7, shows
a greater concentration of hydrocarbon material on the contaminated
side of the target. However, the ratio of hydrocarbon to fluorocarbon decreases at normal incidence, indicating a *thin* contaminating surface layer is present. The ratio of hydrocarbon:fluorocarbon
is quantified in Table I, which also shows fluorocarbon and
oxygen:oxidized-carbon ratios obtained from three fluoropolymer
specimens. The grazing incidence results are high in hydrocarbon
compared to either the normal or integral results, further
indicating the top surface concentration of the hydrocarbon
material.

The ion-etched FEP has a very rough surface structure,
Figure 8. ESCA composition results indicate hydrocarbon in only

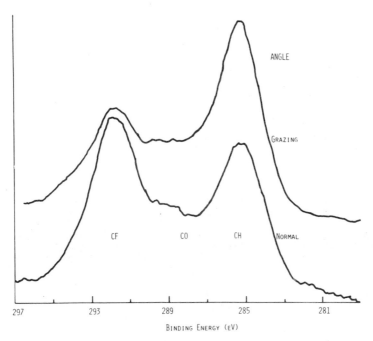

Figure 6. Angle-resolved ESCA photoemission from contaminated
"Teflon" FEP showing significant hydrocarbon in the surface and
sub-surface layers. Dominance of fluorocarbon signal at normal
incidence means the overlying contamination must be <3nm.

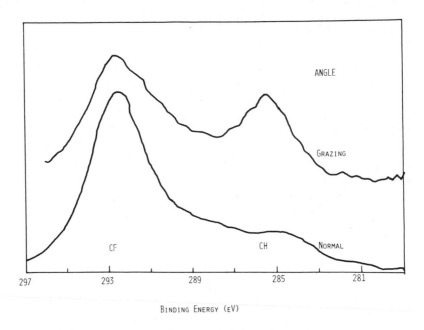

BINDING ENERGY (EV)

Figure 7. Angle-resolved ESCA photoemission from ion-etched
"Teflon" FEP. The top monolayer is partially hydrocarbon in nature,
whereas the subsurface appears entirely fluorocarbon.

Table I. ESCA Peak Intensity Ratios vs Take-off Angle

Sample	ESCA Take-off Angle	$\dfrac{F}{CF}$	$\dfrac{CH}{CF}$	$\dfrac{O}{CO}$
"Teflon" FEP target,	Integral	2.7	1.1	1.6
Side 1	Normal	2.9	0.9	1.8
contaminated	Grazing	2.6	2.0	1.2
"Teflon" FEP target,	Normal	2.6	0.2	–
Side 2	Grazing	4.3	0.7	–
ion-etched				
"Teflon" FEP,	Integral	2.6	–	–
sputter-deposit				

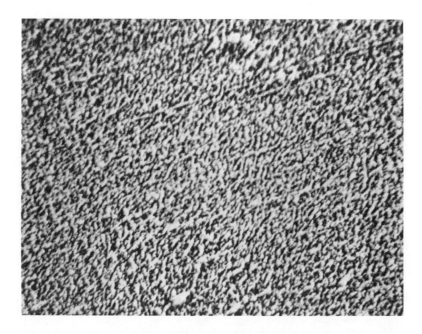

Figure 8. SEM photomicrograph (1000x) of ion-sputtered "Teflon" FEP, for which the ESCA spectra are shown in Figure 7. The surface is rough and pitted to an extreme. Asperities are less than 1μm in diameter and seem to project several micrometers above the surface.

the first atom layer or two (its signal vitually disappears in normal incidence). The hydrocarbon:fluorocarbon ratio increases by a factor of 3.5, and the fluorine:fluorocarbon ratio increases by a factor of 1.7 from normal to grazing incidence. Perhaps perfluoromethyl groups are packed at the surface with a small hydrocarbon fraction.

Surface-Treated Titanium 6-4 Alloy

Morphological and structural details of a variety of surface treatments on Ti 6-4 alloy were obtained by SEM/EDAX and ESCA.[48,49] Further elucidation of these surface structures is shown with AES and ion beam depth profiling in Figures 9, 10, and 11. Figure 9 presents the Auger spectrum obtained from the Ti anodized sample at the end of the depth profiling experiment. All of the constituents of the alloy and oxide surface are visible, including a significant carbon component that arises from the surface treatment and an argon component from implantation during milling. The

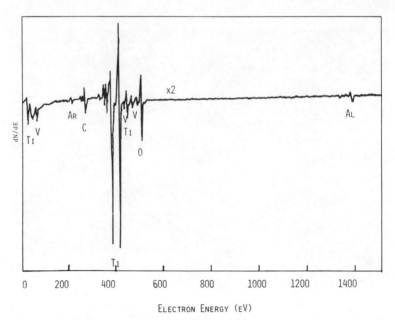

Figure 9. AES spectrum taken at the end of ion profiling through
the oxide layer on anodized titanium 6-4 alloy. The principal
alloying constituents appear, as well as argon implanted during
sputtering.

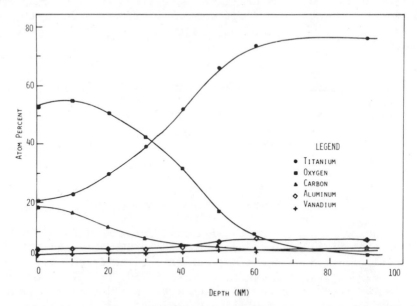

Figure 10. Simultaneous AES/ion-milling depth profile (calibrated
with Ta_2O_5 at 10nm/min.) of anodized titanium 6-4 alloy. An oxide
layer thickness of approximately 60nm and a significant carbon
percentage throughout the oxide layer are obvious.

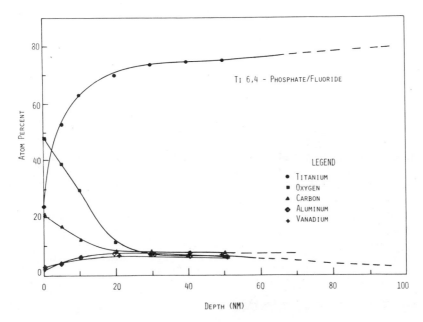

Figure 11.   Simultaneous AES/ion-milling depth profile of
phosphate/fluoride treated Ti 6-4 alloy.   A very thin oxide layer
is apparent, with carbon also penetrating the oxide layer.

contrast between Figures 9 and 10 illustrates that the anodized
Ti 6-4 has an oxide surface layer at least three times as thick as
the oxide layer produced in the phosphate-fluoride treatment.   It
is important to note from both profile experiments that a
significant carbon contamination exists throughout the oxide layer
and is not simply a surface contaminant caused by handling.

CONCLUSION

Modern surface analysis has promoted new understanding of
contamination effects in the related fields of adhesion, corrosion,
and wear.   Successful applications require an understanding of the
fundamentals of each energetic beam technique *and* their synergistic
combination.   Depth profile information is necessary to understand
heterogeneous specimens, as illustrated with grazing angle ESCA
photoemission on fluoropolymers and ion-sputter/Auger analysis of
oxidized titanium 6-4 alloys.

ACKNOWLEDGEMENTS

The support of the NSF Polymer Section (Grant DMR78-05429) and the National Aeronautics and Space Administration (Grants NSG-1124 and NSG-3204) is gratefully acknowledged.

REFERENCES

1. C.J. Powell, Applications of Surface Sci., $\underline{1}$, (2), 143 (1977).
2. R.H. Scott, P.B. DeGroot and J. Caron, Mater. Eval., p.45, Oct. 1977.
3. M. Gettings and J.P. Coad, AERE-Report #8288, Harwell Atomic Energy Reasearch Est., U.K.
4. L. Wiedmann, O. Ganschow and A. Benninghoven, J. Electron Spectrosc., $\underline{13}$, 243, (1978)
5. K. Persy and N. Gurker, J. Electron Spectrosc., $\underline{13}$, 91 (1978).
6. D.E. Williams and L.E. Davis, in "Characterization of Metal and Polymer Surfaces," L.H. Lee, Editor, Vol. 2, p.53, Academic, New York, 1977.
7. A. W. Czanderna, A. Miller, H. Jellinek, H. Kachi, J. Vac. Sci. Technol., $\underline{14}$, 227 (1977).
8. D. M. Young and A. D. Crowell, "Physical Absorption of Gases," Butterworths, London, 1962.
9. J. J. Chessick and A. C. Zettlemoyer, Adv. Catalysis, $\underline{XI}$, 263 (1959).
10. R. E. Johnson, Jr. and R. H. Dettre, in "Surface and Colloid Science," E. Matijevic, Editor, Vol. 2, p. 85, Wiley, New York, 1969.
11. P. C. Hiemenz, "Principles of Colloid and Surface Chemistry," p.453, Marcel Dekker, New York, 1977.
12. F. M. Fowkes and M. A. Mostafa, Ind. Chem. Prod. Res. Dev., 17, 3, (1978).
13. H. R. Anderson, F. M. Fowkes and F. H. Hielscher, J. Polym. Sci., Phys. Edition $\underline{14}$, 879 (1976).
14. P. A. Lee, Phys. Rev., $\underline{23}$, 29 (1978)
15. H. Ishida and J. L. Koenig, J. Colloid Interface Sci., $\underline{64}$, 555 (1978).
16. R. C. Sundahl, J. Vac. Sci., Technol., $\underline{9}$, 181 (1972).
17. D. W. Dwight, J. Colloid Interface Sci., $\underline{59}$, 447 (1977).
18. E. H. Andrews and A. J. Kinloch, J. Polym. Sci.: Symp. No. 46, 1 (1974).
19. R. A. Gledhill and A. J. Kinloch, J. Adhesion, $\underline{6}$, 315 (1974).
20. M. Gettings and A. J. Kinloch, J. Mater. Sci., $\underline{12}$, 2511 (1977).
21. D. T. Clark, in "Handbook of X-Ray and Ultraviolet Photoelectron Spectroscopy," D. Briggs, Editor, Heyden & Son, London, 1977.
22. D. T. Clark and H. R. Thomas, J. Polym. Sci., Chem. Edition $\underline{16}$, 791 (1978).

23. D. T. Clark and H. R. Thomas, J. Polym. Sci., Chem. Edition 15, 2843 (1977).
24. D. T. Clark, A. Dilks and H. R. Thomas, in "Developments in Polymer Degradation, Part I.," p. 87, N. Grassie, Editor, Applied Science Pub. Ltd., London, 1977.
25. D. Briggs and D. M. Brewis, J. Mater. Sci., 12, 2549 (1977).
26. H. Yasuda, H. C. Marsh, E. S. Brandt and C. N. Reilley, J. Polym. Sci., Chem. Edition 15, 991 (1977).
27. D. W. Dwight and W. M. Riggs, J. Colloid Interface Sci., 47, 650 (1974).
28. D. Briggs, D. M. Brewis and M. B. Konieczo, J. Mater. Sci. 11, 127, (1977)
29. P. H. Holloway, "Application of Surface Analysis for Electronic Devices," Pittsburgh Conference on Analytical Chemistry and Applied Spectroscopy, Cleveland, OH, May 12-16, 1978.
30. M. G. Yang, K. M. Koliwad and G. E. McGuire, J. Electrochem. Soc., 122, 675 (1975).
31. P. H. Holloway and D. M. Bushmire, in "Proc. 12th Ann. International Conf. Reliability Physics," p. 180, Las Vegas, 1974.
32. D. W. Bushmire and P. H. Holloway, in "Proc. 1975 International Micro-Electronics Symp.," p. 402. Avail. Dep. NTIS (SAND-75-5600).
33. F. J. Grunthaner, in "ARPA/NBS Workshop IV; Surface Analysis for Silicon Devices," NBS Spec. Publ. 400-23, p. 151, A. G. Lieberman, Editor, 1975.
34. H. L. Marcus, et al., J. Electrochem. Soc., 119, 1348 (1972).
35. A. Hiraki and E. Lugujjo, J. Vac. Sci. Technol., 9, 145 (1977).
36. G. J. Dooley, J. Vac. Sci. Technol., 9, 145 (1972).
37. R. R. Sowell, R. E. Cuthrell, D. M. Mattox and R. D. Bland, J. Vac. Sci. Technol., 11, 474 (1974).
38. T. A. Bush, M. E. Counts and J. P. Wightman, in "Adhesion Science and Technology, Part A." p. 365, L. H. Lee, Editor, Plenum Press, New York, 1975.
39. E. L. Koehler, Corrosion, 33, 209 (1977).
40. A. G. Smith and R. A. Dickie, Ind. Eng. Chem. Prod. Dev., 17, 42 (1978).
41. J. E. Castle, Surface Sci., 68, 583 (1977).
42. J. S. Hammond, J. W. Holubka and R. A. Dickie, American Chemical Soc. Preprints, 39, 506 (1978).
43. J. Ferrante, NASA Technical Note TN D-7933, Washington, D.C., (1975).
44. J. P. Rynd and S. K. Rastogi, Ceramic Bull. 53, (9), 631 (1974).
45. R. A. Chappell and C. T. H. Stoddart, J. Mater. Sci., 12, 2001 (1977).
46. D. H. Buckley and W. A. Brainard, in "Advances in Polymer Friction and Wear," Vol. 5A, p. 315, L. H. Lee, Editor, Plenum Press, N. Y. (1974).
47. D. R. Wheeler, Wear, 47, 243 (1978).

48. W. Chen, D. W. Dwight and J. P. Wightman, "Surface Analysis and Adhesive Bonding III. Titanium 6-4," Pittsburgh Conference an Analytical Chemistry and Applied Spectroscopy, Cleveland, OH, May 12-16, 1978.

49. W. Chen, D. W. Dwight, W. R. Kiang and J. P. Wightman, This proceedings volume, pp. 655-667.

50. D. W. Dwight, in "Characterization of Metal and Polymer Surfaces," L. H. Lee, Editor, Vol. 2, p. 313, Academic Press New York, 1977.

51. A. J. Weigand and B. A. Banks, J. Vac. Sci. Technol., $\underline{14}$, 326 (1977).

52. P. W. Palmberg, J. Electron Spectrosco., $\underline{5}$, 691 (1974).

53. G. L. Lepeyre, R. J. Smith and J. Anderson, J. Vac. Sci. Technol., $\underline{14}$, 384 (1977).

APPLICATIONS OF AUGER ELECTRON SPECTROSCOPY TO

CHARACTERIZE CONTAMINATION

P. A. Lindfors

Physical Electronics Division/Perkin-Elmer
6509 Flying Cloud Drive
Eden Prairie, Minnesota 55344

The fundamentals of Auger electron spectroscopy
(AES) are described. Several examples of the use of
AES to detect contamination are given.

BASIC PHENOMENA OF AUGER ELECTRON SPECTROSCOPY

Auger electron spectroscopy (AES) is a surface sensitive
analytical technique which can identify the elements[1] in the
first few atomic layers[2] on a specimen surface and can provide
estimates as to the concentrations[3] of these elements. The basic
phenomena involved in Auger electron spectroscopy are shown pic-
torially in Figure 1. The specimen to be analyzed is placed in
an ultrahigh vacuum chamber and bombarded with electrons (indicated
by the arrow coming from the right). If these incident electrons
have sufficient energy, they may ionize a core level of a target
atom. This ionization is represented by the vacancy shown at
position A. Conservation of energy requires that the electron
transition from C to A must be accompanied by emission of either
an Auger electron or an x-ray photon. If an Auger electron is
emitted as shown in Figure 1, it will have an energy equal to the
difference in energy levels involved in the transition to fill
the core level vacancy (C to A) less the energy required to raise
the electron from the "C" level to above the vacuum level (where
its kinetic energy could be measured). Because the electron
energy levels in all elements are discrete, Auger electrons have
energies characteristic of their parent atoms. Furthermore, the
Auger electron signals from the elements have been cataloged.[3]

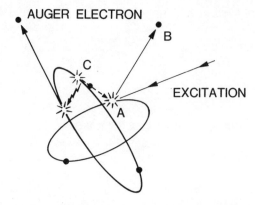

Figure 1.   Basic phenomena involved in Auger electron spectroscopy.

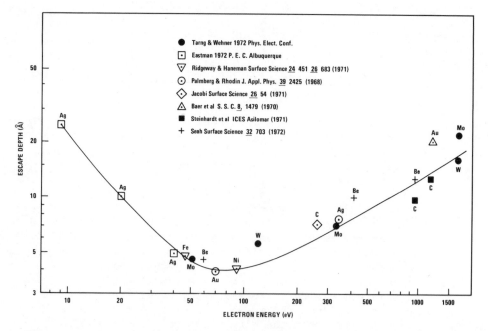

Figure 2.   Escape depth of electrons vs. the energy range normally of interest in Auger electron spectroscopy.

The number of Auger electrons emitted is proportional to the concentration of a particular elemental species in the volume probed. Thus the strength of the Auger electron signal can be used to quantify the AES data.[4]

The surface sensitivity of Auger electron spectroscopy is determined by the mean-free path length of the escaping Auger electron. In Figure 2 is shown a plot of the mean-free path lengths (escape depth) of electrons vs. the energy range normally of interest in Auger electron spectroscopy. Note that the mean-free paths are on the order of 4-20 Å, or a few atomic layers. Auger electrons are created deeper in the specimen, but before they are emitted into the vacuum where they could be measured, they lose sufficient energy such that they are lost in the back-ground. Maximum sensitivity is typically 0.1% atomic concentration.

EXAMPLES OF CHARACTERIZATION OF CONTAMINATION
USING AUGER ELECTRON SPECTROSCOPY (AES)

Example No. 1. Figure 3 shows an AES spectrum from a stained

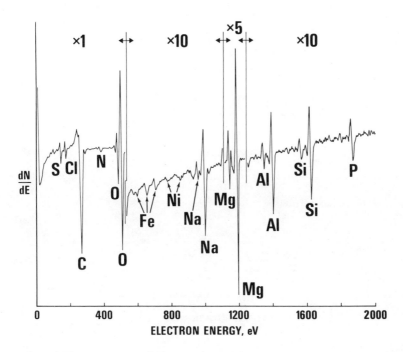

Figure 3.   AES spectrum from a stained area on magnesium alloy.

area on a magnesium alloy. This spectrum is typical, and it will
serve to illustrate that (1) the individual constituents of a
complex specimen can easily be identified, and (2) the traditional
mode of display of AES spectra is dN(E)/dE vs. E, where dN(E)/dE
is the first derivative of the energy distribution describing the
number of electrons emitted from a specimen as a function of their
energy. This derviative signal is derived electronically and the
technique for doing so is described elsewhere.[5]

Example No. 2. The next three examples will illustrate the
use of AES in thin film analysis. Electronically one can detect
the peak-to-peak amplitude of individual Auger electron peaks and
do so on an automatic and sequential basis. Simultaneously one
can probe into a specimen by ion sputtering (ion etching).[6] The
result is a compositional depth profile which is a plot of Auger
signal strength (peak-to-peak amplitude) vs. ion sputtering (ion
etching) time or depth. With the use of appropriate scale factors,
the Auger signal strengths can be converted to atomic percent.[3]
Compositional depth profiles are used extensively to characterize
layers of contamination on surfaces and at interfaces.

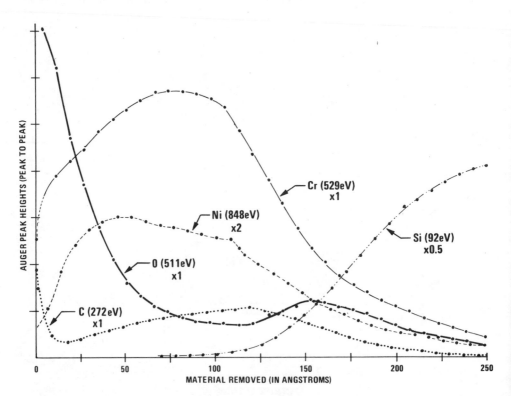

Figure 4. Compositional depth profile of a nichrome film deposited
on silicon (as evaporated).

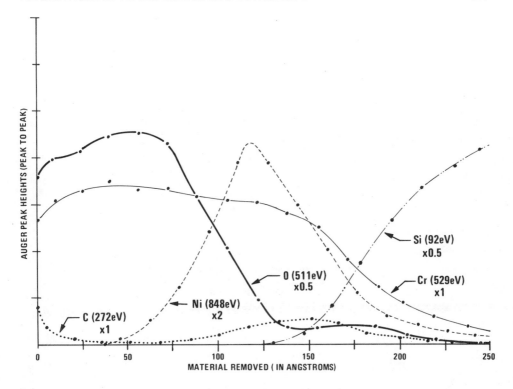

Figure 5. Compositional depth profile of a nichrome file deposited on silicon (after heating at 450°C in air for 30 seconds).

Figure 4 shows a compositional depth profile of a nichrome film deposited on silicon with a thin oxygen rich layer at the surface and some carbon contamination in the film. After heating at 450°C in air for 30 seconds (Figure 5), the film character changed dramatically and this compositional depth profile shows a chromium and oxygen rich layer on the surface with a nickel rich layer adjacent to the silicon ₒsubstrate. A chromium oxide layer on the order of 50-100 Å would probably not be visible and might only be suggested by how it might affect properties of this material in adhesion, electrical resistance, corrosion resistance, optical properties, etc. Not the scale in Angstrom units along the X-axis indicating the depth resolution that is possible.

Example No. 3. The next two examples illustrate detection of contamination at interfaces. Figure 6 shows a compositional depth profile of an alloyed, aluminum/silicon contact. Note that at the aluminum/silicon interface no signal from oxygen impurity was detected. However, in Figure 7 an oxygen signal was detected

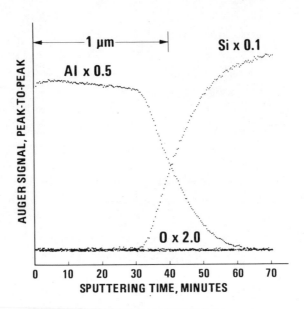

Figure 6.   Compositional depth profile of an alloyed, aluminum/
silicon contact.

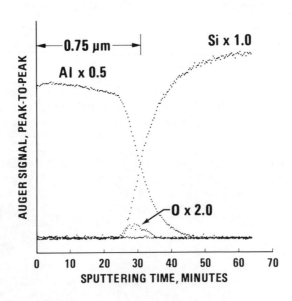

Figure 7.   Compositional depth profile of an alloyed, aluminum/
silicon contact.

at the aluminum/silicon interface, indicating that this second
specimen had been oxidized before the aluminum was deposited.

Example No. 4. Figure 8 illustrates the detection of the
diffusion of gold through a titanium-tungsten layer. The titanium-tungsten layer had been deposited to provide good bonding
between the silicon and gold because silicon and gold do not
readily adhere. Note the gold signal is high near the surface
where it was deposited and also at the titanium-tungsten/silicon
interface where its presence could be detrimental to adhesion.

Example No. 5. This example illustrates the scanning capability of AES and how it might be used to detect localized contamination. A scanning Auger microprobe (SAM) can raster the incident electron beam over a specimen surface and detect Auger
electrons, or secondary electrons, as a function of position.
Figure 9a shows a secondary electron image of an electrical contact where contamination had resulted in high contact resistance.
Figures 9b, c and d show scanning Auger (SAM) images for silicon,
gold and oxygen. White areas in Figures 9b, c and d indicate the
presence of high surface concentrations of a particular element.

Using the SAM technique, very small volumes can be probed for
contamination. SAM systems with incident electron beam diameters
≤2000 Å are commerically available, and the analysis depth is

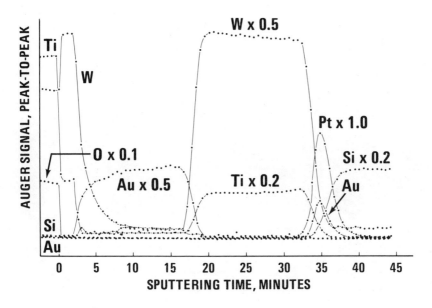

Figure 8. Ti, W, O, Au, Pt, Si multi-layer device.

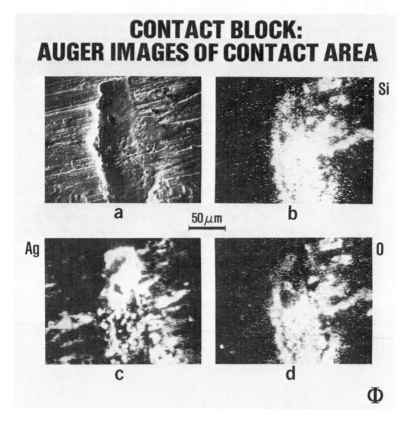

Figure 9. Detection of localized contamination.

limited by the escape depth of the Auger electrons. A combination
of SAM images and compositional depth profiles would produce a
three-dimensional characterization of a specimen.

REFERENCES

1. J. J. Lander, Phys. Rev., 91, 1382 (1953).
2. P.W. Palmberg and T. N. Rhodin, J. Appl. Phys., 39, 2425
    (1968).
3. L. E. Davis, N. C. McDonald, P. W. Palmberg, G. E. Riach and
    R. E. Weber, "Handbook of Auger Electron Spectroscopy", 2nd
    edition, Physical Electronics Industries, Inc., Eden Prairie,
    Minnesota, 1976.
4. R. E. Weber and W. T. Peria, J. Appl. Phys., 38, 4355 (1967).
5. L. A. Harris, J. Appl. Phys., 39, 1419 (1968).
6. P. W. Palmberg, J. Vac. Sci. Tech., 9, 160 (1972).

AUGER AND TEM STUDIES ON THE CONTAMINATION OF CHEMICALLY PREPARED GaAs SUBSTRATE SURFACES

J.S. Vermaak, L.W. Snyman and F.D. Auret

Department of Physics, University of Port
Elizabeth, Port Elizabeth, South Africa

Auger analysis and a gold decorating technique
in conjunction with transmission electron microscopy
(TEM) have been used to systematically determine the
origin, nature and influence of contaminants on (001)
GaAs surfaces which were prepared according to a
variety of chemical cleaning procedures. It was
found that the major contaminants are oxide and carbon. Occasionally Cℓ and K were also observed but
could easily be desorped in vacuum of 10^{-9} Torr at
temperatures higher than 220^{0}C. The oxide could also
be desorbed at temperatures higher than 500^{0}C whilst
the original carbon contamination could not be desorped even at temperatures as high as 600^{0}C. It was
found that carbon contamination as low as 10% (0,3
monolayer) can still influence the epitaxial growth
of Au on GaAs whilst an 0,9 carbon monolayer completely destroyed any epitaxial relationship between
substrate and overgrowth. A chemical cleaning procedure combined with a desorption of the oxide in a
vacuum of 10^{-9} at a temperature higher than 500^{0}C is
proposed to create a well defined stoichiometric
(001) GaAs surface with virtually no contaminants.

INTRODUCTION

The nature and properties of semiconductor surfaces play an
important role in the eventual properties of devices fabricated
from these materials. The reproducible preparation of stoichiometric surfaces that are essentially free from contaminants is
considered an essential first step in producing reliable devices

from compound semiconductor materials.[1,2,3]   Non-stoichiometric
surfaces as well as contaminants can deteriorate the interface
properties of the devices leading to undesirable mechanical, electrical and optical properties.[3]

Compared to the well established silicon technology where the
surface properties of not only elemental Si[4] but also its oxide
are well characterised, relatively little work has been done on
GaAs surfaces. Furthermore, the surface and interface properties
of a compound semiconductor such as GaAs are much more complicated
than that of an elemental semiconductor such as Si.[1,2] This follows because either element in the compound may be deficient to
varying degrees and each may have its own reaction potential with
regard to a particular reagent.[1,2] This can for instance result
in a large number of chemical and structural configurations of the
interface between GaAs and its oxide.[1]

In recent years GaAs is playing an increasingly important
role in the fabrication of a variety of electro-optical devices.
Most of these device structures are fabricated by the process of
either liquid phase epitaxy (LPE) or vapour phase epitaxy (VPE)
and occasionally molecular beam epitaxy (MBE) followed by a vacuum
deposition (VD) of a contact metal or alloy on the surfaces of the
structure. In all these processes a well defined clean and stoichiometric surface is of paramount importance.

Recently Chang et al[1] reported on an anodic oxidation-etch
method to produce well defined stoichiometric GaAs surface with
no detectable excess As or contaminants, except for about 0,5 Å
thick carbon layer. The procedure involves a thick anodic oxide
growth and NH_4OH etch to remove the oxide, followed by a light
anodic oxidation and NH_4OH etch for eliminating the excess As.

This paper deals with an Auger study and gold decorating
technique that were used to systematically determine the nature,
properties and sources of contaminants as well as the stoichiometric nature of GaAs surfaces that were chemically cleaned according to a variety of chemical processes commonly quoted in the
literature.[2,5] A chemical cleaning procedure followed by a vacuum
desorption process is proposed to produce a stoichiometric (001)
GaAs surface with virtually no contaminants.

EXPERIMENTAL   PROCEDURE

Substrate material

All the GaAs substrate material used in this study were obtained from M.C.P. Electronics Limited and consisted of bromine-
methanol polished (001) oriented Si-doped single crystals with a
carrier concentration of $1,7 \times 10^{18}$ carriers/cm^3.

Chemical treatment

The following general cleaning processes with variations indicated in Table I were used to chemically treat the (001) GaAs
surfaces.

Chemical polish. A 0,05% - 0,1% Bromine-Methanol (BM) solution was used to chemically polish the (001) surface in order to
remove any mechanical damage on the surface.

Degreasing. Boiling in hot trichloroethylene for 15 minutes.
Rinsing in hot isopropanol. Drying by evaporating hot isopropanol
from the surface. Thorough rinsing in deionized water ($\rho > 20$ MΩ).

Chemical etching. The surface of the sample was chemically
etched with various etchants as depicted in Table I.

Washing. The free etchant was gradually flooded away with
deionized water. The washing solution was decanted several times
to remove all traces of the etch solution from the surface of the
sample. Care was taken not to expose the surface directly to the
atmosphere and to keep the surface in contact with the water, thus
stabilizing the surface with a thin layer of natural oxide.

Rinsing and drying. The samples were subsequently thoroughly
rinsed in hot isopropanol which removed all traces of the H_2O when
heated slightly in the clean atmosphere of a flow hood.

The following types of surfaces (A to L) were prepared using
the specific chemical treatment as depicted in Table I.

After chemical treatment the substrates were cleaved in two
halves. One half was used to do an Auger analysis on while the
other half was decorated with Au in an ultra high vacuum system
for subsequent analysis in the transmission electron microscope.

Auger analysis

The one half of the GaAs substrate was used to do an Auger
analysis on. The Auger analysis was performed in a Varian LEED/
Auger system equipped with a cylindrical mirror analyser (CMA)
using a 3-keV electron beam. All measurements have been obtained
at normal incidence of the electron beam on the sample which was
positioned normal to the axis of the CMA. Semi-quantitative Auger
analyses were performed using peak to peak heights of Auger spectra obtained in the first derivative mode, yielding the concentration of element i as

$$C_i = (\alpha_i I_i) \left(\sum_j \alpha_j I_j \right)^{-1}$$

Table I. Chemical Preparation of Bromine-Methanol Polished (001) GaAs Substrates

SURFACE	DEGR.	ETCHING PROCEDURE	WASH.	RIN./DRY.
TYPE A	NO	NO	NO	NO
TYPE B	YES	NO	NO	NO
TYPE C[1,2]	YES	$NH_4OH:H_2O_2:H_2O$ (1:1:2) for 45 s	YES	YES
TYPE D[1,2]	YES	$H_2SO_4:H_2O_2:H_2O$ (3:1:1) for 45 s	YES	YES
TYPE E	YES	$HF:HNO_3:H_2O$ (1:3:2) for 45 s	YES	YES
TYPE F	YES	Bromine-Methanol (4:100) for 45 s	YES	YES
TYPE G	YES	$NH_4OH:H_2O_2$ (1:1:2) for 45 s followed by conc. HF stripping	NO	YES
TYPE H	YES	$NH_4OH:H_2O_2:H_2O$ (1:1:2) for 45 s followed by con. HCℓ stripping	NO	YES
TYPE I	YES	$NH_4OH:H_2O_2:H_2O$ (1:1:2) for 45 s followed by con. HCℓ stripping then KOH rinsing	NO	YES
TYPE J	YES	10 - 15% KOH ultrasonic treatment for 5 min.	YES	YES
TYPE K	YES	10 - 15% KOH ultrasonic treatment followed by $H_2SO_4:H_2O_2:H_2O$ (3:1:1) for 1-3 min.	YES	YES
TYPE L	YES	10 - 15% KOH ultrasonic treatment followed by $NH_4OH:H_2O_2:H_2O$ (1:1:2) for 1 min.	YES	YES

1. Chemical treatment of substrates were done in ultrasonically cleaned utensils in standard laboratory atmosphere.
2. Chemical treatment of substrates were done in quartz utensils cleaned with boiling solution of sulphuric acid and sodium dichomate (chromic acid) followed by rinsing with chemically pure KOH and prolonged washing in deionized H_2O ($\rho > 20$ MΩ). Ultrasonic treatment was used during each step while the whole treatment was done in the clean atmosphere of a flow hood.

where the sum is over one peak from each element detected on the surface. I_i and I_j respectively represent the peak to peak heights of elements i and j, while α_i and α_j are their respective inverse sensitivity factors as determined by Chang.[1] Table II depicts the elements monitored in this study, their Auger peak energies used for the analyses, as well as their inverse sensitivity factors normalised to the Si 92 eV. The mean free path λ also included, serves as an indication of the depth from which Auger electrons can be detected. Since this study was done as a function of temperature, and the temperature influences the density of states, which on its turn effects the intensity of the low energy Ga and As peaks (mostly MMV), only their high energy peaks were used.

Having calculated the concentrations C_i of the contaminants present on the surface, the Chang formalism[1], further enabled us to estimate their thicknesses. For a homogeneous surface layer of fractional concentration A ($=C_i$) and mean free path λ on a substrate, its thickness d may be obtained from

$$A = 1 - e^{-d/\lambda}$$

Vacuum deposition

It is known that gold which is vapour deposited on clean (001) and (110) GaAs surfaces grows epitaxially on them.[6,7] Furthermore, this epitaxial growth is very sensitive to any contaminants on the surface.[6] Thus the morphology and structure of gold on the GaAs surface provide a very sensitive technique to determine the presence of contaminants as well as their influence on the epitaxial growth process. The second half of the chemically treated substrate was, therefore, in conjunction with the first half used to determine the presence of contaminated layers and their influence on the epitaxial growth.

This half of the substrate was, after limited exposure to the atmosphere, loaded in a Varian VT 102 ultra-high vacuum system. The vacuum system was of all metal construction with no greased or lubricated joints. The high vacuum was obtained by use of a Vac-

Table II. Elements, Auger Energies and Inverse Sensitivity Factors

Element	Ga	As	Cℓ	K	C	O
Energy (eV)	1067	1228	168	252	273	510
Inverse Sensitivity Factor	3,8	5,3	1,0	1,17	3,0	2,4
Mean free path of inelastic scattering λ (Å)	18	20	12	8	7,5	5

Ion pump and a Titanium sublimation pump combined with three Vac-
Sorb roughing pumps that also contained no organic parts or fix-
tures. To monitor contamination[8] from the stainless steel wall of
the chamber the system was pumped down with and without the bakeout system. Under these conditions a vacuum of better than 10^{-8}
Torr was obtained in less than 2 h without bakeout whilst 10^{-9} -
10^{-10} Torr could be obtained with bakeout.

The samples were mounted on the same stainless steel oven and
could be outgassed before deposition of the Au. The gold was evaporated from a tungsten filament at a rate of 0,2 Å/s to a total
average thickness of 50 Å onto the (001) surfaces of the GaAs held
at 320^{0}C. The evaporation rate and thickness were controlled by
an Edwards quartz oscillating crystal monitor.

The metal films were carbon-backed from a built-in carbon
source to protect them as well as to avoid further contamination
after the specimens were removed from the vacuum.

Transmission electron microscopy

After deposition in the vacuum system the samples were chemi-
cally/mechanically polished from the substrate side to a thickness
of $\sim$ 50 μm by using a 0,5 - 1,0% bromine-methanol solution on a
PAN W polishing pad. Thereafter, the samples were further thinned,
again from the substrate side with a 0,2% bromine-methanol jet
etch until they were transparent for 100 kV electrons. This technique enables one to investigate the morphology, structure and
orientations of both the Au-nuclei still on the substrate as well
as those embedded in the C-overlayer in regions where the substrate
was completely etched away. Both transmission electron microscopy
(TEM) and selected area diffraction (SAD) microscopy were performed
on these samples using a Philips EM 300 electron microscope.

RESULTS AND DISCUSSION

The Auger and gold decorating results are summarized in
Table III. The bright field micrographs (BFM) and selected area
diffraction (SAD) patterns in the last column of Table III refer
to the micrographs in Figure 1.

The desorption of the contaminants as function of temperature
was determined by heating the substrate in a vacuum of 10^{-9} Torr
in the Auger system and continuously monitoring the contaminants
on the surfaces. The general desorption characteristics of the
common contaminants are depicted in Figure 2.

It is clear from the Auger results in Table III that the
dominant contaminants are carbon and oxides. Occasionally chlorine
and potassium were also observed on surfaces that were treated with

Table III. Auger and Gold Decorating Results on Chemically Prepared Surfaces as Described in Table I.

SURF.	AUGER RESULTS: % OF CONTAMINANTS IN SURFACE AREA					
	AT ROOM TEMPERATURE					GROWTH MORPHOLOGY AND STRUCTURE OF AU
TYPE	C	O	Cℓ	K	As/Ga	BFM AND SAD
A	20–30	30–40	–	–	1,23	fig. 1 (a)
B	10	49	–	–	0,75	–
C^1	30–40	15–25	–	–	1,5	–
C^2	15	30	–	–	1,43	–
D^1	10–18	20–30	–	–	1,5	–
D^2	12	27	–	–	1,59	–
E	20	19	–	–	1,59	–
F	20	36	–	–	0,82	–
G	39	7	–	–	1,88	fig. 1 (a)
H	19–37	6–16	2–5	–	1,38–1,88	–
I	6–12	20–40	–	–	1,12–1,38	–
J	0–3	47	–	10–14	0,77	–
K	0–3	22	–	–	1,63	–
L	0–2	28	–	–	1,62	–

HCℓ (type H) and KOH (type J), respectively. In all these cases,
however, the Cℓ and K could easily be removed by heating the substrate in vacuum to approximately 220°C (see Figure 2). This is
in agreement with the results of Uebbing.[9] From the desorption
studies depicted in Figure 2, it is further noted that the oxide
can be removed by heating the substrate to approximately 500 –
550°C for 1 hour in a vacuum at 10^{-9} Torr.

   In all the substrate preparations used in this work the GaAs

Table III (continued)

SURF.	AUGER RESULTS: % OF CONTAMINANTS IN SURFACE AREA					GROWTH MORPHOLOGY AND STRUCTURE OF AU
	OUTGASSING FOR 1 h AT 500^{0}C					
TYPE	C	O	Cℓ	K	As/Ga	BFM AND SAD
A	-	-	-	-	-	-
B	-	-	-	-	-	-
C^1	25–35	0–10	-	-	1,1–1,3	fig. 1 (a)
C^2	13	0	-	-	0,96	fig. 1 (b)
D^1	7–15	0	-	-	1–1,2	fig. 1 (b)
D^2	8	0	-	-	1,2	fig. 1 (b)
E	11	0	-	-	0,69	fig. 1 (b)
F	10	0	-	-	0,64	fig. 1 (b)
G	-	-	-	-	-	-
H	14–22	0–4	-	-	1,05–1,14	fig. 1 (a)
I	5–12	0–7	-	-	1,04	fig. 1 (b)
J	0–3	0	-	-	0,98	fig. 1 (c)
K	0–2	0	-	-	1,08	fig. 1 (c)
L	0–2	0	-	-	1,08	fig. 1 (c)

substrates were etched in acidic and basic aqueous-based systems. This chemical etching of the semiconductor surface involves an oxidation-state change and thus, the formation of oxides as inter-mediates, are to be expected.[1]

Chang et al[1] proposed a mechanism for the chemical action of these etches. Basically in these types of etches it is not the relative solubilities but rather the dissolution rates of the Ga and As oxides that play an important role. In this work all the

a   > 25 % C

b   10–15 % C

c   0 % C

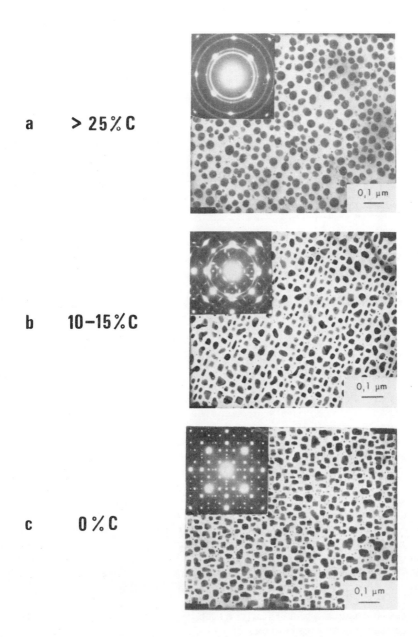

Figure 1. Typical bright field micrographs (BFM) and selected area diffraction patterns (SAD) of Au grown on (001) GaAs surfaces of varying carbon content.

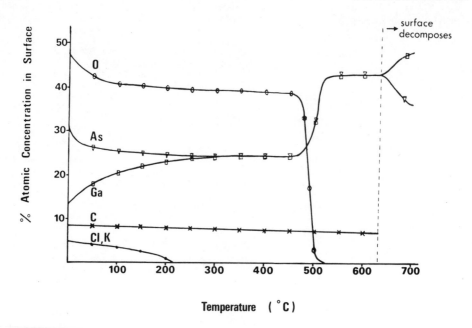

Figure 2.   Desorption of typical contaminants on an (001) GaAs surface.

samples were first Bromine-Methanol (BM) polished. The Br in the BM solution preferentially brominates Ga, thereby inducing migration of Ga to the surface. Upon rinsing, the $GaBr_3$-rich surface is converted to Ga_2O_3, with the subsequent loss of bromine containing species. Surfaces of the type B and F confirm the Ga-rich nature of the surface layer.[1]

The migration of Ga to the surface implies that elemental As is left behind near the interface region between the oxide layer and the unoxidized GaAs substrate. Our results (surfaces of the type C,D,E,G,H,I,K and L) are in agreement with this mechanism as in all these cases the acidic and basic etches dissolved the oxicized GaAs exposing the excess As at the interface.[1] The aqueous solutions then sequentially oxidize and dissolve the As in this region. The fact that all these surfaces (see Table III) have a As/Ga ratio bigger than 1 implies that either insufficient etch times to remove the existing excess As, or etch times that are too long, with the result that the solutions etch through the excess As to the GaAs substrate and in doing so preferentially oxidize and dissolve the Ga atoms leaving an As-rich surface.[1]

Although the Auger technique cannot determine exactly which and in what proportions the Ga and As oxides are present on the

surface, the Auger results in Table III and in Figure 2 would indicate that both are present and furthermore that the surface is
As-rich. The excess As may be due to pile up of elemental As at
the interface.[1] Upon heating the substrate in high vacuum this
excess As evaporates so that the As/Ga ratio tends to one at about
300°C implying equal amounts of As and Ga oxides in the surface
layer. From 300°C to 500°C this thin oxide layer seemed to be
stable but at temperatures higher than 500°C decomposes to leave
an oxide free stoichiometric GaAs surface.

Annealing experiments of GaAs surfaces with 2000 Å thick
amorphous oxide layers revealed[10] a surface oxide crystalline from
350°C upwards. In this work we also observed partial epitaxial
growth of gold on GaAs surfaces that were covered with approximately C_i = 40% oxygen at temperatures higher than 320°C. The epitaxial growth of the gold on the oxide layer implies that the oxide
layer in our case has also recrystalized with a definite epitaxial
allignment to the GaAs substrate. At temperatures lower than 320°C
polycrystalline growth of gold occurs indicating an amorphous or
polycrystalline intermediate layer.

The instability of the thin oxide layer provides an excellent
technique to produce clean GaAs surfaces provided the carbon contamination can be eliminated. From the Auger results in Table III,
it is clear that the C-contamination on all the substrates A to I
is very high. Furthermore, the desorption studies revealed that
the carbon contamination cannot be removed by heating to temperatures even as high as 600°C. The gold decorating results showed
in Figure 1, in conjunction with the Auger results depicted in
Table III, show that carbon contamination bigger than 25% ($\sim 0,9$
monolayers of carbon[1]) resulted in complete polycrystalline growth
(Figure 1 (a)), while carbon contamination as low as 10% (0,3 monolayers of carbon[1]) can still influence the epitaxial growth (Figure 1 (b)). With no C-contamination the gold grows in perfect
epitaxial allignment on the clean stoichiometric (001) GaAs surface. The epitaxial relationship is given by

$$(100)_{Au} \ || \ (100)_{GaAs} \quad \text{and}$$

$$<100>_{Au} \ || \ <100>_{GaAs}$$

in agreement with previous results.[6]

Comparing the results on substrate C^1 and C^2 as well as D^1
and D^2, it is clear that some of the C-contaminants may result
from the chemical solutions, the utensils or the atmosphere. Contamination from the atmosphere was checked by cleaning GaAs surfaces in the Auger system with Ar^+-ion sputtering whereupon atmospheric air was let into the system. On exposing these surfaces to

air at one atmosphere for 20 minutes no additional carbon contamination could be detected after the system was evacuated. The effect of the vacuum environment was also monitored. No carbon contaminants originated from it, not even during the bakeout cycle at $200^{\circ}C$. The cleanliness of the utensils and the chemical solutions are, therefore, the important sources for carbon contamination. The utensil cleaning procedure used for the type C^2 and D^2 surfaces, however, seems to eliminate carbon contamination from these sources and was, therefore, used in all the surface preparations E to L.

In agreement with Chang et al[1] it was also found that carbon contamination was enhanced on surfaces where the oxide layers have been stripped by HCℓ (type H surface) and HF (type G surface). They attribute this to the more reactive nature of As-rich surfaces.

Both the Auger and TEM results on surfaces of the type J, K and L reveal that carbon contamination can to a large extent be eliminated by using the chemical treatment as described in Table I for these surfaces. Furthermore, upon heating these substrates to $\sim 500^{\circ}C$ for 1 hour produce well defined stoichiometric surfaces with virtually no contaminants.

In conclusion, typical chemical "cleaning" procedures depicted in Table I result in carbon contamination of the surface which will influence and deteriorate the interface properties of devices fabricated by LPE, VPE, MBE and VD. As both LPE and VPE growth on GaAs takes place in a H_2-reducing atmosphere at temperatures well above $500^{\circ}C$, it is expected that the oxide layers will not produce any problems in these processes. However, in the metallization of the GaAs for device fabrication, care should be taken to remove the oxide by heating to above $500^{\circ}C$ for at least 1 hour before the metal or alloy evaporation commences at a lower temperature, normally at $350^{\circ}C$.

ACKNOWLEDGEMENT

The financial assistance of the C.S.I.R., Pretoria is gratefully appreciated.

REFERENCES

1. C.C. Chang, P.H. Citrin and B. Schwartz, J. Vac. Sci. Technol. 14, No. 4, 943 (1977).
2. Shiota, K. Motoya, T. Ohmi, N. Miyamato and J. Nishizawa, J. Electrochem. Soc. 124, No. 1, 155 (1977).
3. J.J. Uebbing, J. Appl. Phys. 41, No. 2, 802 (1970).
4. A.S. Grove, "Physics and Technology of Semiconductor Devices", Wiley, New York (1967).
5. B. Schwartz, Critical Reviews in Solid State Sciences 5, 609 (1975).
6. J.S. Vermaak, L.W. Snyman and F.D. Auret, J. Crystal Growth 42, 132 (1977).
7. K. Takeda, T. Hanawa and T. Shimojo, J. Appl. Phys., Suppl. 2, Pt. 1, 589 (1974).
8. T. Oda and T. Sugano, Japan J. Appl. Phys. 15, 1317 (1976).
9. J.J. Uebbing, J. Vac. Sci. Technol. 7, 81 (1970).
10. B.L. Weiss and H.L. Hartnagel, Electronics Lett. 12, 321 (1976).

SURFACE CHARACTERIZATION OF CONTAMINATION

ON ADHESIVE BONDING MATERIALS

J. S. Solomon, University of Dayton
Dayton, Ohio 45469
and
W. L. Baun, Air Force Materials Laboratory
Wright-Patterson AFB, Ohio 45433

The surface composition of adherend materials, as received, is usually entirely different from the bulk and generally unsuitable for adhesive bonding. Consequently, prebonding treatments of adherend surfaces are usually necessary to remove unwanted chemical species and to modify the surface chemistry and topography in order to produce a strong and durable bond. Auger electron spectroscopy, ion scattering spectroscopy, and seconday ion mass spectroscopy were used to characterize adhesive bonding materials as received and at various prebonding stages. The presence of contaminants or undesirable materials has been traced to manufacturing processes, prebonding treatments, and environment. In many instances the presence of contaminants or unwanted materials has been directly related to poor bond performance.

INTRODUCTION

The presence of unwanted chemical species within the interphase[1] region of an adhesive bond becomes an extremely important consideration in the overall performance of bonded materials. This region includes all the material from some point in the bulk adhesive toward and through the boundary between adhesive and adherend to a point in the adherend where the local properties are the same as the bulk properties. Contaminants or unwanted materials within this region, originating from the original adherend surfaces, adherend bulk, and bulk adhesive, can adversely affect bonding mechanisms (i.e. wetting and adhesion) as well as bond performance (i.e. strength and durability). This becomes a very

crucial problem when applying adhesive bonding to aircraft primary
structure fabrication. Table I lists the principle points at
which contaminants or unwanted chemical species can be introduced
into the component materials of adhesively bonded structures.

Initially, many impurities are introduced in the raw materials long before the actual use in adhesive bonding during manufacturing and processing steps. The next possible entry point for
contaminants can occur after the materials are received and prepared for bonding. Prebonding treatments generally are designed
to mechanically or chemically alter the original adherend surfaces.
The various chemical solutions (i.e. acid etches and degreasing
solvents) used in these steps certainly are potential sources for
contamination of the adherend surface. The problem is further
complicated when alloys are chemically treated since the alloy
constituents may be subject to segregation, selective chemical
attack, or smutting. Finally, the environment and handling before
and during the actual bonding can introduce contaminants or even
physically alter the adherend surfaces.

The problem in trying to establish some acceptable level of
impurities in bonding materials is twofold: first, the processing
steps are usually proprietary and therefore the exact chemical
content is not specified; and second, a chemical analysis may not
be useful since impurities such as highly movile sodium may diffuse to the surface during cure cycles resulting in localized high
concentrations. Contamination problems may be minimized by the
proper choice and control of materials and effective bonding pretreatments. However, the only guarantee that the proper choices
are being made is the establishment of a thorough surface characterization program. In our investigations of adhesive bonding
phenomena, surface analysis methods, such as Ion Scattering Spectrometry (ISS), Seconday Ion Mass Spectrometry (SIMS), and Auger

Table I.   Sources of Contaminants or Species Which Could Affect

           Adhesive Bonding and Bond Properties.
           _____

           1.   Raw Materials Processing
                a.   Adherend
                b.   Adhesive

           2.   Prebonding Treatments
                a.   Chemical solution contributions
                b.   Alloy constituents

           3.   Environment
                a.   Storage and handling
                b.   Bonding

Electron Spectrometry (AES) were used to chemically characterize the surface of materials which make up adhesively bonded structures. Not only are these techniques necessary for prebonding studies, but contribute immensely to the overall performance and long term durability studies of bond joints by providing chemical information from failed surfaces.

In this paper examples of surface characterization data from ISS, SIMS, and AES are presented which identify the chemical species introduced to the bonding materials at the various stages outlined in Table I. In many cases a correlation is made with surface contamination and bond performance.

RAW MATERIAL PROCESSING

Adherend

The basic adherend materials used for adhesively bonded structures in aircraft applications are 2024 Al, 7075 Al, 6Al-4V-Ti, and graphite fiber composites. The adhesives are thermosetting epoxies with a Nylon or Dacron mat carrier. Aluminum and titanium undergo many steps in their processing before finally being rolled into sheets of either the pure metal or various alloys. Table II lists some of the prominent impurites that either exist naturally in the original ore or which are introduced during processing, heat forming and fabrication, and heat treatment. In several areas of processing, such as the use of a flux in skimming of the molten bath, it appears that little thought is placed to the possible consequences to adhesive bonding by adding such materials as sodium, potassium, and chlorine into the melt.

Four aluminum alloys studies were 2024, 7075, 7050, and 6061. Their nominal compositions are listed in Table III. Auger spectra in Figures 1 and 2 show the surface compositions of the "as received" alloys to be totally different from the bulk compositions listed in Table III. In fact the surfaces appear to be a magnesium rich oxide which is probably the result of the heat treatment step during final processing. The in-depth elemental profiles, represented in Figure 2 by the Auger dN/dE signal strength as a function of ion sputtering time, show that this oxide layer on 6061 aluminum to be approximately 1000Å in thickness and the magnesium content decreases with oxide thickness. Similar results were found with "as received" sheets of the other three alloys.

In addition to the surface oxide, segregated particles which are rich in the alloy constituents can be found in the bulk as a result of quenching after heat treatment. Figure 3 contains electron micrographs with superimposed X-ray line scans of copper, iron, manganese, and aluminum showing these alloy constituent rich precipitates in 2024 aluminum. These inclusions could be potential

Table II.   Introduction of Impurities in or on Aluminum.

Process	Common Impurity	Mechanism
Melting	Si ore Fe impurity	$Al + SiO_2$ $Al + Fe_2O_3$
Skimming	Na, K, Cl	Flux
Degassing	Cl, Mg, Zn	Flux
Remove Inclusion	Na, Cl, K	Flux
Forming	C	Lubrication
Fabrication	C, S, Cl (soaps, oils, greases)	Lubrication
Heat Treating (Solution heat treatment, 950– 1000°F	KNO_3, $KNO_3/NaNO_3$ $K_2Cr_2O_7$ Silica Sand	Salt Bath Salt Bath Corrosion Inhibitor Fluidized Beds
Precip. heat treats 3.0–500°F	H_2O, O, N, CO_2, CO, SO_2, SO_3, NH, H_2 Mg and others (Diffusion of Alloying Elements through Al clad)	Furnace Atmosphere Excessive Temp

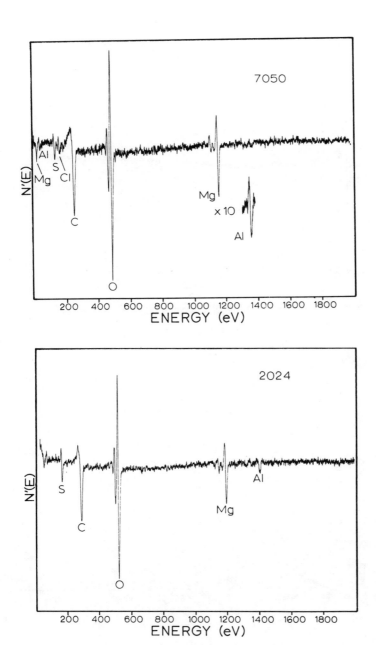

Figure 1.   Auger electron spectra from "as received" 7050 Al and
2024 Al alloy sheets.

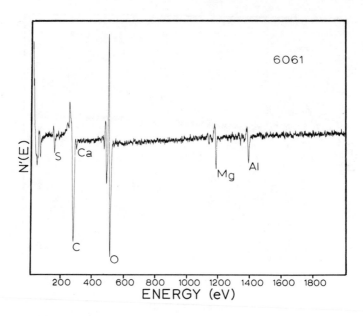

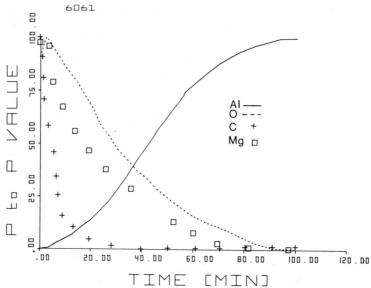

Figure 2. Auger electron spectrum and elemental in-depth elemental profiles for Al, O, C, and Mg from an "as received" 6061 Al alloy sheet.

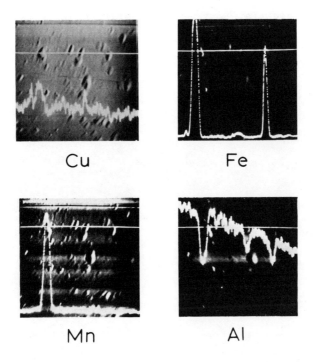

Figure 3. Backscattered electron images at 360X magnification from a polished 2024 Al sheet with superimposed X-ray line scans of Cu, Fe, Mn, and Al.

Table III. Nominal Chemical Composition of Aluminum Alloys[3].

Alloy	Si	Cu	Mn	Mg	Cr	Zn	Zr
2024	---	4.5	0.6	1.5	---	---	----
7050	---	2.3	---	2.25	---	6.2	0.12
7075	---	1.6	---	2.5	0.3	5.6	----
6061	0.6	0.25	---	1.0	0.2	---	----

Table IV.   Potential Sources of Impurities in Epoxy Resins.

Process	Common Impurities	Source
Catalysis	Na, K	Caustic Soda
Prevent emulsification during washing	Na, S	Na H5
Polymerization (special)	Li	Li OH
Intermediate Production	B, F	Lewis acid BF_3
Dehydrohalogenation	Na, Aℓ	Sodium aluminate
Synthesis of ester	Na, K, Cℓ	Na or K salt of acid or acid chloride
Epoxidization	Cℓ	H Cl O C_4 H_9 Cl O

sites for bond failure, especially if they are exposed to a high
humidity environment at which corrosion could be inititated.   They
may also inhibit the effect of pretreatments such as anodization.

Adhesive

The processing of adhesives also is a multistep process in
which a number of impurities, mainly alkali halides, are introduced
as contaminants.   Table IV lists some of the processing steps for
epoxy adhesives and the impurites which can be introduced at specific steps during processing.

Although the epoxies are purified, residual amounts of NaOH
and NaCl remain in the product.   Analysis of four typical formulations (Table V) show that small amounts, from 5-70 parts per

Table V.   Impurity Analysis of Commercial Adhesives (ppm).

	Na	Cl (ionic)	Cl (total)
EA 9628	5	---	1700
FM123	20	---	500
AF55	70	140	680
PL729-3	50	---	---

million (ppm), remain. While the effect of these hydroscopic impurities is unknown, Briggs, et al have shown that small amounts of hydroscopic impurities can drastically affect the water absorption characteristics of elastomers.[5]

PREBONDING TREATMENTS

Metal Alloy Adherends

Bonding "as received" materials such as the aluminum alloys nearly always results in poorly bonded weak structures. Consequently, some type of surface preparation is needed to make the adherend material more bondable. Usually the first step is to clean the surface, but not necessarily make it atomically clean. In some cases, the presence of foreign "atoms" may improve adhesion. Therefore, cleaning a surface for adhesive bonding means the modification of the surface region in a desirable way.[6] Care must be taken using processes that include numerous steps that one step does not negate a previous step, for example, the final step of an aluminum alloy treatment may include preparation of a well defined oxide layer by anodization or other chemical methods. If this surface were then exposed to a long hot water wash, the oxide character would be changed. Consequently, it becomes extremely important that the surface be characterized after each step.

In dealing with aluminum alloys, the first step for bonding preparation is to remove the original magnesium rich oxide surface layer. This can be accomplished by mechanical or chemical methods or both. Listed in Table VI are some of the commercial and non-commercial chemical treatments investigated. These treatments include a simple degreasing to highly reactive alkaline and acid etching. Figure 4 contains Auger spectra along with corresponding in-depth profiles from three chemically treated 6061 alloy specimens. The first was treated with NaOH followed by HNO_3-HF; the second was treated with NH_4HF_2; and the third was treated with the commercial etch solution alodine, a conversion coating mixture of phosphoric acid, chromic acid, and NaF. In each case, the different effects on the surface composition in terms of the alloy constituents and chemical solutions are evident. The corresponding in-depth elemental profiles show the effect of the three chemical treatments on the oxide thickness. The alkaline-acid treatment left the thinnest oxide layer while the bifluoride and chromate conversion coating treatments' net result in terms of oxide removal was negligible.

Similar Auger data was obtained from the other three pretreated alloys. When combined with ISS and SIMS data from the same surfaces a very informative matrix of surface elemental information can be constructed. Table VII is an example of one such

Table VI.   Surface Chemical Treatments for Aluminum Alloys.

ID Number	Description	Treatment
1.	Degrease	Metal sample slurried in acetone, wiped dry, then ultrasonically cleaned in carbon tetrachloride for 5 minutes.
2.	Alkaline	Metal sample submerged in 0.1N sodium hydroxide, room temperature for 2 minutes.
3.	ALK + HNO_3/HF	Treatment #1 followed, then sample submerged in a solution of 170 ml nitric acid, 30 ml hydrofluoric acid, 800 ml distilled water, room temperature for 2 minutes.
4.	HNO_3/HF	Acid solution.
	ALK + FPL	Treatment #1 followed, then sample submerged in a solution of 30 g sodium dichromate, 300 g sulfuric acid, distilled water to make one liter, 50° C for 5 minutes.
5.	FPL	Dichromate solution as described in #4.
6.	NH_4HF_2	Metal sample submerged in solution of ammonium bifluoride (10 g/liter) room temperature for 2 minutes.
7.	Chromate Conversion Coating	Metal sample submerged in solution of 65 g phosphoric acid, 10 g chromic acid, 5 g sodium fluoride, distilled water to one liter, 50°C for 2 minutes.

matrix showing a simple qualitative comparison between the chemical
treatments on 7075 aluminum. The first four columns reflect the
elements present in the bulk alloy and the last three are the
elements found in the various chemical solutions.

Another way to de-oxidize a surface is to use some type of
abrasive such as abrasive paper or grit blast. The advantage of
abrasion in terms of adhesive bonding is the increase in surface

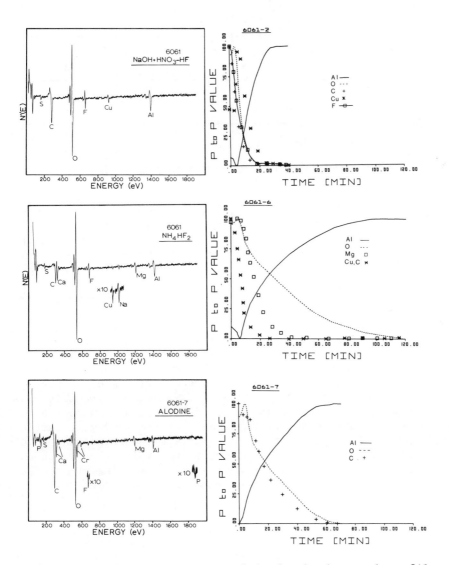

Figure 4. Auger electron spectra and in-depth elemental profiles
from chemically treated 6061 Al alloy sheets.

Table VII. Major element detected at the surface of 7075–T6 aluminum by ISS (I), SIMS (S) and AES (A).

Surface Treatment	Al	Cu	Mg	Zn	Cr	P	F
Degrease	I S A		I S A				
Alkaline	I S A	I S A	I S A	A			
ALK + HNO$_3$/HF	I S A	I S A					I S A
HNO$_3$/HF	I S A	I S A	S				I S A
ALK + FPL	I S A				S A	A	
FPL	I S A				S A	A	
NH$_4$HF$_2$	I S A		I S A				I S A
Alodine	I S A				I S A	A	

roughness and thus an increase in the bond strength simply because
of the increase in the number of sites for the "hook and latch"
effect between adherend and adhesive. Figure 5 contains two positive secondary ion mass spectra from the surfaces of an "as received" and grit blasted (sand) 6Al–4V–Ti alloy. The grit blasting
did remove some of the sodium but left behind an aluminum and
silicon rich surface due to the impact of the sand grit particles.
Their presence on the surface could offset the roughening advantages of grit blasting since they are possible sites for bond
failure, either the result of poor adhesive wetting or just the
fact that the loose grit particles do not adhere very well to the
alloy surface.

Some of these surface pretreatments were used for years to
condition aluminum and titanium alloys for bonding and the resulting bonds were generally good in terms of strength. In aircraft
applications strength as well as long term reliability and durability are important considerations. Consequently, new and better

surface preparations were investigated which would not only result
in stronger bonds, but also offer corrosion protection. As a
result anodization was introduced as a surface treatment since the
growth of the anodic oxide film could be precisely controlled and
the anodized aluminum alloys were more resistant to corrosion since
anodization leaves a surface oxide free of the alloying constituents
such as copper and magnesium if the proper anodization conditions
are chosen.

Again a matrix of surface characterization data was generated to show, for example, the effects of the various electrolytes
and anodization conditions (i.e. voltage, concentration and temperature) on the alloy surface. As in the case of the chemical
treatments, chemical species of the electrolytes are usually detected on anodic oxide surfaces. The Auger spectra from ammomium
chromate and phosphoric acid anodized aluminum shows the presence
of chromium and phosphorous from the respective electrolytes. The
concentration and depth distribution of these species appears to
vary with electrolyte and anodization conditions. Also, the

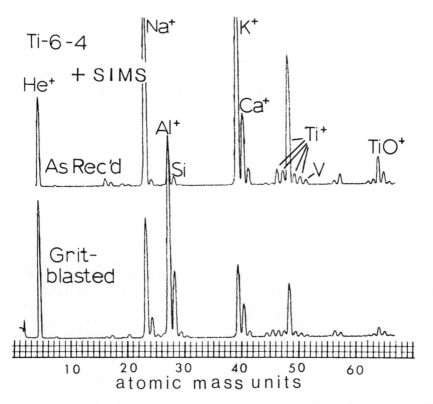

Figure 5. Positive secondary ion mass spectra from "as received"
and grit blasted 6Al-4V-Ti.

chemical states may differ on the surface compared with the bulk oxide.

Figure 6 contains three Auger spectra from three NH_4HF_2 pre-treated 2024 aluminum panels which were anodized at 10V for 15 seconds, 35 seconds, and 60 seconds in 1.0M H_3PO_4. An interesting observation in these spectra is the decreasing amounts of magnesium and phosphorous with anodization time. In-depth profiles of the respective oxide layers show that both magnesium and phorphorous are present only within a few monolayers of the surface and that the oxide thickness increases with time. The in-depth elemental profile of the 35 second anodized 2024 alloy in Figure 7 shows that copper, which is not detected on the surface, is present at the oxide metal interface at higher concentrations than in the bulk which may be the cause of observed defects (pits) on the anodized surface.

During chemical treatments or anodization, **alloys containing** copper, manganese, or silicon are highly susceptible to the formation of smut on their surfaces. The condition usually develops when certain alloys such as 2024 aluminum are etched in sodium hydroxide or other solutions. This is a significant problem in large commercial chemical treatments or anodization baths containing thousands of gallons of chemical solutions which cannot be economically replaced after each use, even though they contain significant amounts of dissolved impurities such as copper. Figure 8 contains SEM micrographs of a copper rich smut material formed on 2024 aluminum by electroless deposition in a chemical etching bath of hot sulfuric-chromic acid for a prolonged time. The micrograph on the right shows what the surface should look like if a shorter etching time is used. One method to remove the smut is to simply physically wipe it off when still wet. Although the wiped surface may appear to be clean, ISS spectra in Figure 9 from a smutted surface which was wiped, still shows excess copper remaining. Since the particles shown in the SEM micrographs are soft and can be smeared into the rough porous cavities, the presence of copper after wiping is not too surprising.

A similar smutting contamination is observed during anodization in which the electrolyte bath contains dissolved copper. The copper may be deposited on the cathode and change the electrical characteristics of the anodized bath. After the current is removed from the electrolyte bath, the copper may go back into solution. If the work piece is then held in the bath for a period without being removed, copper may form on the surface of the anodized film. This copper rich layer then may form a weak boundary layer when the material is adhesively bonded. Figure 10 shows evidence of this contamination effect from an anodizing bath. The cathode piece was aluminum while the anode was 2024 aluminum. The ISS spectra of the aluminum cathode in Figure 10 shows that copper had gone into solution and plated out on the cathode surface while the anodized 2024 aluminum surface was void of copper.

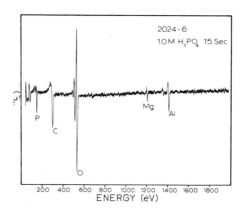

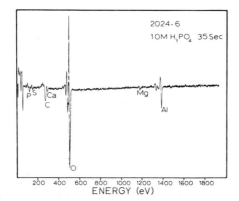

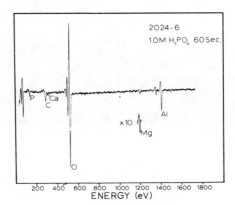

Figure 6. Auger electron spectra from 2024 Al sheets anodized in 1.0M H_3PO_4 for 15, 35, and 60 seconds.

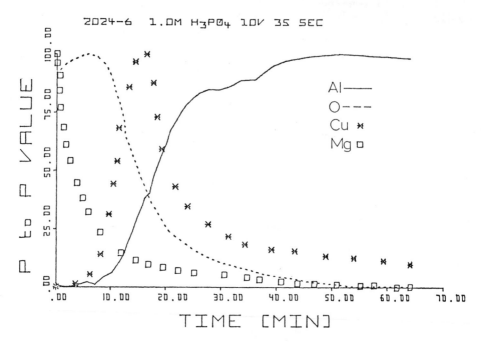

Figure 7. Auger in-depth element profiles for Al, O, Cu, and Mg
from an 1.0M H PO 10V 35 second anodized 2024 Al sheet.

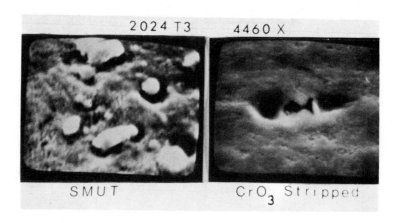

Figure 8. Scanning electron micrographs at 4460 X magnification
of smutted surface and properly etched surface of 2024Al etched
in hot chromic-sulfuric acid.

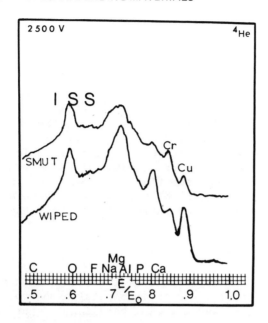

Figure 9. Ion scattering spectra from the smutted surface shown in Figure 8 and from the same surface after wiping.

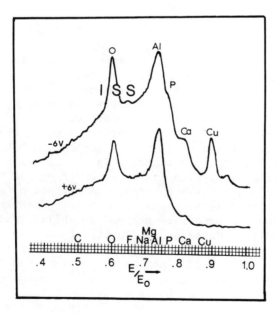

Figure 10. Ion scattering spectra from an Al cathode (-6V) and 2024Al anode (+6V).

Graphite Fibers

The problem of surface chemistry alterations resulting from prebonding treatments is not limited to metal adherends. To enhance adhesion between graphite fibers and a polymer matrix the fibers can also be subjected to pretreatments, such as oxidation. Unfortunately, as Drzal reported, an increase in the amount of material that can be desorbed with moderate thermal treatment, such as curing, can result.[7] Figure 11 contains the Auger spectra from an "as received" untreated fiber and one which underwent a manufacturer's proprietary treatment. The spectrum from the treated fiber shows that the fiber was oxidized. However, during the process it picked up S, N. Ca, and Na. These species have the potential to affect the interaction between the fiber and polymer matrix.

Environment

The third category of potential contamination sources for adhesive bonding materials is environmental exposure. This includes atmospheric exposure and material handling. Impurities chemisorbed or physisorbed, either into the bulk or on the surfaces, from the atmosphere, especially during heat treatments, may drastically influence the mechanical properties of metal adherends. Table VIII shows that the general effect on 2024 aluminum with exposure to SO_2 and water vapor (high humidity or steam) is the decrease in tensile strength and elongation.

After the adherend materials are received and prepared for bonding, handling becomes an important consideration since the panels are usually moved from pretreatment facilities to bonding facilities or stored until needed. Figure 12 shows Auger spectra from phosphoric acid anodized 7075 aluminum, a panel of the anodized 7075 wiped with a paper usually placed between anodized panels for storage, and an anodized panel wiped with a clean cotton glove. There are two noticeable changes in the surfaces of the wiped panels compared to the original. In both wiped panels there is a noticeable increase in the alloying elements magnesium and zinc along with the presence of carbon from residual oil in the paper and glove. An additional effect of wiping, which may not be classified as a contaminant, but nevertheless potentially deleterious to bond strength and durability, was the smearing of the fragile oxide layer, defeating one of the purposes of anodization, namely, to produce a porous oxide with a uniform known thickness. Figure 13 contains the oxygen in-depth concentration profiles of the original anodized 7075 aluminum and two wiped panels. The profiles show a decrease in the oxide film thickness as a result of the wiping. Bonded specimens of the wiped panels failed at the interface of the adherend and adhesive. To determine if the cause of failure was the carbon contamination

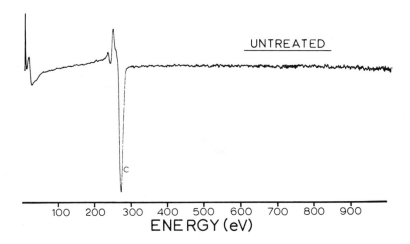

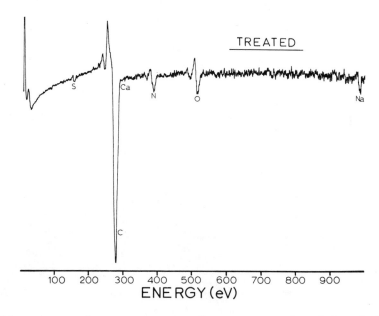

Figure 11. Auger electron spectra from untreated and treated graphite fiber.

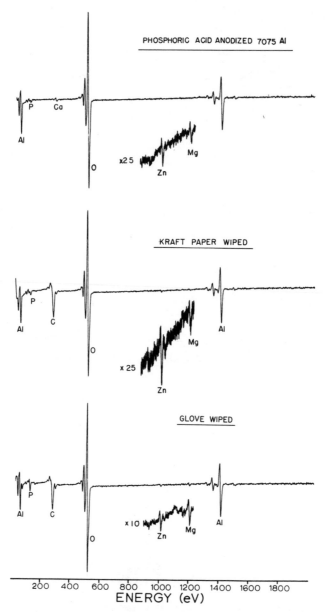

Figure 12.   Auger electron spectra from phosphoric acid anodized 7075 Al, an anodized 7075 panel wiped with Kraft paper and an anodized 7075 panel wiped with a cotton glove.

Table VIII.   Effect of impurities in heat treat atmospheres on mechanical properties of 2024-T4 aluminum alloy sheet.

	Tensile Strength % Decrease	Elongation % Decrease
Dry air containing 0.0002% sulfur dioxide (a)	3	27
Air containing 25% combustion products from natural gas(b)	5	75
Dry air containing 0.0012% sulfur trioxide(c)	7	35
Air containing 0.8% water vapor (40 F dew point) (a)	8	40
Dry air containing 0.007% sulfur dioxide (a)	15	68
Sulfur dioxide (a)	20	64
Air containing 3.4% water vapor (80 F dew point) (a)	25	77
Moist oxygen (a)	29	82
Ammonia (a)	29	82
Water vapor (100% saturated steam) (a)	60	95

(a) Heated at 920 to 930 F for 20 hr.

(b) Heated for 2 hr.

(c) Heated for 30 min.

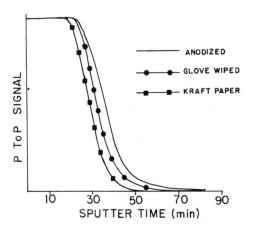

Figure 13.   Oxygen in-depth profiles from phosphoric acid anodized 7075 Al "as is", Kraft paper wiped, and glove wiped.

or smearing of the oxide, a panel was wiped with a piece of degreased aluminum foil and bonded. The resulting structure did
not fail as before which indicates the probable cause of failure
on the bonded glove wiped panels was contamination.

One of the tests for determining bond strength and durability of the adhesive is the wedge test in which a wedge is driven
into the bond joint creating an initial crack within the adhesive.
This is illustrated in Figure 14. The specimen is then exposed to
some environment such as high temperature and humidity and
monitored for crack growth rate and length.

The photograph on the right in Figure 15 shows the exposed
surfaces of an adhesively bonded structure that failed adhesively
during a wedge test in a SO_2 environment. Prior to bonding, a
piece of adhesive tape was placed on the oxide surface at Point A
and then removed. After bonding, this specimen and one which was
not exposed to the tape, were placed in a SO_2 environment to characterize the effects of a high SO_2 atmosphere on the bond using
the wedge test. The specimen that was not exposed to the tape
failed cohesively, while the tape exposed specimen failed cohesively to the tape exposed area where failure became 100% adhesive
and total delamination resulted. The Auger spectra to the left in
Figure 15 show chlorine present on the exposed surface of the adhesive which was in contact with the oxide surface. Analysis of
the adhesive tape shows it to contain lithium chloride. The exposure of this surface to the adhesive tape probably resulted in

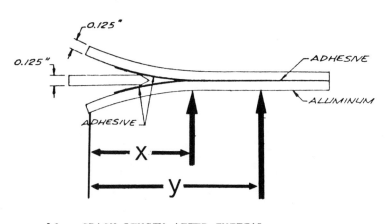

X -- CRACK LENGTH AFTER INITIAL
     WEDGE STRESS

Y -- CRACK LENGTH AFTER EXPOSURE

Figure 14. Illustration of the wedge test for adhesively bonded
structures.

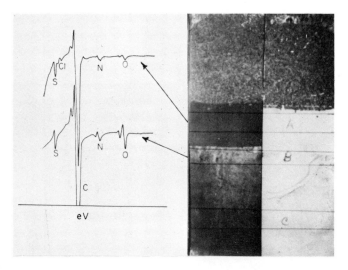

Figure 15. Adhesively bonded anodized 7075Al panels after wedge test with Auger electron spectra from the adhesive side of the failed structure at a point which was in contact with a surface previously exposed to adhesive tape (A) and a point not exposed to the Tape (B).

poor wetting by the bonding adhesive and thus poor bonding. Again, the implication of the results from this example is the potential effects of improper handling of the adherend, especially with bare hands or dirty gloves.

As previously stated, most adhesives used in aircraft applications are thermosetting epoxies with cure temperatures ranging from 250°F to 350°F. At these temperatures surface migration of certain bulk impurities which were introduced during processing was observed. Figure 16 contains positive ion SIMS spectra of an anodized 6Al-4V-Ti substrate at room temperature (156-1) and after it had been heated to 350°F, a typical cure temperature for bonding titanium. Heating has resulted in an increase in sodium and the alloying element aluminum, which probably occurs by grain boundary diffusion to the surface and is quite common in many alloys.

Also, as mentioned earlier, the adhesive can contain a number of impurities and at elevated cure temperatures these impurities can migrate to the interphase region. The positive ion SIMS spectra from a room temperature and 250°F cured epoxy in Figure 17 show a very large increase in sodium and potassium at the 250°F cure temperature. Analysis of such adherend and adhesive materials which were bonded and subsequently failed at high humidity and elevated temperatures indicate early crack propagation at the adhesive-

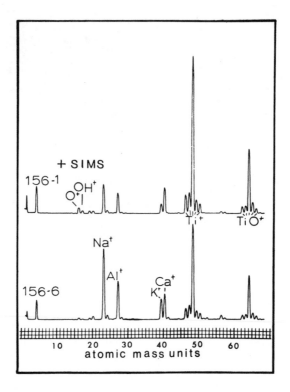

Figure 16.   Positive secondary ion mass spectra from 6Al–4V–Ti at
room temperature (156–1) and after being treated to 350°F (156–6).

oxide interface. Large amounts of impurity ions such as sodium and
potassium are usually present in this region. At elevated cure
temperatures the very mobile ions such as lithium, sodium, and
potassium become concentrated at the interphase. Then, if the conditions for bond failure under water attack are those for hydroly-
sis, the diffusion of alkali ions to the interphase region would
increase the osmotic potential and enhance the destructive ingress
of water at the interphase.

Finally, the bonding environment itself can be a source of
contamination. The large panels bonded for aircraft applications
are cured in autoclave chambers.[8] The specimens to be bonded are
usually covered with a sealed bag which is evacuated with a roughing pump. The backstreaming gases from the pump and the outgassing
components from the bag during the cure cycle are high in carbon,
chlorine, and sulfur. These contaminants can be introduced to the
interphase region of the bond joint at or near the structures'
edges.

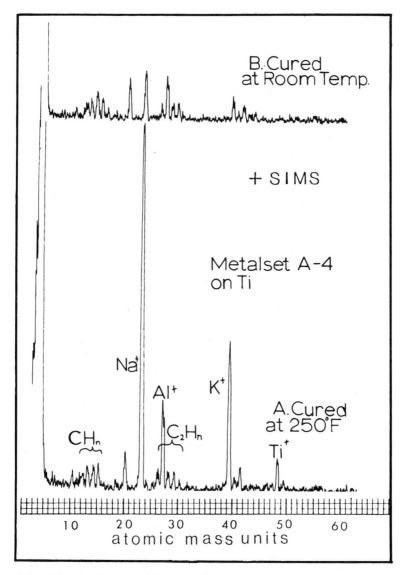

Figure 17.   Positive secondary ion mass spectra from room temperature cured and 250°F cured epoxy adhesive.

Conclusions

        In conclusion, the surface composition of candidate materials
for adhesively bonded structures could be significantly different
from the bulk due to contamination introduced during raw material
processing, bonding pretreatments, and environmental exposure.

Variations of composition and the appearance of certain elements
at the interphase can be all important in the bondability and long‛
term durability of adhesive bonds. A major research problem which
tends to keep adhesive bonding in the realm of a "black magic" art
instead of an exact science is the lack of mechanical and environmental tests which are sensitive to small variations in surface
chemistry. Consequently, further research is needed to design
tests whose statistical variations are small enough to show the
effects of controlled surface composition changes as well as the
presence of unexpected contamination.

References

1.   L. T. Drzal, "Summary of the Workshop Held on the Role
      of the Polymer Substrate Interphase in Structural Ad-
      hesion", Air Force Materials Laboratory Technical Re-
      port AFML-TR-77-129, July 1977.
2.   K. R. VanHarn, Editor "Aluminum, Fabrication and Finish-
      ing", Vol, 3, American Society for Metals, Metals Park,
      Ohio, 1967.
3.   "Adhesive Bonding Alcoa Aluminum", Aluminum Company of
      America, Pittsburgh, 1967.
4.   H. Lee and K. Neville, "Handbook of Epoxy Resins",
      McGraw-Hill Book Company, New York, 1967.
5.   G. L. Briggs, D. C. Edwards, and E.B. Story, Rubber
      Chem. Technol. $\underline{36}$, 621 (1963).
6.   D.M. Mattox, "Surface Cleaning in Thin Film Technology,"
      SAND 74-0344, Sandia Laboratories, Jan. 1975.
7.   L. T. Drzal, Carbon $\underline{15}$, 129 (1977).
8.   W. H. Gutmann, "Concise Guide to Structural Adhesives",
      Reinhold Publishing Company, New York, 37, 1961.

APPLICATION OF ISS/SIMS IN CHARACTERIZING THIN LAYERS (~10nm)

OF SURFACE CONTAMINANTS

G. R. Sparrow, Analytical Systems, 3M Co., St. Paul, Minn. and Steven R. Smith, Engineering and Design Center, Case Western Reserve University, Cleveland, Ohio.

During research, development and final production, a product is constantly subjected to numerous forms of contamination. Usually such contamination appears on the outer surface of a device and can dictate much of the performance and chemistry that subsequently occurs. Often even thin layers of contaminants less than 10Å can drastically alter desired results, especially in operations involving adhesion, paint coating, lubrication or catalysis.

Not only are designed experiments and systematic routine analytical techniques necessary to optimize product quality and manufacturing efficiency but also complete characterization of a surface contaminant and its origination must be made.

Modern technology makes routine surface characterization very applicable and practical to many types of products presently being manufactured. This paper discusses how surface contaminants can be systematically categorized, and what their sources may be. Practical application of Ion Scattering Spectroscopy (ISS) and Secondary Ion Mass Spectrometry (SIMS) to detection and characterization of contaminants is discussed for a wide range of different products.

INTRODUCTION

What is the difference between a good product and a bad
product? Not a trivial question! Unfortunately, poor product
performance is often correlated with surface contamination. Such
contamination can slowly increase in extent or frequency over
periods of months or years, resulting in gradual decrease in
product performance, or drastically increased manufacturing costs,
failures, and waste. Or, all too often it can complicate technical
development, thus shelving valuable technical ideas.

A rapidly growing concern is being directed at understanding
and controlling surface contamination. During the course of
complete evaluation and description of a material, the following
questions can be routinely asked:

Is the surface contaminated?

What is the chemical and physical nature of the surface?

Where are the different chemical components, especially
contaminants, located on the surface laterally and with
respect to depth.

How can the surface contaminants be controlled?

Obviously the control of surface integrity requires the
ability to chemically and physically characterize it. During the
past few years, several analytical tools have been developed to
provide detailed chemical information about a material surface.
Before describing these techniques it is necessary to define the
term surface. In general, a surface may be described as that
portion of a material that is most directly affected by or affects
the external space around it. Chemically it includes a diffuse,
weakly adsorbed layer of gaseous material, a thin layer of
chemisorbed atoms, and a reacted, treated, or contaminated layer
which is almost always different than the underlying bulk material.
Physically it can include microtopographic variations from a few Å
to hundreds of microns, defects, inclusions, and an infinite number
of texture features. Strictly speaking, it is difficult to define
the term "surface" on an absolute scale since the concept "surface"
is directly associated with the FUNCTION and desired PERFORMANCE of
a material. This can include as little as the outer monolayer of a
material such as in paint and adhesive coatings technology or can
extend several microns into a material such as in abrasive and
machining operations. A complete characterization of the surface
should include a coordinated investigation of both the physical and
chemical aspects. Classical techniques such as x-ray analysis,
spark source mass spectrometry, emission spectrometry, atomic
absorption, neutron activation, etc. are useful for

characterization of bulk impurities but generally provide essentially no real surface information. The physical characterization of a surface is most commonly done by scanning electron microscopy and is described elsewhere. [1,2]

CONTAMINATION CLASSIFICATION

The term "contamination" is often a diffuse, general description applied to hopefully suggest why some device has failed. However, a more systematic, objective approach requires specific answers to the questions:

1. What type of contaminant is occurring?

2. What is its source?

3. What are its effects?

In this text contaminants are classified according to several parameters including:

> Physical/chemical type
> Source
> Distribution
> Frequency of occurence

The type of contaminant may be organic, inorganic, trace, bulk, surface segregated, etc.; and its distribution may be uniform throughout the matrix, heterogeneously isolated in surface islands or in depth or it may be highly surface enriched. Its frequency of occurrence may be periodically correlated with processing time, material lot numbers, personnel activities or apparently random.

Sources of contamination appear to be categorized according to four basic stages of manufacturing:

I.   Contaminants Originating from Raw Material

II. Contaminants Resulting from Cleaning and, Preparation and Processing

III. Contaminants due to Distribution, Packaging, or Storage

IV. Contaminants due to Incorporation of Foreign Materials.

Modern analytical techniques can provide routine surface characterization at each stage of Research and Development as well as Processing and Manufacturing. This paper discusses the techniques available for such characterization and some of their applications to identification of surface contaminants for

solutions of problems. Ironically, contamination is not always
associated with poor product performance but may be an unknown
ingredient necessary for superior device performance. This implies
the general description of surface contamination: A material that
is not intentionally deposited on a surface or not necessarily
known to be present. Furthermore, it may be at a very trace level
or it may comprise the majority of the surface.

CONTAMINANT DETECTION AND IDENTIFICATION

Modern surface analytical instrumentation includes numerous
techniques among which AES, ESCA, ISS, and SIMS are by far the most
common. AES, (Auger electron spectroscopy), and ESCA, (Electron
Spectroscopy for Chemical Analysis) are techniques involving energy
analysis of electrons emitted from surfaces subjected to electron
and x-ray radiation respectively. They will not be discussed here
but generally include extensive surface information relating to the
outer 10 to 50Å of a specimen. Auger is especially useful where
high spatial resolution (5μm) is necessary and ESCA is
particularly useful for providing information concerning the
specific chemical bonding of surface atoms. Further discussion on
surface analysis methods can be found elsewhere.[3]

The specimen surface often dictates the entire chemical and
physical characteristics observed in a product. If the desired
surface is contaminated in any manner, even by only a single layer
or less of atoms, its properties can be drastically altered.
Complete chemical analysis of a specimen surface should provide the
following information:

> Identification of elements
> Concentration of elements
> Variation of concentration with surface depth
> Distribution across the surface
> Chemical nature of the bonding.

The dashed curves in Figure 1 illustrate how surface
information may relate to desired product performance. Curve A
illustrates the usefulness of surface information obtained from a
product or process in which generally only the outer few
monolayers may dictate performance. Note that surface information
obtained is of less and less value the deeper it is obtained from
the surface. This is especially true of many problems relating to
adhesion, paintability, lubrication, friction, fracturability,
photo properties, etc. In these cases it is essential to obtain
maximum information from the very outer surface layers.

Curve B illustrates usefulness of information obtained from
surfaces where the performance is not necessary dependent on just
the outer few monolayers but perhaps on tens or hundreds of

monolayers. This can occur in systems involving abrasives,
corrosion, reaction, physical and electrical properties and
appearance, etc. In these cases, analytical information
integrated over the first 50 or even several hundred Å may provide
sufficient information to be useful in solving a particular
problem.

Also illustrated on Figure 1 are shown the nominal expected
surface depths from which the analytical signal of each technique
originates. Note that x-ray analysis includes signals originating
over depths on the order of microns. Usually such information

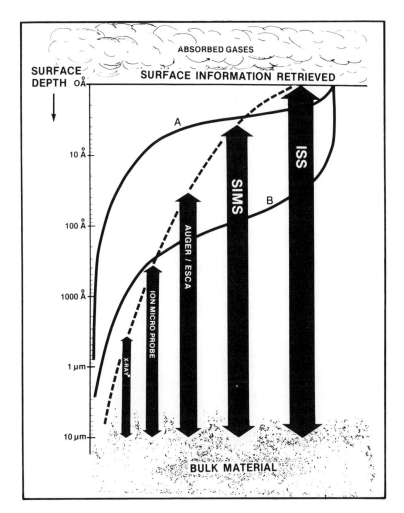

Figure 1. Amount of usefulness of surface information retrieved as
a function of depth from the surface. Relative comparison of
various techniques.

must be considered as due to bulk material and not due to real
surface character. At the extreme opposite are data obtained from
ISS which originate predominantly only from the very outer layer
of surface atoms. In this paper only ISS and SIMS are described.

ION SCATTERING SPECTROSCOPY (ISS)

ISS is relatively simple in principle and application. A
monoenergetic beam of positive ions such as He or Ne is directed
at the specimen surface. The low energy ions (0.5 to 5kV) are
scattered from atoms on the surface with an energy loss consistent
with a simple binary scattering event. The energy of ions
reflected back from the specimen at a particular angle is directly
related to the mass of the probe ion and the surface atom. Simply
recording the number of scattered ions vs their energy results in
a spectrum in which each surface element gives rise to a unique
peak.

Figure 2 illustrates an Ion Scattering Spectrometer analyzer
and a typical spectrum. In the most common commercial ISS system
the ion source is located coaxially within two concentric
cylinders which act as an energy analyzer of the ions scattered at
an angle of 138° from the specimen surface. Such a device
provides very high detection sensitivity, on the order of 10,000

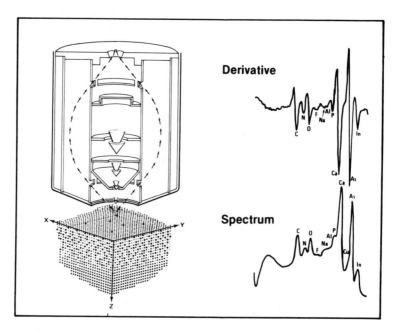

Figure 2. Ion Scattering System and a typical spectrum with its
derivative.

counts per second per nA of ion beam current. Although a single
scan of less than 1 second can routinely be used to identify most
major surface elements present, the data recorded in the following
studies were acquired by signal averaging several one second scans
on a data processor and subsequently storing them automatically on
magnetic tape.

ISS is especially sensitive to high mass elements such as
metals above Ti and the signal is directly related to the atomic
concentrations at the surface. It is essentially independent of
matrix effects thus permitting convenient direct representation of
approximate actual surface concentrations. Actual quantitative
analysis is accomplished by normalizing the measured peak
intensities using measured elemental sensitivities such as
reported by Erickson.[4] Detection limits on the order of a few ppm
can be realized for high mass elements such as As or Sb in a Si
matrix.

SECONDARY ION MASS SPECTROSCOPY (SIMS)

As the ion beam impinges the specimen surface, surface
material is sputtered away. Some of this material includes
positive and negative ions. Extracting these ions from the target
area using an appropriate energy filter and focussing them into a
mass spectrometer permits their mass analysis. Although the SIMS
process is thus the technique of analyzing the mass and intensity
of ions sputtered from a specimen surface, two distinctly
different types of instrumentation are employed. In the earliest
versions a very intense ion beam of high energy (10 to 15 kev) was
used in conjunction with a magnetic sector type mass spectrometer.
This resulted in very rapid sputter removal of surface material,
very high detection sensitivity, and normally provided information
from a surface depth of well over 200Å. This domain should more
approximately be termed Ion Microprobe Analysis.

The most commonly used technique today involves use of low
energy, (500 to 5000 eV) low density ion beams and quadrupole mass
spectrometers. The term SIMS used throughout this work will refer
to such low energy techniques.

Figure 3 illustrates a typical positive SIMS spectrum
obtained during analysis of a CdS/CuS solar cell material. Note
that all elements including hydrogen are directly detectable as
well as trace elements. Also commonly observed are clusters of
elements correlating with the chemical bonding nature of the
surface components. In specific cases, these fragments can lead
to a very positive identification of a specific molecular
structure.[5,6]

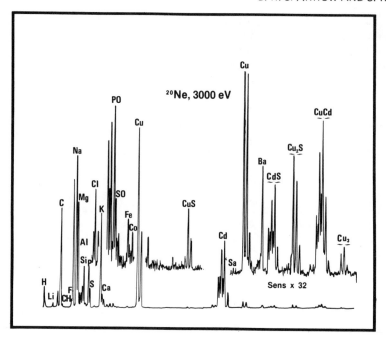

Figure 3. Typical SIMS spectrum.   CdS/CuS Solar Cell.

Unfortunately in SIMS the relative instrumental sensitivities to the elements can vary over a range of about 5 orders of magnitude. Furthermore, the ion yield (sensitivity) is very dependent upon the matrix in which a particle element is included. Thus actual surface concentrations may be very different than observed ion intensities in a given spectrum. Although techniques exist for obtaining semi-quantitative SIMS results,[7, 8, 9] they generally involve extensive reference materials and calculations or are restricted to specific types of samples.[10]

The latter work by Sparrow shows that: in general, the relative sensitivities in routine SIMS are highest for the most electropositive elements (lowest ionization potential) such as the alkalis, and lowest for the electronegative and noble elements. Reversing the ion energy filter electronics permits mass analysis of the negative ions. Electronegative elements such as C, O, and halides are especially amenable to analysis by negative SIMS. This also includes ion clusters which are stable anions such as CN^-, SO^-, etc. or anion-stabilized species such as oxides MO^-, where M is another element.

CHARACTERIZATION OF SURFACE CONTAMINANTS: APPLICATIONS

The following few examples of contamination detection and characterization are classified according to the stage of product development that contributed primarily to the problem.

Stage I   Raw material as received.

One of the most common production problems involves variation of incoming raw material. Unless routine, documented, incoming Q.C. methods are used, poor quality material may go unnoticed thus causing problems during manufacture or even with the ultimate consumer. In fact good knowledge of the actual contaminants and their surface concentrations on raw materials can lead to much more efficient and economical operations during processing and production.

The following three examples illustrate the minor surface changes that can drastically affect final results. In most cases the surface contaminants went essentially undetected by conventional analytical procedures.

Figure 4A illustrates both ISS and SIMS spectra for one of the most common types of raw material problems; the existence of high levels of impurities commonly associated with certain types of metal coils such as Al. In this particular problem certain lots of 1024 Al sheeting completely wore out tool and die parts used to fabricate assembly line parts. Although all lots contained essentially the same types of surface contaminants the defective coils contained very high levels of Na, Mg, Si, Ca, and F to much greater depths. In comparison, spectrum B, of the chemically cleaned material, was acceptable. Yet it was prohibitive to clean all coils. Routine Q.C. analysis of a few samples from each new coil would have helped eliminate extensive costs for tool and die replacements as well as production shutdowns.

Figure 5 illustrates how trace level impurities can affect material performance. The 1000 ppm of Pb observed in one lot of ZnO powder was the only difference detected among 4 similar materials. This particular lot of material exhibited inferior photo and electrical properties. Other analytical techniques showed no detectable differences.

Figure 6 illustrates substantial levels of Mo on the surface of silicon wafers destined for use in integrated circuits. The source of Mo was unexplained but subsequent to analysis it was discovered that some of the annealing ovens contained Mo rods.

Without much surface characterization, there are few actions that can be taken. Knowledge of raw material quality permits a

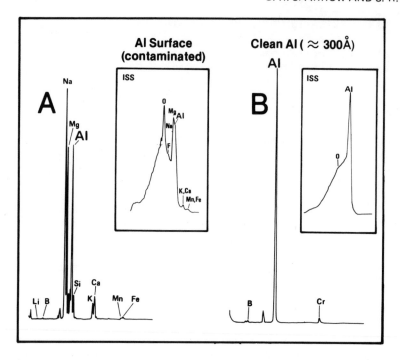

Figure 4. SIMS spectrum of abrasive, contaminated Al and chemically cleaned Al.

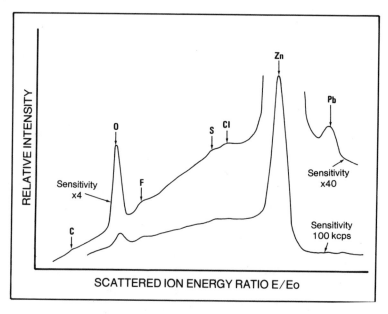

Figure 5. ISS spectrum of ZnO powder containing 10ppm Pb bulk.

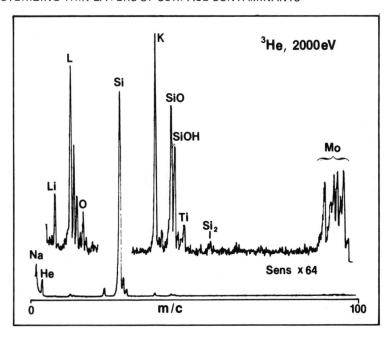

Figure 6. SIMS spectrum contaminated Si water.

laboratory to decide upon cleaning and processing techniques subsequent to using, rejecting, finding alternate material sources, or changing product specifications or uses.

Stage II   Cleaning and Processing

Proper cleaning and preparation of raw materials prior to use constitutes the most difficult and extensive aspect of many production problems. Without a good understanding of the chemical nature of the starting material, efficient and effective material preparation and cleaning procedures cannot be developed very rapidly. Certainly volumes can be written concerning techniques involving chemical preparation of materials. The following examples illustrate how ISS/SIMS can be applied in a systematic manner to directing the development of these techniques. Contamination specific to preparation of surfaces for bonding is discussed elsewhere.[11]

Figure 7 shows one expanded ISS spectrum and other ISS spectra obtained during a depth profile of the surface of Al sheeting chemically treated in preparation for fabrication of composite aircraft structural members. The source of each contaminant is attributable to the treatment or history of the Al. For example

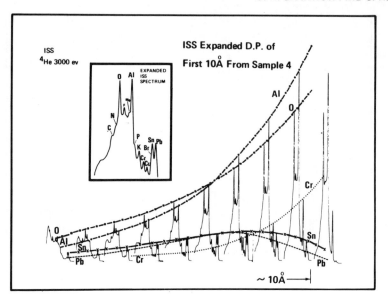

Figure 7. ISS spectrum of outer 5 Å of Al chemically cleaned for adhesion.

Mg, and F are common impurities of Al sheeting. C, and N are normally associated with environmental organic material or handling. The P, K, and Cr occur due to use of a $K_2 Cr_2 O_7$, PO_4 etching solution. The Cu is an alloying element within the Al. The Sn, Pb, and Br were unexpected contaminants found as trace impurities in the $K_2 Cr_2 O_7$ reagent.[11]

However, the elemental concentrations with respect to depth are also necessary in order to determine the thickness of this contaminated layer. Figure 8 shows the concentration of these elements within the outer 25 Å of the surface. The ultra high depth resolution (UHDR) analysis of surfaces can lead to very detailed understanding of how processing affects a surface. The insert illustrates how ISS analysis was also used to accurately measure the thickness of the anodized oxide film.

Often different relative concentrations of elements can drastically alter performance. Figure 9 illustrates the O to Si ratio for three SiO_2 wafers exposed to different conditions during growth of the oxide layer. Note that in the outer 80 Å O/Si ratios varying by less than 3% were readily monitored.

Figure 10 illustrates the ISS depth profile of a silica wafer found to be defective. A detailed UHDR ISS depth profile illustrates that the surface of this wafer is badly contaminated

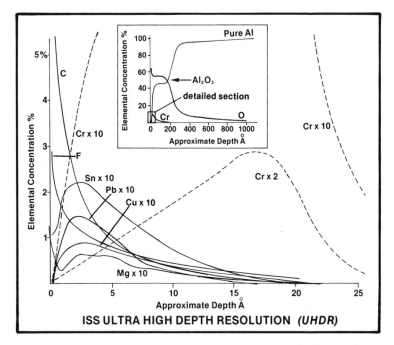

Figure 8. Concentration depth profile of treated Al surface illustrating detail of outer 25Å.

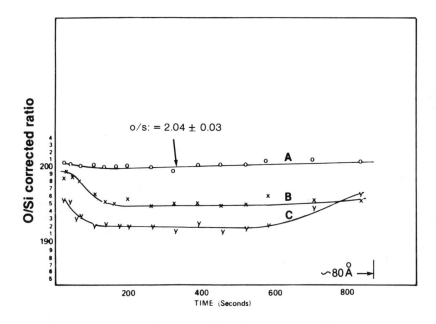

Figure 9. O/Si ratio obtained by ISS analysis of three amorphous SiO₂ films.

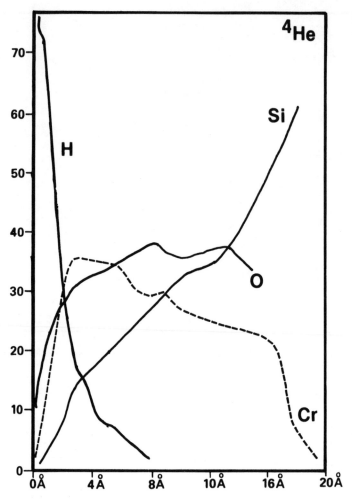

Figure 10. Ultra High Depth Resolution (UHDR) profile of Cr
contaminated Si water.

with a thin layer very rich in Cr and other impurities. In fact
the surface is predominantly Cr. Yet due to a deeper source of
signal origination (escape depth) other techniques such as ESCA
showed only 3% Cr due to dilution of spectral intensities by
sublayers of Si.

   One of the most overlooked key steps to an effective surface
cleaning or processing operation is the rinsing. Figure 11
compares two Al surfaces prepared in the same manner but in one
case curve B, there was a 60 second delay between the cleaning
solution and a distilled water rinse. Air oxidation or exposure

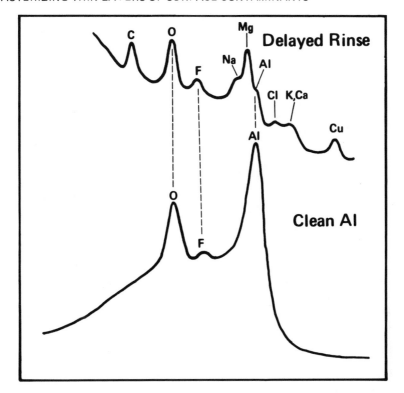

Figure 11. ISS spectra of two cleaned Al surfaces showing the effects of a 60 second delay in rinsing (B).

no doubt can contribute substantially to formation of nonsoluble oxide species which become segregated on the surface.

Figure 12 illustrates the results obtained from an extensive systematic research program directed at improving Al adhesion to Kapton, a polyimide. Note in the Figure the ISS spectra of the outer surface of each polymer A through D and its surface after sputtering away approximately 30 Å. Each surface represents a different cleaning procedure for the same polymer including some laboratory preparations and some production line products. Depth profiles throughout the first 25Å of each surface for 10 samples indicated a completely different chemical nature for all the samples.

Stage III Distribution and Handling.

Upon completion of rigorous chemical processing procedures, materials can be recontaminated through improper storage conditions, environment, poor containers, or simply by handling.

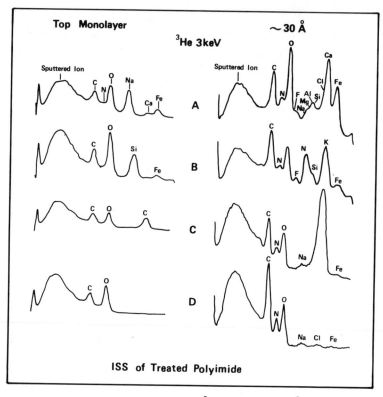

Figure 12. ISS spectra of the top 5Å and from 25Å deep for 4 samples of polyimide surface cleaned by different techniques.

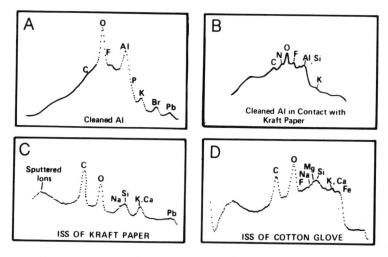

Figure 13. ISS spectra of clean Al surface and the same sample after contact with Kraft paper which resulted in loss of 30-40% adhesion strength.

The most critically affected surfaces involve those specifically
designed for maximum adhesive bonding. Even a single monolayer of
contaminant can tie up surface bonding sites thus drastically
reducing product adhesive strength. Figure 13 illustrates the ISS
spectra obtained from very clean Al, two common production
materials including kraft paper, and white cotton gloves and the
surface of the same clean Al merely placed in contact with the
Kraft paper. Spectral stripping techniques utilizing a digital
data processor very strongly indicates the contaminant deposited
on the cleaned Al was transfer of organic material from Kraft
paper to the Al. The cause of this was probably due to the fact
that after processing and cleaning the Al sheeting was placed on
Kraft paper. The result of such improper storage was the
reduction of bond strength by 30 to 40%.[11]

Stage IV Utilization

     A major problem relating to surface contamination is that of
corrosion. This may occur as actual removal of desirable surface
components or as chemical change such as oxidation. The
development and use of solar materials is an extremely important
area of interest in todays energy-conscious environment. Studies
of glass substrate surfaces as well as the coating material used
in solar cells is essential to development of this technology.
Figure 14 illustrates ISS analysis of several treated glass
surfaces. Note that most reagents, either acidic or basic
solutions, tend to preferentially remove the surface alkali
elements such as Na, K, Ca. This can alter subsequent adhesion of
coated films. Furthermore, the dissolved species can frequently
redeposit as surface contaminants on other materials in contact
with the same solution.

     Fracture of devices subjected to long term stress or
elevated temperature can result in severe product problems. Often
this type of fracture is related to a migration of bulk impurities
to grain boundaries within the material. As such it may be highly
concentrated at grain boundaries, as high as a full monolayer of

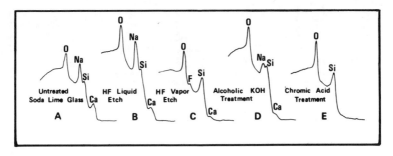

Figure 14. ISS spectra of treated glass surfaces.

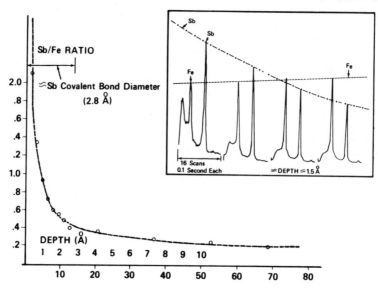

Figure 15. ISS analysis of grain boundary fracture in Sb doped steel.

material yet occurring at trace levels within the bulk material. Figure 15 illustrates the ISS surface analysis of a steel pin fractured along grain boundaries due to migration of Sb. The bulk concentration of Sb was 600 ppm.

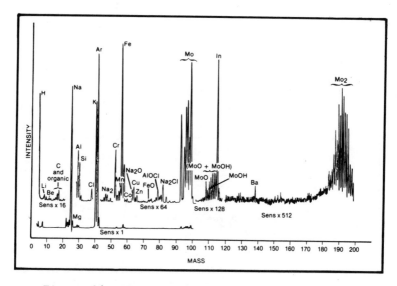

Figure 16. SIMS spectrum of contaminated Mo.

Stage V Foreign Material

At nearly any stage of production or utilization, material extraneous to the bulk composition of a device can be incorporated onto a surface. Figure 16 illustrates the content of such contamination for a pure Mo metal exposed to laboratory and environmental atmospheres for long periods of time. Note again that interpretation of such SIMS spectra must include considerations of relative SIMS sensitivities to the various elements. For example in that spectrum, the Na and K signals are very high yet their concentrations are very low due to the fact that these elements exhibit much higher relative sensitivities than Mo.

SUMMARY

The need for a well defined, systematic, routine program of surface analysis of product material throughout various stages of analysis is obvious. The techniques of ISS and SIMS are illustrated by practical applications to common problems arising during research, development, and manufacturing. The major features of ISS include its ultrahigh surface sensitivity to the outer atomic monolayer, its independence from sample matrix effects, its routine use for insulating materials, and its precision for successive measurements. SIMS features include extremely high detection sensitivity, its ability to provide chemical bonding information, and its detectability of all elements. Where feasible these techniques combined with others, such as SIMS on the SEM,[12] are becoming essential in the development of new products and optimizing product quality and cost.

REFERENCES

1. G. B. Larrabee, in "The Characterization of Solid Surfaces,"
    Scanning Electron Microscopy, IITRJ, Vol. I, pp. 39-51, 1977.
2. J. I. Goldstein and H. Yakowitz, Editors,"Practical Scanning
    Electron Microscopy," Plenum, New York, NY, 1975.
3. A. W. Czanderna, Editor, "Methods of Surface Analyses,"
    Elsevier, Amsterdam, The Netherlands, 1975.
4. R. E. Ericson and J. T. McKinney, "Quantitative Surface
    Analysis by Ion Scattering Spectroscopy," Federation of
    Analytical Chemistry and Spectroscopy Society, 1977.
5. H. W. Werner and H. A. M. DeGrefe, Rad. Effects, 18, 269
    (1973).
6. Benninghoven, W. Sichtermann, and S. Storp, Thin Solid Films,
    28, 59 (1975).
7. C. A. Andersen, Int. J. Mass. Spectra Ion Phys., 2, 61 (1969).
8. J. M. Schroer, T. N. Rhodin and R. G. Bradley, Surface Sci.,
    34, 571 (1973)
9. W. H. Gries and F. G. Radenauer, Int. J. Mass Spectro. Ion
    Phys. 18, 111 (1975).
10. G. R. Sparrow, "Quantitative SIMS Approximations for General
    Applications in Surface Analysis," American Society for Mass
    Spectrometry, 1977.
11. G. R. Sparrow, "Characterization of Cleaned and Prepared
    Bonding Surfaces by ISS/SIMS," Society of Manufacturing
    Engineers, 1977.
12. J. Leys and R. Ruscica, Ind. Res/Dev., 20,   (1978).

REDUCTION OF CONTAMINATION ON TITANIUM SURFACES STUDIED BY ESCA

W. Chen[*], D. W. Dwight[+], W. R. Kiang[+] and J. P. Wightman[*]

Virginia Polytechnic Institute and State University

Blacksburg, Virginia   24061

X-ray photoelectron spectroscopy (ESCA) was used to determine compositions of some titanium and Ti 6-4 alloy surfaces before and after a variety of cleaning pretreatments. Although ubiquitous carbon contaminating layers were present, trace elements characteristic of specific etching solutions were detected. When the same specimens were analyzed in the AEI ES-100 and the duPont 650 photoelectron spectrometers, excellent agreement was obtained in binding energy values, but peak intensities were significantly different. Two types of argon plasmas (external to the spectrometer) were rapid and effective to reduce carbon contaminants to a minimum, promoting identification of trace elements associated with the substrate.

INTRODUCTION

Titanium 6-4 alloy is a highly stabilized, alpha-beta phase alloy, using 6% aluminum as the alpha stabilizer and 4% vanadium as the beta stabilizer, imparting toughness and strength at temperatures up to 627 K (750°F) (1). The general corrosion resistance of titanium alloys is superior to many common engineering metals, because of its natural, tenacious, self-healing oxide film, usually developed in the environment of water (2). However, in joining titanium alloys by welding, adhesive bonding, or other

---

[*]Department of Chemistry
[+]Department of Materials Engineering

techniques, it is difficult to clean and prepare good bonding surfaces. A variety of specialized surface treatments have been developed over the past decade (3), and increasingly sophisticated
measurements are being used to characterize the surfaces involved.

The use of alkaline and phosphate/fluoride surface treatments
on Ti 6-4 were characterized in a review of surface analysis applied to several adherends. Differences between treatments were
seen by Scanning Electron Microscopy (SEM) and in surface elemental
composition as determined by Electron Spectroscopy for Chemical
Analysis (ESCA) (4). Other surface preparations for titanium have
been described, but without detailed analysis (5). In a study of
a proprietary structural adhesive (HT 424) with aluminum 2024-T3
and Ti 6-4, ellipsometry and surface potential difference measurements identified substrates which would result in poor adhesive
bonding (6). The use of Auger Electron Spectroscopy (AES) with
ion-sputtering depth profiling, electron microscopy and diffraction
and x-ray diffraction showed that the locus-of-failure alternated
between the surface oxide layer, the interface and primer layer,
and into the bulk adhesive. Phosphate/fluoride and Turco surface
treatments gave similar bond strength and failure loci. Only a
nitric acid/fluoride treatment produced low bond strengths, and
this treatment left large copper concentrations in the oxide film.

Determination of time-to-failure (durability) at 333 K (140°F)
and 95% relativity humidity at various loads, indicated stress-
durability almost an order of magnitude greater with phosphate/
fluoride treatment that with alkaline-cleaned alloys. Outdoor
exposure in both stressed and unstressed adhesive joints provided
a similar comparison. By non-destructive thermoholography the
rearrangement of titanium dioxide from anatase to rutile crystal
structure was detected under the bond. Electron diffraction determined surface structures that were stabilized with ions that
promote anatase formation (7). Similar observations have been
made by other workers who comment that ∿ 10% volume change accompanies the phase change, and may adversely affect bond durability
(8).

Nylon-supported FM400 (modified epoxy) adhesive was used in
conjunction with a corrosion inhibiting primer in a bond durability
study of six different treatments on Ti 6-4 alloy. Lifetimes de-
termined at $1.03 \times 10^{7} \text{Nm}^{-2}$ (1500 psi) stress, 334 K (160°F) and
100% relative humidity. Samples of lowest lifetime showed the
greatest amount of interfacial failure, and longer-lasting surface
treatments (lifetimes varied from 15 to over 1000 hr) produced
evidence of more cohesive failure in the adhesive layer. The
phosphate/fluoride surface treatment showed the lowest lifetime
while several other acid treatments (pre-treated with base) produced lifetimes averaging around 500 hr (9). AES/ESCA/SIMS were
used simultaneously to study the oxidation of titanium in the mono-

layer range (10). Successive stages of oxidation led to significant changes, first in the AES, then in the SIMS signals, and finally to a chemical shift in XPS. Also a number of treatments of titanium 6-4 alloy and pure titanium were studied with x-ray and electron diffraction, and no anatase form was found (11). Another study found oxidation of the metal was twice as fast as the alloy in water vapor at high temperature and pressure, and only rutile was found (12).

We concluded that the literature contains useful information on titanium, its alloys, surface treatments and adhesive bonding, but specific details of chemical and morphological structures that result from several commercial processes are not available or the data are conflicting. Thus we undertook a systematic study of many common treatments steps using scanning electron microscopy (SEM) with energy dispersive analysis of x-ray fluorescence (EDAX) and x-ray photoelectron spectroscopy (ESCA). As a first stage in accessing the relative durability of the different surface treatments, changes occasioned by aging at 505 K (450°F) were monitored.

In general, our results showed that surface chemical composition was relatively constant, while profound changes occurred in morphology. The SEM results have been published elsewhere (13), but can be summarized briefly as follows: Ti 6-4 coupons had totally different morphology on opposite sides before grit blasting; that destroyed the original structures, leaving the surface heavily worked and fragmented and covered with minute fracture debris. Aluminum content significantly increased (uniformly and no Al_2O_3 grit blast particles appeared). Each of the four chemical surface treatments produced unique surface morphology, and only the two acid treatments bore any resemblance to each other. A qualitative ranking: Turco>Pasa-Jell>phosphate/fluoride>anodize was made by considering both (1) degree of roughness, and (2) degree of change in surface morphology after a 10 hr, 505 K (450°F) exposure. Oxidation diminished the size of surface structures generally, and favored the alpha phase at the expense of beta.

Initial ESCA results on those samples indicated that 25-75% of the photoemission derived from a contaminating carbon overlayer. Plasmas have been demonstrated more effective than liquid reagents in reducing carbon contamination on rhodium and iron-cobalt alloy (14). Water contact angles and AES data were correlated in determining the amount of residual contamination after each type of cleaning. AES was used to monitor the relative efficacy of cleaning boron-doped silicon (111) semiconductor wafers by several liquid reagents, plasmas, and ion-beam sputtering (15). Again, inert gas plasmas were superior. Therefore we employed argon plasmas to reduce the carbon contamination prior to ESCA analysis.

The results reported below bear upon three points: (1) Semi-quantitative analysis of Ti 6-4 alloy surfaces before and after four commercial surface treatments and thermo-oxidative aging, (2) Comparison between results obtained in two photoelectron spectro- meters (AEI ES-100 and duPont 650), and (3) Time study of both custom and commercial plasma cleaning on pure titanium metal.

EXPERIMENTAL

Materials. Lap shear coupons of Ti 6-4 were provided by D. Progar, NASA Langley Research Center, both before and after grit blasting. Pure titanium foil was obtained from the Alfa Division of Ventron Corporation, Danvers, MA 01923. Anodizing was performed by the Boeing Company using a proprietary process. Procedures followed for the three chemical treatments are listed in Tables I-III. Also samples after each of the four treatments were exposed for 10 and 100 hours at 505 K (450°F) in air.

Plasma Cleaning. Both custom and commercial apparatus were used to reduce the carbon contamination on the titanium alloy coupons and pure foils. A "Plasmod" (Tegal Industries, 360 Wharf St., Richmond, CA 94804) with argon gas at 1 torr employed a crystal-controlled 13.6 MHz discharge at about 50 watts. The custom apparatus used argon at 5 torr and a Tesla coil ($\sim$5 MHz) discharge. In this setup specimens were electrically grounded while they were not in the Plasmod. The effects of the custom discharge were studied both before and after mounting the specimens on ESCA probes.

Table I. Phosphate/Fluoride Treatment

1. Solvent wipe - methylethyl ketone.
2. Alkaline clean - immerse in SPREX AN-9, 30.1 g/ℓ, 353 K (80°C) for 15 min.
3. Rinse - deionized water at room temperature.
4. Pickle - immerse for 2 min. at room temperature in solution containing 350 g/ℓ of 70% nitric acid and 31 g/ℓ 48% HF.
5. Rinse - deionized water at room temperature.
6. Phosphate/fluoride treatment - Soak for 2 min. at room temper- ature in solution containing 50.3 g/ℓ of tri sodium phosphate (Na_3PO_4); 20.5 g/ℓ of potassium fluoride (KF); and 29.1 g/ℓ of 48% hydrofluoric acid (HF).
7. Rinse - deionized water at room temperature.
8. Hot water soak - deionized water at 338 K (65°C) for 15 min.
9. Final rinse - deionized water at room temperature.
10. Dry - air at room temperature.

Table II. Pasa-Jell 107 Treatment

1. Solvent wipe - methylethyl ketone.
2. Alkaline clean - immerse in SPREX AN-9, 30.1 g/ℓ, 353 K (80°C) for 15 min.
3. Rinse - deionized water at room temperature.
4. Pickle - immerse for 5 min. at room temperature in solution containing 15g nitric acid (HNO_3) 15% by weight; 3g hydrofluoric (HF) acid 3% by weight; and 82g deionized water.
5. Rinse - deionized water at room temperature.
6. Pasa-Jell 107 Paste - Apply to the titanium surface with an acid resistant brush covering the entire surface by cross brushing.
7. Dry - for 20 min.
8. Rinse - deionized water at room temperature.
9. Dry - air at room temperature.

ESCA Analysis.   Spectra on some samples were obtained on an AEI ES-100 photoelectron spectrometer with an aluminum anode ($K\alpha_{1,2}$ = 1486.6 ev) and digital data acquisition.   Rectangular (5x20 mm) samples were mounted on the sample probes using double-sided adhesive tape.   In addition, ESCA spectra on all samples were obtained on a duPont 650 spectrometer with a magnesium anode ($K\alpha_{1,2}$ = 1253.6 eV) and analog display on an X-Y recorder.   The carbon 1s level at 284.0 eV was used to evaluate the work function of both spectrometers (16).   Circular (d = 6.4 mm) samples were

Table III. Turco 5578 Treatment

1. Solvent wipe - methylethyl ketone.
2. Alkaline clean - immerse in Turco 5578, 37.6 g/ℓ, 343-353 K (70-80°C) for 5 min.
3. Rinse - deionized water at room temperature.
4. Etch - immerse in Turco-5578, 419 g/ℓ, 353-373 K (80-100°C) for 10 min.
5. Rinse - deionized water at room temperature.
6. Rinse - deionized water at 333-343 K for 2 min.
7. Dry - air at room temperature.

mounted on the sample probes using double sided adhesive tape.

Wide-scan spectra for each sample were obtained using the duPont spectrometer, attempting to detect all surface components present in significant amounts. The duPont spectrometer is espe- cially suited for rapid analysis; a wide scan (0-700 eV) can be obtained in about one hour. Narrow scan spectra were obtained to establish precisely the binding energy and intensity of each peak noted in the wide scan.

Elemental assignments for each peak were based on standard binding energy tables (16). Further analysis of the ESCA results is possible using the measured intensities (I_i) in counts/sec and tabulated photoelectric cross sections (σ_i) (17). Using several assumptions the following equation was derived to approximate the atomic fraction (AF_i) of a given surface species (i):

$$AF_i = \frac{I_i/\sigma_i}{\Sigma(I_i/\sigma_i)}$$

It should be emphasized that these calculations are semi- quantitative at best, and primarily serve to reduce the data to comprehensible tables.

RESULTS AND DISCUSSION

The ESCA results discussed below represent two separate con- tamination studies. The first relates to four common chemical treatments designed to clean and etch Ti 6-4 for bonding. Inter- ference of carbon contamination with ESCA analysis prompted the second study of argon plasmas to reduce that contamination.

Pretreatment and Thermal Aging

The ESCA results for these samples in Table IV show that the grit-blasted sample has the highest aluminum content, in agreement with EDAX results (13). The large carbon content indicates organic contamination from the grit-blast operation.

Anodize. Organic contamination from environmental exposure and sample handling prior to analysis, as well as deposition during analysis, invariably gave rise to a large carbon 1s photopeak. Indicative of tetravalent oxide, the major titanium $2p_{1/2}$ and $2p_{3/2}$ oxide photopeaks were observed at 464.0 and 457.7 eV, respectively. The separation of the two peaks by 5.8 $\pm$ 0.4 eV is in excellent agreement with a previously published separation of 6.0 eV (16). Oxygen 1s photopeaks (529.6 $\pm$ 0.4 eV) were observed for each sample: a relatively thick oxide layer for the anodized sample was deter-

Table IV. ESCA Peak Parameters for Ti 6-4 Before and After Chemical Surface Treatments

Peak Assignment	As-received BE	%	Anodized BE	%	Phosphate/Fluoride BE	%	Pasa-Jell BE	%	Turco BE	%
F 1s	684.2	2	684.1	—	684.2	3	684.7	1	685.0	2
Cr $2p_{3/2}$	—	—	—	—	—	—	576.3	2	—	—
O 1s	530.0	24	529.9	13	529.5	41	529.6	42	529.4	32
V $2p_{3/2}$	514.2	1	515.0	1	515.5	1	514.6	1	515.7	1
Ti $2p_{3/2}$	457.9	6	457.9	7	457.0	17	457.6	14	457.9	10
N 1s	—	—	399.2	1	—	—	—	—	—	—
Ca $2p_{3/2}$	346.6	1	—	—	—	—	—	—	—	—
K 2p	—	—	—	—	287.9	1	—	—	—	—
C 1s	(284.0)	59	(284.0)	77	(284.0)	35	(284.0)	36	(284.0)	53
Cl 2p	—	—	197.6	1	197.3	1	—	—	—	—
P $2p_{3/2}$	—	—	—	—	132.6	1	—	—	—	—
Al 2s	118.5	8	118.2	1	118.2	1	118.0	2	119.1	2
Si 2p	—	—	—	—	—	—	103.2	2	—	—

BE = Binding Energy (eV)

% = Atomic Percent, from normalized peak intensity.

mined by ion milling/AES profiling (18). The O/Ti ratio of 1.9 is close to the expected value for TiO_2. Vanadium is a 4% bulk component of Ti 6-4, but its surface concentration was generally less than one percent, and the stoichiometric ratio of Al/V is greater than 1.5 determined from the bulk composition. A small nitrogen ls photopeak at 399.4 $\pm$ 0.2 eV was noted. Fluorine was present as a fluoride with ls binding energy of (684.4 $\pm$ 0.4 eV). Chloride ion was detected by a very weak 2p photopeak at 197.9 $\pm$ 0.3 eV.

Phosphate/Fluoride. The ESCA results (Table IV) show traces of phosphorous, fluorine, potassium and chlorine from the surface treatment process, retained in the surface through the rinsing step.

Pasa-Jell. The results show that chromium and silicon (in the Pasa-Jell 107 paste) are residual surface components after this treatment.

Turco. The results in this case indicate minimum residual contamination. Fluorine is the only element not associated with the substrate, and its origin is unclear.

Thermal Aging. A small amount of nitrogen was detected in each of the samples after aging for 10 hours at 505 K (450°F); otherwise, there were no significant changes in the ESCA spectra of any of the samples after 100 hr of thermal aging. The binding energy values above agree closely with preliminary results on similar specimens reported earlier (19).

It should be re-emphasized that the calculated surface compositions are accurate to $\pm$ 15% at best, and care should be exercised in interpreting the results. On the other hand, most analytical techniques provide bulk elemental analyses, while the ESCA technique probes the composition of the surface layer (<10nm) only, where attention is focused in many problems. For example, in adhesion the definition of the interfacial region is tantamount to understanding bond failure. The ESCA technique has shown the presence of different trace elements in the interfacial region after specific chemical pretreatments. Some of these trace elements could serve as catalysts in polymer degradation and/or substrate corrosion, and thereby influence bond durability.

Comparison of AEI and DuPont Spectrometers

Since ESCA was collected for some of the anodized samples on both the AEI and duPont spectrometers, it is of interest to compare the results. A tabulation of average binding energies and intensities for Ti, Cr, F, O and C obtained from both spectrometers on three different anodized samples is given in Table V. There

Table V. ESCA Parameters for Anodized
Ti 6-4 in two Spectrometers

Peak	AEI ES-100		duPont 650	
Assignment	BE (eV)	%	BE (eV)	%
F 1s	684.4+0.1	4	684.5+0.2	1
Cr	577.2	1	—	—
0 1s	529.7+0.2	37	529.8+0.2	19
Ti (IV) $2p_{3/2}$	458.0+0.3	15	458.1+0.3	7
C 1s	(284.0)	43	(284.0)	73
C/Ti	2.9		10.0	

BE = Binding Energy (eV)   % = Atomic Percent, from
                                      normalized peak intensity

is remarkable consistency in the binding energies; also peak
shapes were similar as gauged by a comparison of the peak widths
at half-height. Chromium was detected in the AEI ES-100 but not
in the duPont 650. In this case where the chromium surface concentration was <1%, the repetitive scan, digital acquisition used
with the AEI ES-100 offered an advantage over the analog mode.
The duPont multichannel analyzer was not in operation during this
work, but appears necessary for analysis where the signal-to-
background ratio is small. The C/Ti ratios are 2.9 and 10.4 for
the AEI and duPont spectrometers, respectively. This may be
caused by variations in either sample contamination, x-ray beam
effects, or instrument geometry.

Significant difference in peak intensities were noted with
different spectrometers. The duPont 650 signals were typically
30 times greater than those obtained with the AEI ES-100. The AEI
ES-100 uses a slit between the sample and analyzer reducing transmission compared to the duPont 650; differences in intensity were
expected. This results in faster analysis in the duPont spectrometer. For example, actual analysis time for the five principal
elements was 33 minutes in the duPont 650 spectrometer and 210
minutes in the AEI ES-100 spectrometer. Sample contamination
occurred to a greater extent in the duPont spectrometer as gauged
by the larger C/Ti ratio. Recent studies on the build up of contamination during ESCA measurements shows that the vacuum level
and instrumental factors governing neutralization of hole site
are important factors (20). Peak intensity differences between
our AEI ES-100 and duPont 650 spectrometers could be due to similar
factors.

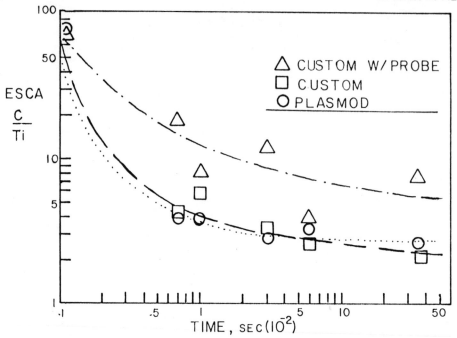

Figure 1.   The decrease in C/Ti atomic ratio (determined from ESCA peak heights) as a function of exposure time to Ar$^+$ plasmas generated by three methods convenient for subsequent ESCA analysis.

Comparison of Plasma Techniques for Cleaning Titanium

The ESCA results suggest that the removal of a carbonaceous overlayer might enhance the intensity of substrate photopeaks, and permit detection of trace elements associated with the titanium surface. Without in situ argon ion cleaning, the question is, can an argon plasma external to the spectrometer significantly reduce the carbon contamination? Three timed series of ESCA studies were made on pure titanium foil, with Ar$^+$ plasmas generated under different conditions. Figure 1 shows the C/Ti ratio vs time of plasma exposure. It is clear that most of the cleaning action is accomplished within one minute. Furthermore, the ESCA probe seems to lead to additional residual contaminants, as can be seen in Table VI.

Custom rf discharge with sample on grounded probe. The ESCA results obtained in this case detected calcium, nitrogen, potassium, sulfur and zinc after argon cleaning, each at less than 1%. It is likely that these trace elements resulted from sputtered constituents of the sample probe and the glass walls of the plasma apparatus. The C/Ti ratio for the as-received Ti was 67, and it decreased significantly to 19 after 70 sec. exposure to the argon plasma.   There is no further significant decrease in the carbon signal even after a 1 hr. exposure.   Since the sample was mounted on a grounded probe, removal of carbon contamination is due to the action of both electrons and argon ions.

Custom rf discharge with sample grounded (no probe).   When the Ti sample was grounded, but not mounted on a sample probe, only calcium was observed as a trace contaminant after argon plasma cleaning.

Plasmod.   The custom and PLASMOD rf discharge appear equally effective in removal of the carbon contamination from titanium surfaces.

SUMMARY

Collectively the data suggest that the major carbonaceous contamination of metal oxides occurs prior to analysis, rather than deposition in the spectrometer.   It is obviously desirable to have

Table VI. ESCA Parameters for Pure Titanium Foil
Exposed to Argon Discharges for 70 sec.

Element	As-Received BE	%	Custom (probe) BE	%	Custom (no probe) BE	%	Plasmod BE	%
O	529.4	5	529.4	10	529.3	31	529.3	32
Ti	457.5	1	457.6	4	457.3	13	457.7	13
N	——	—	400.2	1	——	—	399.0	1
Ca	——	—	346.8	1	346.6	1	346.5	1
C	(284.0)	88	(284.0)	66	(284.0)	53	(284.0)	50
Cl	197.9	2	197.7	7	——	—	197.6	1
Al	——	—	——	—	——	—	118.5	1
Si	102.2	3	101.7	3	101.4	2	101.4	1

BE = Binding Energy (eV)   % = Atomic Percent, from
                                                normalized peak in-
                                                tensities.

in situ cleaning, but in lieu of this capability, external argon plasma cleaning can reduce carbon contamination significantly, leading to improved detection of trace constituents on the substrate of interest.

Binding energy values obtained with two different designs of commercial photoelectron spectrometers were in excellent agreement, but the relative intensities of photopeaks showed no correlation.

Trace contaminants, characteristic of different processes used in the pretreatment of titanium 6-4 alloy remain in the surface and are detected readily by ESCA.

ACKNOWLEDGEMENTS

Support for this work under NASA Grant NSG-1124 including a graduate research assistantship for one of us (WC) and the assistance of Mr. D. J. Progar at the NASA Langley Research Center are acknowledged gratefully.

REFERENCES

1. R. F. Muraca and J. S. Wittick, "Materials Data Handbook on Titanium 6Al-4V," NASA Tech Brief B73-10372, October 1973.
2. ASM Committee on Metallography of Titanium and Titanium Alloys, Metals Handbook, 7, 321 (1972).
3. R. H. Shoemaker, Titanium Sci. Technol. 4, 2401 (1973).
4. W. C. Hamilton, Appl. Polym. Symp. 13, 105 (1972).
5. E. D. Newell and G. Carrillo, Materials and Processes for the 1970's, SAMPE Conference, 1973, p. 131.
6. T. Smith and D. H. Kaelble, Report No. 74-73, Air Force Materials Laboratory,Wright-Patterson Air Force Base, Ohio(1974).
7. R. F. Wegman and M. J. Bodnar, SAMPE Quarterly 5, 28 (1973).
8. R. Dalard, C. Montella and J. C. Sohm, Surf. Technol. 4, 367 (1976).
9. D. M. Stifel in "New Industries and Applications for Advanced Materials Technology," SAMPE Pubs. Azuza, CA., 1974, p. 75.
10. A. Benninghoven, H. Bispinck, O. Ganschow and L. Wiedmann, Appl. Phys. Lett. 31, 341 (1977).
11. K. W. Allen and H. S. Alsalim, J. Adhesion 6, 299 (1974).
12. A. Motte, C. Coddet, P. Sarrazin, M. Azzopardi, and J. Besson, Oxid. Metals 10, 113 (1976).
13. W. Chen, D. W. Dwight and J. P. Wightman in Symposium on Surface Analysis, Pittsburgh Conference on Analytical Chemistry and Applied Spectroscopy, Cleveland, OH, May 12-16, 1978. ASTM Publication, in press.

14. D. F. O'Kane and K. L. Mittal, J. Vac. Sci. Technol. 11, 567 (1974).

15. M. G. Yang, K. M. Koliwad and G. E. McGuire, J. Electrochem. Soc. 122, 675 (1975).

16. K. Siegbahn, C. Nordling, A. Fahlman, R. Nordberg, K. Hamrin, J. Hedman, G. Johansson, T. Bergmark, S. Karlsson, I. Lindgren, B. Lindberg, "ESCA Atomic, Molecular and Solid State Structure Studied by Means of Electron Spectroscopy," Almqvist and Wiksells, Uppsala, 1967.

17. J. H. Schofield, J. Electron Spectrosc. 8, 129 (1976).

18. D. W. Dwight and J. P. Wightman, This proceedings volume.

19. T. A. Bush, M. E. Counts and J. P. Wightman in "Adhesion Science and Technology," L. H. Lee, Editor, p. 365, Plenum, New York, 1975.

20. E. S. Brandt, D. F. Untereker, C. N. Reilley, and R. W. Murray, J. Electron Spectrosc. 14, 113 (1978).

AN ESCA STUDY OF SURFACE CONTAMINANTS ON GLASS SUBSTRATES FOR CELL ADHESION

B.D. Ratner*, J.J. Rosen°, A.S. Hoffman*, and L.H. Scharpen†
*Department of Chemical Engineering and Center for Bioengineering
°Department of Pathology and Center for Bioengineering
University of Washington, Seattle, Washington 98195
†Surface Science Laboratories, Palo Alto, CA 94303

Glass surfaces used as substrates for cell culture techniques must be reproducibly cleaned of surface organic contaminants. ESCA, which examines the uppermost 10-50Å of glass surfaces, has been used to ascertain the efficiency of various cleaning procedures by analyzing the nature and quantity of deposited foreign substances on cleaned glass.

Glass coverslips were found by ESCA to have significant amounts of organic carbon at their surface. Oxidizing acidic cleaning agents and organic cleaning solutions reduced the levels of surface carbon from the "as received" condition. Argon etching could produce surfaces with extremely low C/Si ratios. Such surfaces rapidly adsorbed carbonaceous material when reexposed to air. U.V.-ozone cleaning was highly effective in removing organic carbon from pre-cleaned glass.

ESCA data from glass specimens with various levels of organic carbon contamination suggest a model in which the initial monolayer of adsorbed organic material at the glass surface is polar in nature while subsequent monolayers are hydrocarbon in nature.

Swiss 3T3 cells were used to investigate the biological effects of residual organic surface films. Significant differences were not observed between attachment to glass plates cleaned by a variety of techniques.

INTRODUCTION

Glass surfaces used as substrates for cell culture techniques must be reproducibly cleaned of surface organic contaminants. Trace material microanalysis, polymer adhesion and microelectronic devices also require clean glass surfaces. A substantial body of literature exists on the cleaning of glass surfaces[1,2]. Analysis of the effectiveness of the various cleaning techniques was, until recently, performed by indirect methods. Three general categories of techniques were used for such analysis:

1) the contact angle of liquids on glass
2) the frictional resistance of the glass surface
3) the adhesion of other substances to the glass surface

Results obtained with these techniques were difficult to interpret, particularly with respect to quantitatively relating the contaminant to the parameter being measured. Also, these techniques do not identify the nature of the contaminant.

More recently, sophisticated analytical techniques have been applied to the problem of studying glass surfaces. A number of studies using these techniques are summarized in Table I.

Electron spectroscopy for chemical analysis (ESCA) has a number of significant advantages for studying surface contamination layers on glass. The specific advantages of ESCA for this application are:

1) All elements except hydrogen can be studied.
2) Information about the oxidation states and molecular environments of the various elements can be ascertained.
3) The technique can be used for quantitative analysis.
4) ESCA is relatively non-destructive.
5) Depth profiling can be achieved by a number of non-destructive techniques[16,17].
6) The homogeneity of surface layers can be assessed[17,18].

Many reviews have been published describing the ESCA technique and its applications[19,20].

In this study, organic carbon contaminant films on glass were analyzed by ESCA. It was demonstrated that very low ratios of carbon to silicon can be produced by argon etching and U.V.-ozone treatment of the glass surfaces. Acid cleaning and organic cleaning solutions, though not as effective in removing organic carbon as the argon treatment, can produce relatively clean surfaces.

Cell interaction with glass has already been shown to be

Table I. Surface Studies on Glass, Silicon, and Quartz Using
Contemporary Analytical Instrumentation

Analytical Technique(s)	Investigator(s)	Cleaning Studied	Reference
ISS, SIMS	Sparrow & Mismash	Yes	3
Auger Spectroscopy	Rynd and Rastoyi	No	4
	Pantano and Hench	Yes	5
	Yang, et al.	Yes	6
	Vig	Yes	7
Infrared Reflection Spectroscopy, AES	Hench	Yes	8
Electron Energy-Loss Spectrum	Wei	No	9
Ellipsometry	Vedam and Malin	Yes	10
Ion Microprobe	Heyndrychx	No	11
ESCA	Bryson, et al.	Yes	12
	Hickson	No	13
LEED	Jona	Yes	14
C-V Characteristics of MOS Structures	Kesavan & Andhare	Yes	15

affected by the method of substrate preparation[21]. In this study
indications have been obtained that the rate and extent of 3T3
cell adhesion is not related to the amount of organic carbon present on glass surfaces over the range studied.

EXPERIMENTAL

The glass specimens used in these experiments were Gold Seal
#2 microscope coverslips and Diamond Brand #2 microscope cover-
slips. For ESCA analysis, Gold Seal coverslips were cut into 10mm
x 13mm sections. Diamond Brand coverslips were analyzed uncut as
12mm disks.

Specimens treated with Sparkleen, Isopanasol, Ivory Soap,
RBS-35, chromic acid or No-Chromix were cleaned using the following technique. The coverslips were individually placed in glass
scintillation vials which were then filled with appropriate cleaning solutions. The cleaning solutions used are listed in Table II
along with their concentration and source of supply. The vials
were capped and the contents placed in an ultrasonic cleaner for
5 minutes. Visible residues were removed from the vials by rinsing
with water purified by ion exchange and then reverse osmosis. The
vials containing the glass plates in water were then sonicated
again for 5 minutes. Thorough water rinses and sonications in
water were repeated two more times. Then the final rinse water

B. D. RATNER ET AL.

Table II. Cleaning Solutions

	Manufacturer	Concentration (in H_2O)
Sparkleen	Fisher Scientific Co.	4 g/l.
Isopanasol	C.R. Callen Corp.	1:64
Ivory Soap	Proctor & Gamble, Inc.	0.1%
RBS-35	Pierce Chemical Co.	1:50
Chromic Acid (Chromerge)	Manostat Co.	--*
No-Chromix	Godax Laboratories, Inc.	--*

*Mixed as directed into concentrated H_2SO_4.

was poured from the vials, and the vials containing the glass specimens were dried under vacuum over anhydrous magnesium perchlorate. Control experiments for the effectiveness of the sonication procedure were run under identical conditions except that the ultrasonic cleaner was not switched on. Another control experiment used a simple water soak (1 hour) and rinse without other cleaning agents or sonication.

Acid washing involved placing coverslips into a 1:1 mixture of hot (80°C) reagent grade nitric and sulfuric acid for 5 minutes. Immediately after this treatment, the coverslips were placed into glass vials and rinsed with sonication as described for the other treatments.

Sodium hydroxide cleaned glass was prepared by first soaking the glass for one hour in 0.1N NaOH, and then for one hour in 0.1N HCl. Three final water rinses with sonication were used to wash away remaining HCl.

The U.V.-ozone treatment protocol has been described in detail in a paper by Vig and LeBus[22]. Specimens were placed three inches from the U.V. source for varying periods of time.

Argon etching (∿5Å/min) was performed in the vacuum chamber of the ESCA spectrometer.

ESCA spectra were taken on a Hewlett Packard Model 5950B ESCA system. A 0.8 kwatt monochromatized X-ray beam from an aluminum anode was used for all spectra. An emission from an electron flood gun was used to neutralize static charge buildup on the nonconducting glass surfaces.

The ratio of organic carbon to silicon, determined using normalized ESCA peak intensities, was calculated to indicate the level of surface carbon contamination. The bulk glass was assumed to contain no organic carbon. Normalized ESCA peak intensities were

obtained by dividing the integrated number of photoelectron counts
in the peak area by the number of scans and then dividing again by
photoionization cross sections for the various elements for Al K_α
excitation. Photoionization cross sections used were carbon (1s)
= 1.000, silicon (2p) = 0.817, and potassium (2p3/2) = 3.97. Overlapping peaks were resolved in the C_{1s} spectra using a Dupont 310
curve resolver. A C_{1s} peak width of 1.3 eV, measured for polyethylene, was used for the curve resolution.

The assay for cell adhesion kinetics utilized clean glass substrates as control surfaces and as supports for various polymer and
protein coatings of interest in this laboratory. The following
procedure was used for cell attachment experiments. Swiss 3T3
mouse cells are radiolabelled with 0.25 μCi/ml of H^3-thymidine for
24 hours and then grown in label-free media for an additional 24
hours prior to the attachment assay. Cells were gently dissociated
from their growth dishes with EGTA and inoculated into multiwell
dishes containing the test substrates. After times from 5 minutes
to 120 minutes, the substrates were removed from the wells, dip-
rinsed in phosphate-buffered saline containing divalent cations
(100 mg/l $CaCl_2$ and 100 mg/l $MgSO_4 \cdot 7 H_2O$) and the percent of
adhered cells determined by a scintillation counter.

RESULTS

ESCA survey scans examining all elements in the surface region
of the glass were performed on an uncleaned (as received) Gold Seal
coverslip and on an Isopanasol cleaned coverslip. Results are reported in Table III. Many elements increased or decreased in concentration indicating that substantial alterations in the surface
chemistry of the glass occurred upon washing. Carbon, silicon,
and oxygen comprised approximately 90 atom % of the surface region
of both specimens. Upon cleaning, their relative abundances changed
significantly. Since oxygen is present in both the organic carbon
contaminating film and in the glass, its level was not used in this
study to monitor the condition of the surface.

Typical high resolution ESCA spectra of glass for the regions
270-290 eV (carbon and potassium) and 85-105 eV (silicon) are shown
in Figures 1 and 2, respectively. The results from these measurements for various cleaning techniques are summarized in Table IV
in terms of the elemental ratio of carbon to silicon at the glass
surfaces.

ESCA spectra for uncleaned glass, Isopanasol cleaned glass and
HNO_3-H_2SO_4 cleaned glass are shown in Figure 3. The deconvoluted
component peaks of these ESCA spectra are also shown and their
relative contributions to the total C_{1s} signal are indicated. The
peaks at 285.0 eV are indicative of hydrocarbon-type material.
Those at $\sim$286.4 eV are probably hydroxyl or ether-type compounds.

Table III. Relative Abundance of Elements at the Surface of Cleaned
            and Uncleaned Glass

Relative Abundance (Atom %)

Element Detected	Uncleaned Glass[a]	Isopanasol Cleaned Glass[b]	Change in Concentration Upon Cleaning
O	36.1	50.2	+
Na	2.5	0.5	−
Zn	0.7	0.3	−
Ti	2.5	0.5	−
N	1.6	1.0	−
K	2.1	1.2	−
C	32.3	16.0	−
P	1.0	1.2	+
Si	19.5	27.1	+
Al	1.5	1.8	+
Cu	−	0.1	+
Ba or Co	−	0.1-0.2	+

a - Gold Seal #2 coverslip - as received.
b - Gold Seal #2 coverslip after washing with Isopanasol
    as described in the Experimental section of this paper.

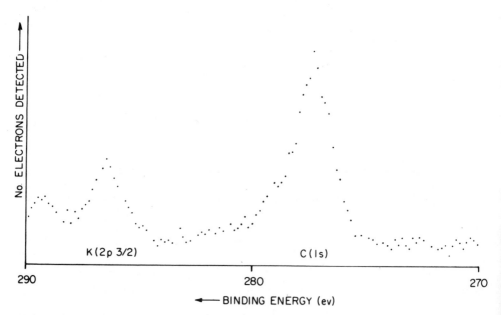

Figure 1. ESCA C(1s) spectrum of organic carbon contamination film
           on glass.

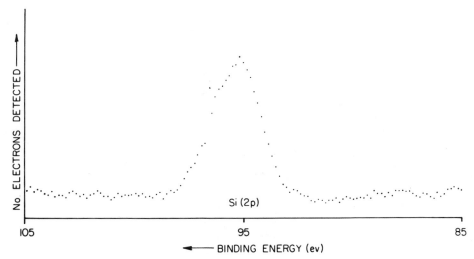

Figure 2.   ESCA Si(2p) spectrum of glass.

Table IV.   C/Si Ratio for Glass Specimens Cleaned by Various Methods.

Treatment	C/Si Atom Ratio	# of Measurements
Sparkleen	0.50±0.05	5
Isopanasol	0.51±0.17	8
Isopanasol (15 min. sonication)	0.42	1
Isopanasol (hot)[a]	0.60	1
Ivory Soap	0.81	2
RBS-35	0.52	2
Chromic Acid	0.37	2
Chromic Acid (hot)[a]	0.33	1
No-Chromix	0.29	1
HNO_3-H_2SO_4 (hot)	0.38	2
NaOH-HCl	0.47	1
U.V.-ozone(uncleaned glass)	0.52[b]	1
U.V.-ozone(Isopanasol cleaned glass)	0.19[b]	1
Isopanasol-no sonication	0.39±0.10	3
Water soak & rinse - no sonication	0.81	1
Uncleaned glass	1.76±0.36	4

a. Solution added to vial at 90°C. At the end of the 5 minute
    sonication, the solution temperature was 50°C.
b. Analyses of these specimens were performed immediately after
    removal from the U.V. source. All other specimens were stored
    in air in clean vials 3-4 days prior to analysis.

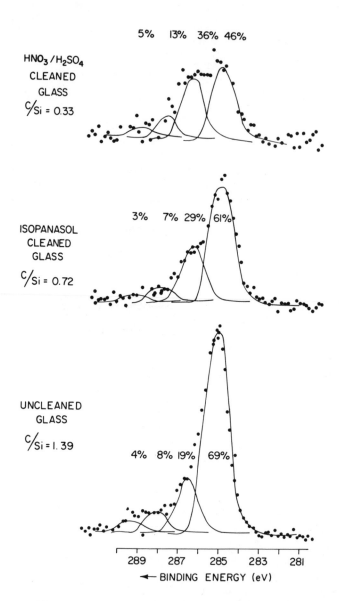

Figure 3. C_{1s} ESCA spectra for $HNO_3-H_2SO_4$ cleaned glass, Isopanasol cleaned glass and uncleaned glass. Relative
contributions of each of the subpeaks to each total C_{1s}
spectrum are indicated.

Peaks at ∿288.0 eV and ∿289.3 eV may be related to ketone and acid/ ester type compounds respectively. These peak assignments assume that the carbon compounds at the surface consist primarily of carbon, hydrogen and oxygen. This assumption is supported by the low levels of other elements which might form compounds with carbon and which could cause substantial chemical shifts (see Table III). These results indicate that cleaning the glass surfaces primarily reduces the hydrocarbon portion of the organic carbon contamination film.

Experiments involving argon etching in the ESCA unit are shown in Figure 4. After 30 minutes of argon etching, the sample was removed from the vacuum chamber of the ESCA unit and allowed to interact with the laboratory atmosphere for 10 minutes. It was then placed back into the ESCA unit and remeasured.

Since the C/Si ratio more than tripled after only 10 minutes of exposure to laboratory air, this experiment was redone with longer exposure times to air. Results are shown in Table V.

The C/Si ratios and C/K ratios for some of the specimens incorporated into Table IV were compared (Figure 5). For C/Si ratios

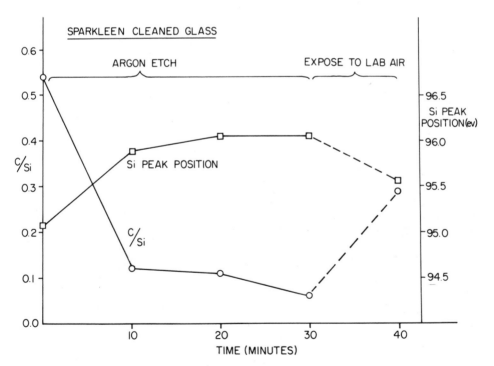

Figure 4. C/Si ratio and Si peak position as a function of time of argon etching of the glass surface.

Table V.   Recontamination of Argon Etched Specimen.

Treatment	C/Si
Glass – Sparkleen Cleaned	0.43
Argon Etch – 30 min.	0.03
Expose to air – 10 min.	0.13
Expose to air – 16 hours	0.35

below 1.2, a linear coefficient of determination (r^2) of 0.73 was obtained. If all points are counted, $r^2 = 0.52$.

Kinetic attachment experiments for Swiss 3T3 mouse cells to various treated and untreated glass surfaces have been performed.

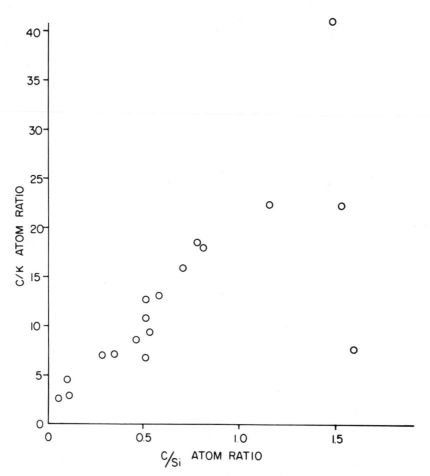

Figure 5.   Relationship between C/Si ratio and C/K ratio for glass specimens cleaned by various techniques.

Plots of cell attachment as a function of time in the presence and
absence of 10% calf serum to uncleaned glass and glass cleaned with
Isopanasol are shown in Figures 6 and 7. Also, the results from
four Swiss 3T3 cell attachment experiments at 120 minutes interaction time are presented in Table VI. The data in Figures 6 and 7
and in Table VI indicate that the extent of organic carbon contamination on glass does not measurably affect 3T3 cell attachment
over the range of contamination levels studied.

DISCUSSION

The lowest C/Si ratios observed in these experiments were for
glass which had been argon etched within the ESCA unit for 30 minutes (Figure 4, Table V). These values provide a functional definition of a clean glass surface. However, the properties and/or
chemistry of the glass surface may have been altered by this extreme treatment. Evidence for alteration of surface chemistry is
seen in the position of the silicon ESCA peak (Figure 4) which is
shifted ∿1 eV upon argon etching. This 1 eV shift to a higher
binding energy was also observed in an earlier experiment involving
argon etching of glass. The C_{1s} peak shifts ∿0.5 eV with the same
treatment. Changes in the chemical composition of silicon surfaces
which have been subjected to argon ion bombardment have been report-

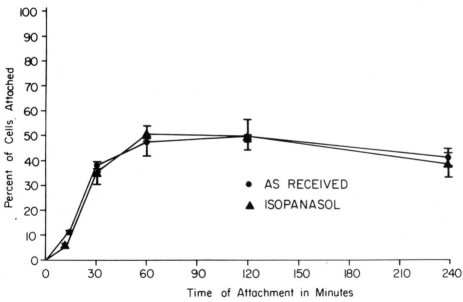

Figure 6. Effect of cleaning technique on the time dependence of
           attachment of Swiss 3T3 cells to glass in the presence
           of 10% calf serum.

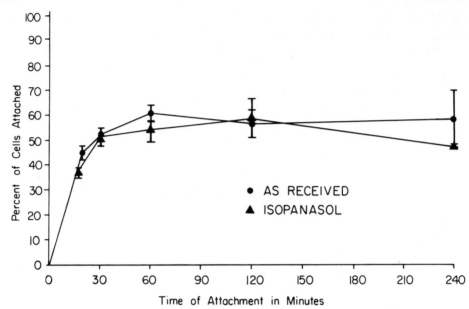

Figure 7. Effect of cleaning technique on the time dependence of
attachment of Swiss 3T3 cells to glass in the absence
of serum.

Table VI. Cell Attachment After 120 Minutes to Glass Surfaces.

Expt. #	Cleaning Technique	Percent of Cells Attached
1[a]	As received	56.3 ± 14.3
	Argon Etch[c]	65.2 ± 5.4
	Isopanasol	74.6 ± 2.2
2[b]	As received	60.2 ± 15.0
	Argon Etch[c]	60.6 ± 14.0
	Isopanasol	72.2 ± 12.9
3[a]	As received	60.1 ± 8.7
	Isopanasol	68.9 ± 11.0
	H_2SO_4/HNO_3 (hot)	65.8 ± 5.7
4[b]	As received	54.2 ± 4.7
	Isopanasol	61.4 ± 3.5
	H_2SO_4/HNO_3 (hot)	52.1 ± 4.4

a - attachment experiment performed in absence of serum.
b - attachment experiment performed in presence of serum.
c - stored in air after argon etch.

ed[9]. A change in the work function of the surface due to dehydration or altering the organic carbon film may provide an alternate
explanation for the observed Si peak shifts.

The results in Table V indicate that argon etched glass specimens must be stored under high vacuum in order to remain relatively
free of surface carbon contamination. The rapid recontamination
rate implies that this method of cleaning glass surfaces would not
be suitable for general laboratory application.

Commercial, non-acidic laboratory cleaning solutions all performed comparably and produced moderately clean glass surfaces.
The replicate C/Si ratios for Sparkleen (0.50+0.05) and for Isopanasol (0.51+0.17) as listed in Table V provide an estimate of the
reproducibility that can be expected using these techniques. Isopanasol and RBS-35, both liquid cleaners consisting of a mixture of
surfactants and chelating agents, were not superior to Sparkleen
detergent. Longer sonication times or elevated temperatures did
not improve the performance of Isopanasol. Ivory Soap may be
slightly less efficient than the other laboratory cleaning agents
in removing organic carbon contamination from glass. Surprisingly,
sonication did not improve the cleaning effectiveness of Isopanasol.
The absence of crevasse-like features, or irregularities on the
smooth glass surface may have rendered the sonication step unneces-
sary.

All of the acidic cleaning solutions were found to be comparable in their ability to remove organic carbon contaminants from
glass. Since chromium contamination has been reported to be a
problem with chromic acid cleaning solutions[23], the presence of
chromium on chromic acid cleaned glass was investigated by ESCA.
In three samples studied, either very little or no chromium was
observed. Similar observations were made by Pantano and Hench[5].
Since ESCA is a surface oriented technique, chromium which has
diffused into the glass and could potentially leach out, might not
be detected. These observations on chromium, therefore, are not
conclusive.

The sodium hydroxide cleaning would be expected to etch the
surface of the glass itself thereby removing any organic carbon
film. The C/Si value of 0.47 may represent recontamination of a
very clean surface.

The C/Si value for the uncleaned glass plate exposed to the
U.V.-ozone treatment decreased from ∿1.6 to 0.52. A glass plate
previously cleaned using Isopanasol decreased in C/Si ratio from
0.72 to 0.19 upon U.V.-ozone treatment. Vig and LeBus have suggested that for surfaces contaminated with substantial amounts of
organic carbon, the U.V. radiation may crosslink the film forming

an intractable layer[22]. It is possible that such a layer was formed
on the uncleaned glass plate.

It has been reported that the U.V.-ozone cleaning technique
removes essentially all carbon from the surface of quartz and glass.
The low C/Si ratios observed in these experiments may represent
contamination which occurred in the interval (<5 minutes) between
switching off the U.V. lamp and placing the sample in the ESCA
unit. One glass specimen which was U.V.-ozone cleaned, observed by
ESCA, and then reexposed to laboratory air for 13.5 hours showed
a 4x increase in the C/Si ratio in this time period.

The ESCA data shown in Figure 3 have shed considerable light
on the mechanisms of both contamination and cleaning. Ten atom
percent carbon at the surface of a smooth glass specimen represents
approximately one adsorbed monolayer. Therefore, almost all the
specimens studied in these experiments have between one and four
monolayers of organic carbon on their surface. Thinner contaminant
films (lower C/Si values) show proportionately more carbon compounds containing groups capable of participating in polar interactions and less hydrocarbon-type compounds. For the uncleaned
glass specimen which probably contains four or more monolayers of
organic carbon material at its surface, a large fraction of the
film is hydrocarbon in nature. A model which can explain these
observations is illustrated schematically in Figure 8. The "clean"
glass surface is highly polar in nature. It would preferentially
tend to adsorb organic compounds which could participate in hydrogen bonded interactions. Typical functional groups which might
enter into such bonding include hydroxyl groups, ether oxygens,
organic acids and esters. Once the polar portions of these molecules
were tied up in hydrogen bonding interactions with the glass, their
non-polar portions would be oriented outwards. These could interact
by Van der Waals bonding forces with hydrocarbon contaminants. All
further contamination layers would tend to be hydrocarbon in nature.
Washing could readily remove these hydrocarbon layers which would
be held to the surface by relatively weak forces. The H-bonded
first monolayer would be difficult to remove. If it were removed,
there would be a strong driving force for it to reform. The multi-
layer structure of this organic carbon contaminant film could be
investigated using angle resolved ESCA studies[16]. As the angle of
the sample with respect to the detector was increased, one would
expect to observe more hydrocarbon and less of the polar compounds
if this model were correct.

The model proposed in Figure 8 could also explain the results
obtained using the U.V.-ozone cleaning procedure. Carbonyl containing compounds are much more likely to interact with U.V. radiation and with ozone than hydrocarbon compounds. Therefore, on
pre-cleaned glass, the U.V.-ozone treatment is extremely effective

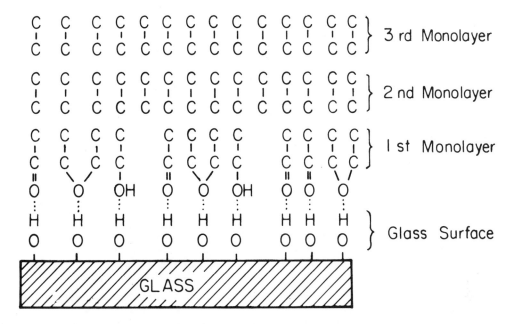

Figure 8.   A schematic representation of the organic carbon film on
glass. C–C–OH, C–C=O and C–C–O–C–C are polar-type organic compounds. C–C represents hydrocarbon-type organic
compounds. Dotted lines indicate hydrogen bonding interaction.

in eliminating the tightly bound monolayer. However, if an overlayer
of hydrocarbon is present, this layer may crosslink or block the
interaction of the U.V. radiation and ozone with the polar monolayer.

Potassium is a common component of many glasses. The potassium
concentration at the surface of the glass is observed simultaneously
with carbon in the 270–290 eV range. A correlation was observed
between the C/Si ratio and the C/K ratio (Figure 5). Although potassium data is easily obtained simultaneously with carbon data
(Figure 1), the C/K value is not as reliable an indicator of surface
contamination as the C/Si ratio since potassium is present in very
small quantities compared to silicon and since it can diffuse or be
displaced from the surface.

The absence of an effect of the various cleaning techniques on
the attachment of 3T3 cells to glass was surprising. Other investigators have reported effects of glass surface treatment on cell
attachment. For example, Rappaport et al. studied the attachment
of HeLa cells to glass and found that the total negative charge
of the glass or the amount of sodium in the glass could affect this
process[24]. Related effects of glass surface charge were also found

on L cells and monkey kidney cells. Smith et al. reported large differences in the adhesive strength of 3T3 cells to glass before and after glow discharge[21].

Even the lowest level of organic carbon contamination investigated in this study represents a monolayer of organic carbon on glass. Further adsorption of additional organic carbon layers does not appear to alter 3T3 cell adhesion even though the composition of the upper monolayers may differ from that of the initial monolayer. Since even the initial monolayer would leave the exposed surface predominantly hydrocarbon in nature, this might explain why cleaned glass surfaces show similar attachment behavior to those which were used as received.

This insensitivity to organic carbon contaminants is in marked contrast to the effect of adsorbed monolayers of proteins to the glass surface. The addition of 10% calf serum to the attachment media significantly delays the initiation of the rapid attachment phase. This may be due to more specific interactions between cell surface proteins and substrate bound proteins[25].

Monolayer protein films (~ 0.5 $\mu g/cm^2$) generally depress cell adhesion[25-27]. Certain adsorbed proteins enhance cell adhesion,[26,28] but these effects seem to be biospecific, rather than the non-specific adhesion decrease seen with most proteins which have been examined. The difference between the non-specific bioadhesive properties of most protein adsorbed layers, and the bioadhesive behavior of the hydrocarbon contaminant films is striking.

Preliminary experiments in our laboratory have indicated that the adhesion of human erythrocytes to glass, as measured in a thin channel flow cell similar to that used by Mohandas et al.,[29] may be sensitive to the degree of organic surface contamination[30]. Therefore, different cell types may respond in different ways (i.e., with different adhesive strengths) to surface contamination films and generalizations on the effect of these contamination films on cellular interaction may not be possible.

CONCLUSIONS

Organic carbon contamination on glass coverslips can be reduced by a factor of four or greater using various laboratory cleaning solutions. Strong acidic oxidizing cleaning solutions are perhaps slightly more effective than other laboratory cleaning solutions. Argon etching can reduce organic carbon at the surface to very low levels. However, an extremely reactive surface is produced which rapidly becomes coated with an organic carbon contaminant film. U.V.-ozone treatment of glass surfaces was found to be highly effective in reducing surface contamination.

The ESCA spectra of glass plates covered with various amounts of adsorbed organic carbon suggest a model in which the initial contamination monolayer is polar in nature and hydrogen bonded to the glass. Additional contamination layers are hydrocarbon in nature. These additional layers are easily removed by most washing procedures. Removal of the hydrogen bonded layer necessitates much more vigorous cleaning procedures (e.g., exposure to U.V.-ozone).

A question of extreme importance is that of recontamination rates. The C/Si values of 0.3-0.7 seen for most cleaning techniques may represent stable, moderately clean surfaces. If surfaces were examined immediately after cleaning, or stored under high vacuum, larger differences in the efficacy of the various cleaning methods have been noted. However, the results obtained here might represent those which could be expected for glassware in an actual laboratory situation, i.e., glassware cleaned and stored in laboratory air for a period of days.

The amount of organic carbon contaminant film on glass did not influence 3T3 cell attachment. However, preliminary data with washed erythrocytes indicates effects on attachment related to the method used to clean the glass surfaces. The sensitivity of cells to contaminant films during the attachment process may vary significantly from cell type to cell type. This might be manifested in the strength of adhesion rather than the rate or extent of cell attachment[31]. This parameter is difficult to evaluate using the dip-shear rinse technique.

The ESCA technique has been shown to be of immense value in studying organic carbon contamination on glass. Future experiments will attempt to clarify the nature of the contaminant films, recontamination rates, changes occurring in the glass surface due to cleaning, and the relationship between organic carbon contamination films and cell adhesion.

ACKNOWLEDGEMENT

This work was supported by NIH grant #HL19419. The authors wish to thank Dr. Thomas Horbett for his preliminary erythrocyte adhesion data and helpful suggestions and Barbara Pedersen and Michael Schway for their excellent technical assistance.

REFERENCES

1. L. Holland, "The Properties of Glass Surfaces", Chapt. 5, John Wiley & Sons, Inc., New York, 1964.
2. F.M. Ernsberger, Ann. Rev. Mater. Sci., 2, 529 (1972).
3. G.R. Sparrow and H.E. Mismash, in "Quantitative Surface Analysis of Materials" (ASTM Special Technical Publication #643),

N.S. McIntyre, Editor, p. 164, ASTM, Philadelphia, PA, 1977.

4. J.P. Rynd and A.K. Rastogi, Ceramic Bull., 53, 631 (1974).
5. C.G. Pantano and L.L. Hench, J. Testing Evaluation, 5, 66 (1977).
6. M.G. Yang, K.M. Koliwad and G.E. McGuire, J. Electrochem. Soc., 122, 675 (1975).
7. J.R. Vig, "Surface Studies for Quartz Resonators", Army Electronics Command Report AD-785 513, Sept., 1974, available through National Technical Information Service, Springfield, VA 22151.
8. L.L. Hench and E.C. Ethridge in "Surface Contamination: Its Genesis, Detection and Control," K.L. Mittal, Editor, Vol. 1, pp. 313-326, Plenum Press, New York, 1979.
9. P.S.P. Wei, Appl. Physics Letters, 17, 398 (1970).
10. K. Vedam and M. Malin, Mater. Res. Bull., 9, 1503 (1974).
11. P. Heyndrychx, Glastechnische Berichte, 44, 543 (1971).
12. C.E. Bryson, L.H. Scharpen and P. Zajicek, This proceedings volume.
13. K. Hickson, Glastechnische Berichte, 44, 537 (1971).
14. F. Jona, Appl. Physics Letters, 6, 205, (1965).
15. R. Kesavan and P.N. Andhare, Indian J. Technol., 13, 403 (1975).
16. C.S. Fadley, Progress in Solid State Chem., 11, 265 (1976).
17. D.T. Clark and H.R. Thomas, J. Polym. Sci., Polym. Chem. Ed., 14, 1671 (1976).
18. D.T. Clark, A. Dilks, D. Shuttleworth and H.R. Thomas, J. Electron Spectros. and Related Phenom., 14, 247 (1978).
19. T.A. Carlson, "Photoelectron and Auger Spectroscopy", Plenum Press, New York, 1975.
20. D.T. Clark, in "Polymer Surfaces", D.T. Clark and W.J. Feast, Editors, Ch. 16, John Wiley & Sons, Inc., New York, 1978.
21. L. Smith, D. Hill, J. Hibbs, S.W. Kim, J. Andrade and D. Lyman, ACS Polymer Preprints, 16(2), 186 (1975).
22. J.R. Vig and J.W. LeBus, IEEE Trans. on Parts, Hybrids and Packaging, PHP-12(4), 365 (1976).
23. E.G. Butler and W. Johnston, Science, 120, 543 (1954).
24. C. Rappaport, J.P. Poole and H.P. Rappaport, Expt. Cell Res., 20, 465 (1960).
25. F. Grinnell, Intern. Rev. Cytol., 53, 65 (1978).
26. B.D. Ratner, T.A. Horbett, A.S. Hoffman and S.D. Hauschka, J. Biomed. Mater. Res., 9, 407 (1975).
27. A.C. Taylor, Expt. Cell. Res. Suppl., 8, 154 (1961).
28. F. Grinnell, D.G. Hays and D. Minter, Expt. Cell Res., 110, 175 (1977).
29. N. Mohandas, R.M. Hochmuth and E.E. Spaeth, J. Biomed. Mater. Res. 8, 119 (1974).
30. T. Horbett, Personal Communication (1978).
31. J. Waldberger, M.S. Thesis, University of Washington, Seattle, WA (1978).

AN ESCA ANALYSIS OF SEVERAL SURFACE CLEANING TECHNIQUES

C. E. Bryson, L. H. Scharpen and P. L. Zajicek

Surface Science Laboratories

4151 Middlefield Road, Palo Alto, California 94303

Several cleaning techniques for materials used in a manufacturing process demanding high surface cleanliness were evaluated using ESCA and UV spectrophotometry. The techniques examined included UV-ozone cleaning, argon ion milling and several wet chemical techniques. The UV-ozone was emphasized since it appears to be moderately gentle and partially selective to hydrocarbons. We also compared the recontamination rates of the cleaned samples in room air, and in 10^{-9} Torr vacuum of the ESCA spectrometer. The relative recontamination with hydrocarbon picked up by clean surfaces when rinsed with reagent grade solvents was examined.

INTRODUCTION

In the manufacturing of atomic clocks, the cleanliness of the components that are used to make the cesium beam apparatus is of utmost importance. It has been found that the long-term stability of the electron multiplier, the ion pump and the cesium source all depend on a clean environment. The most critical is the electron multiplier. The data reported here resulted from part of the effort spent in developing a better cleaning procedure for the cesium beam tubes (C.B.T.) manufacturing process.

Two needs dominated the definition of the requirements imposed on this cleaning procedure. One was the need to provide a wide margin of cleaning ability against unexpected contaminants. A survey of the possible contaminants from a number of the component vendors revealed little could be excluded. In addition to the expected fingerprints and machining oils, unusual items like

687

chewing tobacco (used) and sawdust were encountered. The other need was the capacity to process the volume of parts needed without saturation of the degreasing solvents and cross-contamination.

The primary tool used for evaluating the techniques considered was X-ray photoelectron spectroscopy (XPS or ESCA). This was supplemented in one case with U.V.-visible spectrophotometry. This latter technique was used for selecting the fluids to be used in a degreasing operation. ESCA was used for evaluating the effects of the degreasing, acid cleaning, H_2 firing and U.V.-ozone cleaning.[1] The overall result is, of course, measured by the life and stability obtained with the cesium beam tubes. The following specific examples are situations where the above mentioned measurement techniques yielded rapid and meaningful results for a production environment.

A)   Degreaser fluid selection and contamination monitoring
B)   Oxide removal
C)   Quartz resonator cleaning--U.V.-ozone

MEASUREMENT APPARATUS AND PROTOCOLS

The ESCA spectra were taken on a Hewlett-Packard 5950A and 5950B spectrometer located at Surface Science Laboratories in Palo Alto. The U.V.-visible absorption spectra were taken on a Cary 14 interfaced with a Data General Nova 840 located in the Chemistry Department at Stanford University. Most of the cleaning procedures were carried out at Hewlett-Packard in the Santa Clara facility. The problem of transporting a sample between these facilities (approximately 12 miles) was solved by using the sorbtive properties of aluminum oxide.[2] The samples were wrapped in aluminum foil freshly removed from a roll. This appears to have been sufficient. Essentially the same results were obtained on comparing the spectra from the cleanest samples we obtained--quartz crystals cleaned with U.V.-ozone both after cleaning at Hewlett-Packard and after cleaning at Surface Science Laboratories. The time between the removal of the samples from the cleaner and the insertion in the ESCA vacuum chamber was four hours for the samples cleaned at Hewlett-Packard and 45 seconds for those cleaned at Surface Science Laboratories.

The composition of the material on the surfaces of the samples was calculated from the areas under the peaks of the ESCA spectrum. The areas were normalized using the theoretical photo-ionization cross sections.[3] Corrections for variations in analyzer transmission and electron escape depth with energy were neglected since these cancel within the certainty to which they are known. This last approximation along with the general uncertainties in the cross sections and area measurements limit the absolute accuracy of composition calculations to approximately 20% of value

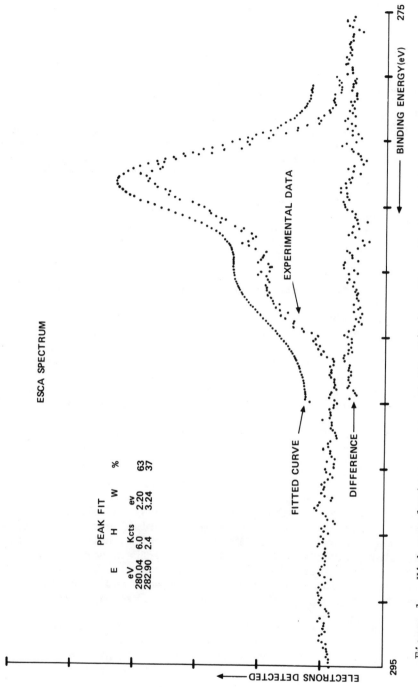

Figure 1. High resolution spectrum of the C(1s) region of the photoelectron spectrum.

for each element. Repeatability from sample to sample for similar samples is approximately 2% for strong signals. Table II lists the compositions found for the surfaces of several aluminum samples. A good measure of the sample to sample variation is the aluminum to oxygen ratio which is 0.51 with a standard deviation of 0.02. The variation in carbon to aluminum ratio reflects the variation in surface contamination on the samples.

Where reported, the quantity of oxidized carbon was determined from high-resolution spectra such as shown in Figure 1. The experimental data was fit with a sum of two Gaussian curves using a least squares routine. The ratio of oxidized carbon to unoxidized carbon was taken to be the ratio of the area under the two peaks, the peak near 281 eV corresponds to unoxidized carbon and the broader peak near 283 eV corresponds to the oxidized carbon. The term "oxidized carbon" is used as opposed to a specific oxide because the width of the 283 eV peak is too wide to be a single oxide.

DEGREASERS

All of the parts used in this manufacturing process are first degreased in two commercial vapor degreasers, first in one and then the other. The use of two degreasers in tandem provides three advantages:

A) The possibility of taking advantage of complementary properties of two solvents
B) Provision of a reserve margin of capacity is provided since the second degreaser is used only on relatively clean parts
C) Less ill effect due to operator mistake, increased chance of at least one step being done properly

The fluid used for the first degreaser, "Freon" TMS, is an azeotrope of methanol in trichlorotrifluoroethane.[4] The fluid used in the second degreaser is trichlorotrifluoroethane, "Freon" TF. Both degreasers are equipped with a spray and ultrasonic cleaner. These fluids were tested in comparison with 1,1,1,-trichloroethane "Freon" and TMC, an azeotrope of methylene-chloride in trichlorotrifluoroethane. Table I lists the absorbance measured at 200 nm for several sets of quartz flats. These flats had all been previously cleaned in nitric acid and hydrofluoric acid and then cleaned with U.V.-ozone. The absorbance obtained on the cleaned flats was measured and stored digitally. The flats were then contaminated with fingerprints and water-soluble machining oils. Two were left uncontaminated to serve as a reference and a control. The samples were wiped as clean as possible with tissue and vapor degreased as described above. They were not immersed in the solvents. The spectra were then recorded and digitally corrected with the spectra of the clean flats. This

Table I. U.V. Absorbtion of Cleaning Residues.

Solvent Used	Substance @ 200 nm		
1,1,1,-Trichloroethane	.026	±	.002
"Freon" TF alone	.030	±	.002
"Freon" TMS alone	.040	±	.002
"Freon" TMC alone	.050	±	.002
"Freon" TMS followed by	.019	±	.002
TF Control	.001	±	.002

corrected absorbance is listed in Table I. A clean flat was used in the reference beam in both cases.

When repeated with non-water soluble oils, the results were similar except the TMS treatment produced a lower absorbance than the 1,1,1,-trichloroethane. Since most hydrocarbons absorb strongly at 200 nm, absorbance is a measure of cleanliness.[5] For a strong absorber with an absorbtivity of 10^5 an absorbance of .02 corresponds to a film thickness of 20 Angstrom.

To establish the level of residues left by the "Freon" TMS, TF combination and to spot check the performance of the degreasers during production, pieces of aluminum foil were cleaned as they were taken off a fresh roll and then analyzed with ESCA. Table II lists the surface composition found for some of these samples. The samples for Case A were cleaned at the end of the day during

Table II. ESCA Analysis of Solvent Residues on Aluminum Foil.

Case A

Sample	Al	C	O	F	C/Al	O/Al
Fresh	27	18	53	1.3	.67	1.96
"Freon" TMS & TF	25	26	47	.4	1.04	1.92

Case B

	Al	C	O	F	C/Al	O/Al
Fresh	26	24	48	.5	.92	1.85
"Freon" TMS & TF	28	12	58	1.1	.43	2.08
"Freon" TF	24	28	47	.4	1.17	1.96
"Freon" TMS	27	16	54	2.0	.59	2.00

which an exceptionally large number of very dirty parts were cleaned. The samples for Case B were obtained after the degreasers had run all night, thereby cleaning the solvents by distillation.

For comparison a set of 10 samples of aluminum foil off the same roll was analyzed without any cleaning. The analysis yielded a mean C/Al ratio of .63 ± .23; by discarding two high numbers the mean dropped to .52 ± .12.

OXIDE REMOVAL

Oxides were removed primarily by acid etching. Various acids are used for different metals. The principal problems encountered were rinsing and drying. Until a Quick Dump[6] system was installed, large quantities of water were required to adequately remove all traces of the acids used. The drying problem which was mostly a matter of eliminating spots and recontamination was solved by using a "Freon" TF vapor dryer preceded by a propanol rinse. Table III shows the atom percent concentration of the elements found on the surface of two samples dried in two ways. Both samples were OFHC copper that had been cleaned in 1 molar HCl. They were rinsed in an ultrasonic water bath and the Quick Dump. Both were then rinsed in propanol with ultrasonics for 15 seconds. Sample I was then blown dry with pure nitrogen. Sample II was dried with a "Freon" TF vapor dryer.

In production, the vapor dryer is giving uniform results with virtually no spotting.

QUARTZ RESONATOR CLEANING

U.V.-ozone (U.V.O.) cleaning has been shown to be a very promising technique for cleaning quartz crystal resonators.[7] In preparation for introducing this technique into a production environment, we did an evaluation to compare U.V.O. to conventional chemical cleaning and argon ion etching. This work is summarized below.

Two pieces of equipment were used for the U.V.-ozone cleaning reported on in this paper. One consists of a serpentine quartz Hg lamp mounted below a stainless steel rotary platform.

Table III. ESCA Comparison of Drying Techniques.

Sample	Cu	C	O	Cl
I	35	38	26	1.2
II	37	24	39	.3

Both the lamp and platform are enclosed in a stainless steel
chamber. The crystals that are to be cleaned are placed in
circular trays which are in turn placed in holes in the rotary
platform. In operation, the platform revolves at about one R.P.M.
The crystals are directly over the lamp approximately 20% of the
time. The distance from the lamp to the crystals is about 2 cm.
This geometry provides a slower cleaning rate than might be
achieved with more optimum geometry but does allow a large number
of crystals to be processed at one time. A minimum exposure time
has not been determined; however, an exposure time of 24 hours has
been found to be satisfactory and was used for this work.

The other U.V.-ozone cleaner used is a small portable
commercial unit.[8] This unit, as delivered, had a plastic filter
holder and a tray lined with conductive foam. In this configuration it left a 4.5% carbon surface and after extended exposure,
approximately 3 hours, Sn, N, F, and S were observed to build up
on the quartz surface. Removing the filter holder and lining the
sample tray with aluminum foil solved this problem. This cleaner
was used for the recontamination study and carbon removal rate
study reported below. For the recontamination study, the exposure time was 20 minutes with the sample 1.5 cm from the lamp.
For the removal rate study the exposure was at the same distance
but in one minute intervals.

The argon ion etch cleaning was accomplished in the preparation chamber of the 5950A spectrometer using a 3M SIMS rastered
ion gun. The ion beam used was 0.1 mm in diameter rastered over
a 7 x 15 mm area. The ESCA spectra were taken from a 5 x 1 mm
area centered in the rastered area. The preparation chamber is
directly attached to the sample chamber of the ESCA spectrometer
allowing the sample to be analyzed immediately after etching
without exposure to room atmosphere. The preparation chamber
base pressure is approximately 10^{-7} torr. The base pressure of
the sample chamber is approximately 10^{-9} torr. Samples prepared
with argon ion etching serve as a control to establish the
minimum carbon level that can be achieved when measuring with
ESCA. This minimum due to recontamination is approximately 1.5%
for quartz and does vary depending on the base pressure of the
sample chamber.

Table IV lists the concentration of the elements found on
several quartz samples. Samples 1 and 2 were taken from normal
production cleaning prior to a final etch. This cleaning consisted of a thorough degreasing followed by several acid and
solvent rinses. Sample 3 was taken from the same batch as sample
2 after receiving a final etch and rinsing. For sample 4, a piece
of quartz had received additional chemical cleaning followed by
U.V.O. This result was typical of what we found on many trials.
The data listed as Sample 5 were obtained by taking the same

Table IV.   Cleaning Residues on Quartz.

Sample #	1	2	3	4	5
Si	36	28	25	33	32
O	58	51	51	65	63
C	6	13	17	2	2
N	.3	–	–	–	–
Cl	–	.7	1.1	–	–
Ca	.07	–	–	–	–
Cu	.09	–	–	–	–
Al	–	2.2	2.5	–	–
Zn	.05	.7	.3	–	–
F	<.04	4.4	3.9	–	3

piece and recleaning with U.V.O. but placing the sample in a "Teflon" holder.

These results show that there is considerable variation in the results obtained with the production cleaning process, little contamination from the final etch, and the U.V.O. with proper fixturing does a superior job.

An attempt was made to determine if the more vigorous methods of cleaning such as U.V.O. and argon ion etching (A.I.E.) led to different recontamination levels. The samples were analyzed and then recontaminated by rinsing in reagent grade propanol or undergoing a 24 hour exposure to air in a polystyrene box. The quartz cleaned well with either U.V.O. or A.I.E., i.e., to a level of carbon of about 2%. On recontamination, the carbon level went to near the value found before U.V.O. or A.I.E., that is about 15%. No significant differences were noted.

The rate carbonous material is removed from the surface of quartz with U.V.O. and the subsequent changes in surface chemistry were examined as follows: The sample was first cleaned to a carbon level of approximately 2.5% and then contaminated with a roughing pump oil. The excess oil was wiped off vigorously with a lint-free cloth. The sample was then analyzed and exposed to the U.V.-ozone environment for one minute. This sequence was repeated three times and then the exposure time was extended to two minutes. The data were analyzed as described in Section III and are plotted vs exposure time in Figure 2. The oxidized carbon to unoxidized carbon ratio increased rapidly and remained nearly constant until the total carbon level dropped to approximately 5%. The ratio then started dropping as the total carbon level dropped. This indicated that oxidization plays a major role in the cleaning process especially for multi monolayer levels.

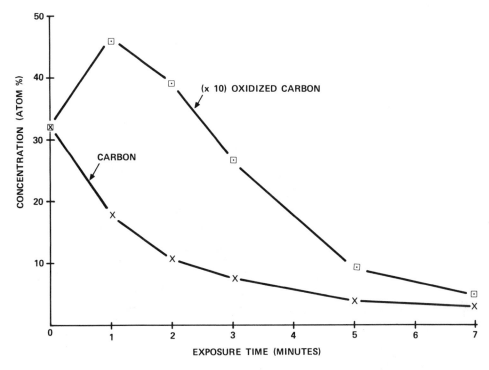

Figure 2.   Carbon removal from a quartz surface with U.V.-ozone.

During a similar experiment, the steam test was performed
on the sample both prior to and after ESCA analysis. If one takes
a 10% carbon level as determined by ESCA to correspond to a single
monolayer, then the results of the steam test agree well with the
definitions given in Reference 7. Briefly, fogging of the sample
occurred for contamination level near one monolayer, i.e., 7%
carbon and greater. Orange peel was observed from approximately
5% down to 2% carbon at which level poor fringes were starting
to form, again agreeing with the earlier results. Good fringes
thought to correspond to contamination levels less than .1 monolayer were observed before inserting the sample into the spectrometer but only poor fringes were observed on removing the sample
and re-testing with the steam test. This indicates that the 2%
carbon level limit we observed with the ESCA is the result of
recontamination due to the ESCA vacuum system. More work needs to
be done to extend the ESCA measurements to levels below 0.2 monolayer for carbon.

CONCLUSIONS

The cleaning procedures implemented have proved adequate
for the task of cleaning C.B.T. parts. Accelerated aging tests
specific to the life limiting mechanisms that result from contamination indicate a greatly increased life can be expected. ESCA
has proven to be a useful tool in determining procedures and in
checking them.

ACKNOWLEDGMENTS

We would like to thank Robert Linder and Ruth Records of
the Chemistry Department at Stanford University for the U.V.-
visible spectra.

REFERENCES

1. John R. Vig, "Surface Contamination, Its Genesis, Detection,
    and Control," K.L. Mittal, Editor, Vol. 1, pp. 235-254, Plenum
    Press, New York, 1979.
2. M. L. White, in "Proc. 27th Annu. Symp. Frequency Control,"
    U.S. Army Electronics Command, Ft. Monmouth, N.J.,
    AD 771042, pp. 79-88, 1973.
3. Compiled by: R.N. King, University of Utah, 2062 MEB,
    Salt Lake City, Utah 84112.
4. TF, TMS, and TMC are registered trademarks of E. I. du Pont
    de Nemours and Co., Wilmington, Delaware.
5. H.H. Jaffe and Milton Orchin, "Theory and Applications of
    Ultraviolet Spectroscopy," John Wiley and Son, New York, 1962.
6. Manufactured by Santa Clara Plastic, Santa Clara, California.
7. John R. Vig, and John W. LeBus, IEEE Trans. PHP, PHP-12,
    No. 4., December 1976.
8. Model S-528, Manufactured by Ultra-Violet Products, Inc.,
    San Gabriel, California.

QUANTITATIVE TECHNIQUES FOR MONITORING SURFACE CONTAMINATION

Tennyson Smith

Rockwell International Science Center

Thousand Oaks, California 91360

Phosphoric acid anodized A17075-T6 samples have been deliberately contaminated to varying controlled levels for the purpose of checking their detection with nondestructive surface tools. Essentially all types of contamination, surface damage from handling and process errors can be adequately detected by ellipsometry. Surface potential difference (SPD), photoelectron emission (PEE), and water contact angle were successful to varying degrees. A computer operated mapping facility is demonstrated to be successful for automatic nondestructive inspection (NDI).

INTRODUCTION

It was our purpose to develop a nondestructive inspection (NDI) technique that can be used, just prior to layup, to detect contamination on surfaces. In this report we focus on contamination of phosphoric acid anodized aluminum alloys. A broad interpretation of contamination includes surface damage due to handling and processing errors during surface preparation as well as organic contamination from various sources.

The type of organic contamination that will be present in a bonding facility will depend upon the precautions taken, ranging all the way from outdoor operations to ultra clean room conditions. The author visited some Rockwell adhesive bonding facilities and found adhesive bonding performed in normal factory environment and semi-clean room environments. Semi-clean room is somewhere between factory conditions and ultra-clean rooms used for semiconductor

697

Table IA.   Contamination Sources in Factory Environments

        Industrial smog
        Cigarette smoke
        Oil and grease associated with machinery
        Ink
        Food remnants from lunch boxes (orange peel, banana
             peel, bread, coffee, etc.)
        Human breath, perspiration and natural body oils
             (fingerprints, women's cosmetics, etc.)
        Clothing lint
        Plasticizers from plastics
        Process solutions

Table IB   Contamination Sources in Semi-Clean Room Facilities

        Industrial smog
        Clothing lint
        Human breath, perspiration and natural body oils
        Plasticizers from plastics
        Process solutions

integrated circuitry facilities. Table 1A is a list of contamination sources found in the factory environments, Table 1B is for a
semi-clean room.

     Contamination related problems are divided into four cate-
gories; processing errors, handling damage, human contamination
and possible contamination from smog aerosol particles. In the
case of phosphoric acid anodized aluminum, process errors include
incorrect anodize voltage or time, contamination from bath
solutions and delay in removing from phosphoric acid before rinsing.
Handling damage refers to mechanical damage to the anodic layer by
handling with clean cotton gloves, kraft paper or kimwipes. Human
contamination is indicated in Table 1.

     Representative constituents of smog[1] include fatty acids,
alcohols, alkenes, etc. dissolved in low molecular weight hydrocarbons and water. Direct measurement of aerosol liquid water
content in smog ranged from less than 10% to more than 50% by
weight depending on humidity.

     The problem is to develop NDI techniques for the various types
of contamination. A technique should be one that can be used to
inspect all areas of surface treated panels or parts. However, the
need for sophisticated equipment that can scan curved or shaped
parts depends on the critical nature of the parts and the number of
parts to be inspected. In many factory situations, inspection of
control samples may be sufficient. This study is limited to flat
panels or parts but the techniques described in this report can be

adapted to the inspection of shaped or curved parts provided the
radii of curvature are not too small.

The solution to the problem lies in the development of NDI
tools that can detect deviations from an acceptable surface condition. The first step involves characterization of an acceptable
surface, i.e., to identify the boundaries of signals from surface
tools for which the surface is acceptable. The second step involves
establishing which tools can detect deviations from the acceptable
surface. The third step involves automatic scanning of surfaces
with the surface tools to identify regions that deviate from the
acceptance band (are considered contaminated). The type of readout
depends on the circumstance and desired spatial resolution. A map
of the part, showing contaminated regions, might be desirable. On
the other hand, automatic scanning, with a light or sound warning
if a given minimum area deviates from the acceptance band, might be
desired. This study is restricted to automatic mapping with a two
dimensional contamination plot and/or averaging the deviation from
the acceptance band over selected areas.

EXPERIMENTAL

Surface Tools

The surface techniques that are conducive to rapid scanning of
surfaces are: ellipsometry, surface potential difference (SPD),
photoelectron emission (PEE) and these tools have been described
previously[2],[3]. In the case of ellipsometry, a new automatic scanning ellipsometer has been developed[4]. This ellipsometer has the
advantage of no moving parts so that the rate of scanning surfaces
is only limited by the rate of motion of the sample under the
sensor. Maps were also made by automatically placing drops of water
on the surface and automatically checking if the drop would wet the
surface.   This was done by a probe that followed the drop dispenser.
If the drop did not spread, the probe would detect its presence by
making electrical contact with the sample as the probe touched the
drop.

Computerized Mapping

There may be no need for a computer in the finally chosen NDI
instrument but we are using a Data General "Eclipse" MDL S/200 minicomputer to control automatic scanning of the sample under the
sensing heads, for data acquisition and data processing. In order
to develop a field instrument, our approach is to first establish a
signal range or band within which the sample surface is proper.
This is done by mapping properly prepared samples and having the
computer report the average signal value, the mean deviation from
the average value and the maximum and minimum values. After storing
the actual signal values for each sample position, we can print out

the array and observe the number map; or, we can make a map with the
X-Y plotter. The map is made by dividing the chosen area on the
chart paper into an array of smaller areas and plotting within each
small area a number of dots. The number of dots is proportional to
the signal amplitude in the corresponding map position. To simulate
field use, the computer is programmed to suppress any chosen band
of the signal and plot deviations from the band. If the band is
chosen as zero, all values that deviate from the mean are plotted.
To reveal contamination in greater contrast, the bandwidth is increased so that dots that correspond to proper surface are suppressed.
Choice of the proper bandwidth in the factory will depend on the relationship between contamination level and joint strengths and
durability. These aspects are not considered in this paper. Figure 1
shows the instrument head mounting. The sample is automatically
moved under the detector heads.

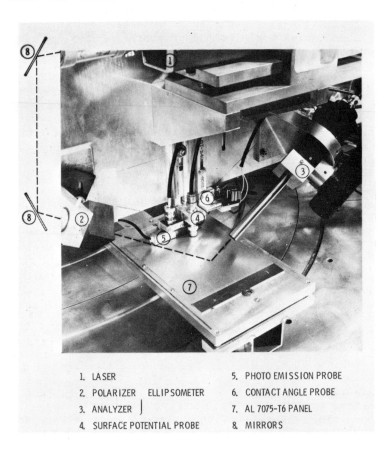

1. LASER	5. PHOTO EMISSION PROBE
2. POLARIZER ELLIPSOMETER	6. CONTACT ANGLE PROBE
3. ANALYZER	7. AL 7075-T6 PANEL
4. SURFACE POTENTIAL PROBE	8. MIRRORS

Figure 1   Computer operated contamination mapping facility.

EXPERIMENTAL RESULTS

Sensitivity to Contamination

Processing Errors: The sensitivity of the surface tools to processing errors of three types were considered, i.e., wrong anodizing voltages, times and post anodizing exposure to phosphoric acid.
Table 2 gives the ellipsometric, SPD and PEE results. It should be
noted that for Table 2, the manual ellipsometric data were taken at
an angle of incidence $\phi = 60^o$ rather than 70°, for comparison with
the automated ellipsometer which was set at 60°. Due to surface
roughness, the automated ellipsometer (operates on light intensity)
yields different Δ and ψ values than for the manual (light null)
technique. Nevertheless, both techniques yield values related to
film thickness as observed for hydroxide dissolution into the phosphoric acid (lower portion of Table 2). The water contact angle
θH_2O for freshly anodized aluminum is near zero whereas that for the
aged (contaminated) aluminum, $\theta H_2O \sim 50^o$. In each case the SPD decreases approximately linearly with film thickness. SEM pictures
indicate that at a given voltage the hydroxide film on A12024-T3 is
thinner (and perhaps more dense) than for that on A17075-T6.

FPL etched (sulfuric acid-dichomate) aluminum alloys have a film
thickness of about 100Å and PEE of about 300 (pA). Anodic films of
1400Å or greater (Table 2) yield about 60 (pA). The constancy of
PEE with thickness (for anodic films) indicates that photoemission

Table II. Sensivity of Surface Tools to Processing Errors on
A17075-T6.

Anodic Parameters			Ellipsometry, $\lambda = 6328$, $\phi = 60$				SPD	PEE
			Automated		Manual			
Potential	Anodize Time	Post Anodize Time	Δ	ψ	Δ	ψ	Esti-mated Thick.		
(volts)	(min)	(min)	(deg)	(deg)	(deg)	(deg)	Å	Volts	(pA)
Anodize								Fresh Aged 6 Mo.	
0	0	0					100	1.0	300
5	20	0	160	45	152	45.5	1400	0.5 0.3	60
10	20	0	158	40	153	40.8	3300	0.3 0.2	60
15	20	0	162	40	157	46.1	4650	0.2 0.2	60
20	20	0	145	45	135	42.3	5900	0.06 0.1	60
25	20	0	166	39	170	41.0	7200	-0.1 -.2	60
H_3PO_4 Exposure									
10	20	0	167	44	170	42.0	1472	0.2	40
10	20	2.5	155	44	152	45.0	1375	0.5	50
10	20	5.0	144	40	122	41.0	225	0.7	200
10	20	7.5	144	38	124	38.7	225	0.7	1200
10	20	10.0	141	35	117	36.0	270	0.7	1600

may be from the outer region of the hydroxide film rather than from
the substrate metal. However, this is unlikely because alumina
does not emit until the photoenergy is of the order of 9 eV as compared to 5 eV used in these measurements. Another possibility is
that the hydroxide is essentially transparent to electrons in the
porous outer film but are attenuated by the inner barrier layer.

Data at the bottom of Table 2 reveal that exposure to H_3PO_4
after anodizing decreases the film thickness about 100Å in the first
2.5 minutes and about 1150Å in the next 2.5 minutes. Further exposure leaves about 225Å of film but roughens the surface with etch
time. The increased roughness is indicated by the decrease in ψ
and the increase in PEE.

These results demonstrate that the ellipsometer and SPD are
sensitive to film thickness but that PEE and water contact angle are
not. For small changes in film thickness (~ 200Å) Δ changes by about
1 deg/10Å. The SPD changes by about -1 mV/10Å for A17075-T6 and
-2.4 mV/10Å for A12024-T3.

Handling Damage: It was discovered at McDonnel Douglas Corp. by
M. Danforth, that rubbing a phosphoric acid anodized A17075-T6
surface with a clean cotton glove or Kraft paper would degrade the
surface with respect to bond endurance, as measured by the wedge
test[5].

The effect of rubbing the anodized surface with cotton, Kraft
paper, or kimwipe paper tissue, is about the same, i.e., increase
Δ, decrease ψ, increase SPD, and increase θH_2O, with little change
in photoelectric emission (PEE). To see if rubbing with material
that was contamination free would affect surface parameters, the
surface was rubbed with quartz wool. The cleanliness of the quartz
wool was checked by allowing single wool fibers to touch water. The
fibers were immediately wetted indicating they were clean. The
surface property trends are the same for rubbing with contamination
free material as for all the other materials, indicating that the
changes are not due to contamination but due to some mechanical
disturbance of the hydroxide film.

To deliberately contaminate a surface with a low energy material,
samples were rubbed with a clean Teflon bar. Δ decreased by $10°$
rather than increase, as for cotton glove, etc., SPD increased from
0.13 to 0.3, PEE decreased from 120 (pA) to 80 (pA) and θH_2O increased from 0 to $50°$. It was also thought that if contamination
particles were deposited by rubbing with the different materials,
that heating the sample might cause the contamination to spread. A
sample was heated with a hot air gun after rubbing with Kraft paper.
The only property to change appreciably was SPD from 1.1 to 0.5 volts
Again, it was concluded that contamination is not the cause of change
in surface properties.

To check the effect of pressure, without rubbing against the sample, aluminum foil was placed between the sample and the cotton glove and the foil was rubbed with hard pressure. This procedure has little effect on the surface parameters, except that in some cases SPD increases. The foil was FPL etched and checked for zero contact angle before use. An area of the sample was also pressed using clean platinum foil, no changes in surface properties occurred. Samples were pressed with foil that was not properly cleaned. In these cases, SPD, PEE and θH_2O increase. A sample was rubbed with aluminum foil. The increase of PEE from 80 (pA) to 14000 (pA) indicates the large amount of abrasion exposing fresh aluminum with very thin oxide.

Organic Contamination: To establish the sensitivity of the surface tools to the presence of organic contamination, measurements were made on samples dipped through monolayers on water and samples exposed to aerosols of petane containing large molecules in solution. These two types of sources simulate contamination during the removal of panels from the phosphoric acid solution after anodizing, and exposure to smog, respectively.

Samples were dipped into pentane that contained 50µg of stearic acid per ml. One sample was repeatedly dipped for six sec and measurements made after the sample had dried after each dip. Another sample was left in for 36 sec, dried and measured. The changes in Δ, ψ, SPD and θH_2O are given in Figure 2. The refractive index of stearic acid is 1.66 and the approximate change in $\delta\Delta$ per Angstrom is calculated to be $\sim 1.4°/10\text{Å}$. The lower plot of $-\delta\Delta$ vs. time in Figure 2 indicates that for repeated dipping, about two monolayers (56Å) have adsorbed in 25 sec and that further dipping adds little to this. The open circle for the sample that remained in the solution for 36 sec indicates that only a monolayer was deposited. Therefore, the process of repeated removing the sample from solution causes more acid to be adsorbed. However, the open circle for the sample after 144 sec more is at about the same position as for repeated dipping. The contact angle increases dramatically even for partial monolayers and reaches 110° at two or three monolayers. The change in SPD is approximately linear with increasing contamination thickness, rather than decrease as for growth of the hydroxide film. The SPD increases by approximately 10 mV/10Å for growth of the hydroxide film. The SPD increases by approximately 10 mV/10Å for growth of the anodic hydroxide.

One exposure to aerosol containing anthracene deposits about 10Å and doubles the carbon peak for Auger electron spectroscopy (AES) (with respect to the control) but leaves the surface wettable. One exposure to aerosol with stearic acid leaves about 22Å which increases θH_2O to about 36° and is detected by AES (carbon peak, 1.5 v. 0.4 for the control). A sample that had been degreased but

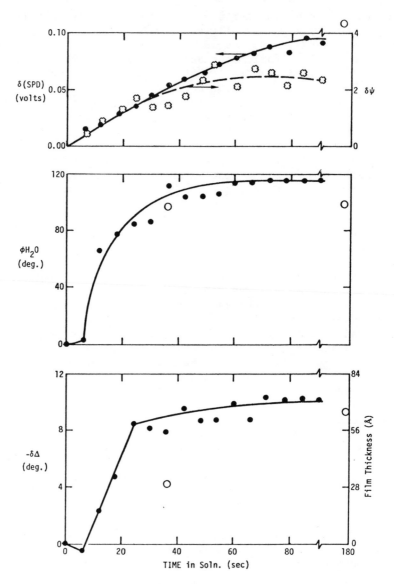

Figure 2.   Plot of changes in SPD, θ H$_2$O, Δ and ψ for stearic
            acid/Al.

not anodized left SPD high, $\theta H_2O \sim 58^o$ and the carbon peak at ~ 2.7.
The PEE is about five times as large as for anodized samples due to
a much thinner oxide layer.

Aging in Laboratory Air: All of the surface tools indicate
growth of contamination during aging of anodized samples in laboratory air, but at the extremely low rate of about $1\text{\AA}$/day or a monolayer ($25\text{\AA}$) in about 25 days. This can be compared to contamination
of FPL etched Al2024-T3, of about 1 monolayer in 10 hrs. Auger
electron spectroscopy confirms the contamination to be carbonacious
and that the rate is approximately $1\text{\AA}$/day.

NDI, Automated Computer Mapping

Aluminum panels (4'x5') were anodized and mapped with the automated mapping facility to obtain the acceptance band for each surface
tool. The panels were then deliberately contaminated or damaged and
remapped, with the acceptance band suppressed, to reveal the contamination or damage. Only one type of contamination is reported
here.

Human Contamination: We received 1'x1'x0.033" production
anodized panels of Al2024-T3 from McDonnel Douglas. The uniformity
of the anodiz film was checked by mapping two panels with ellipsometry, SPD, PEE and water contact angle. All of the positions on
the panels were wettable ($\theta H_2O \sim 5^o$). The anodic films were estimated
to be about 3700 to $3800\text{\AA}$ with index of refraction about 1.3 (i.e.,
about 20% porous). The average values of Δ were 121.6±1.2,
127.2±1.3, and 116.1±1.3. These values correspond approximately to
an average variation of $14\text{\AA}$ over one side of a panel, $35\text{\AA}$ from one
side of a panel to the other and $100\text{\AA}$ from one panel to another. The
average surface potential difference (SPD) values proved to be
0.27±0.03, 0.29±0.04, 0.26±0.03 volts for measurements on one day
and 0.15±0.02 and 0.16±0.02 volts for measurements on the next day.
This change is due to drift of the reference electrode. The photoelectron emission values average 23±2, 23±1, 20±2 (pA) on one day
and 26±2 on the next. There was little difference between values of
SPD and PEE from one side to the other and from one panel to the
other. The uniformity in each case was within ±10% of the average
value.

A panel was divided into 2"x2" squares for deliberate contamination. Every other square was contaminated, according to the
pattern in Figure 3. Contamination included coffee (wetted and
dried), soda pop (wetted and dried), smudge with clean cotton glove,
fingerprints, 3 in 1 oil, hand lotion, lipstick, ink, human cough,
cigarette smoke (after filtering through a human lung) and smog
constituents docosane, hexadecylamine and stearic acid in pentane
aerosol. The Δ plot in Figure 3b reveals all contamination but
cigarette smoke and the cough. The ψ plot (Figure 3c) reveals all

contamination but stearic acid, cigarette smoke and the cough. The
SPD plot (Figure 3d) reveals all contamination except the cough and
just barely the cigarette smoke. A water contact angle plot revealed the stearic acid, hexadecylamine, docosane, lipstick, hand
lotion, 3 in 1 oil and fingerprints, but not the cigarette smoke,
cough, ink, cotton smudge, soda pop or coffee.

Finally, the test of the automated NDI techniques with ellipsometry and SPD is shown in Table 3 for production panels from
McDonnel Douglas. The average value of Δ ,ψ and SPD computed by the
computer for 100 sq. inches of each panel is recorded in Table 3.
The acceptance bands for Δ , ψ and SPD have been chosen as
$141 < \Delta < 144$, $41 < \psi < 43$ and $.2 < SPD < .4$. Panel 4 was rotated
90° and replotted to see the effect of the grain direction. The
average value of Δ increased by 2° while ψ and SPD changed very
little. The change in Δ is due to the difference in light scattering with the direction of roughness. The difference is small enough
to be neglected in terms of the acceptance band. It would be useful
in practice to scan all panels with the grain in the same direction
but probably not necessary.

Clean Cotton Glove Smudge		Soda Pop		Coffee
	3 in 1 oil		Fingerprints	
Ink		Lipstick		Hand Lotion
	Cigarette Smoke		Cough	
Stearic Acid		hexadecylamine		docosane

Figure 3.   Computer plot of Δ ,ψ and SPD for 1'x1' production
            panels with contamination due to human error.

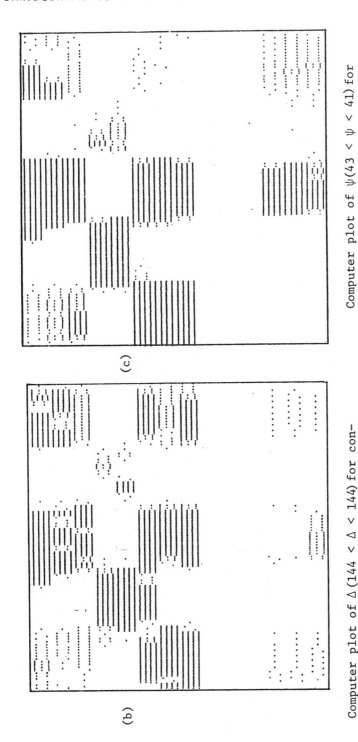

(c)

Computer plot of ψ (43 < ψ < 41) for contaminated panel 4 from McDonnel Douglas

(b)

Computer plot of Δ(144 < Δ < 144) for contaminated panel 4 from McDonnel Douglas.

Figure 3 (cont.)

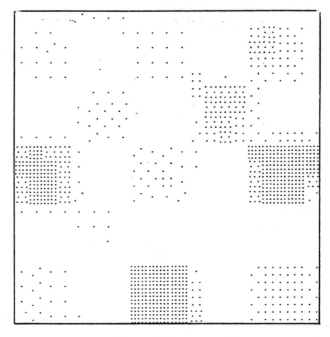

Figure 3(d). Computer plot of SPD (.28 < SPD < .48) for contaminated panel 4 from McDonnel Douglas.

Table III. Quality Assurance Table for Production Panels from
McDonnel Douglas

Acceptance Band 141 < Δ < 144,   41 < ψ < 43,   0.2 < SPD < .4.
λ = 6328 Å, Angle of Incidence 60°, Automated Ellipsometer.

Panel	Automated NDI Average Values			Pass Accpt. Band	Re-inspect	Visual Inspection
	Δ	ψ	SPD			
Md 1	142	42.5	0.30	yes		OK
3	144	42.4	0.32	yes		OK
4	142	42.0	0.33	yes		OK
Rotate 4,90°	144	42.1	0.32	yes		OK
5	146	40.1	0.35	no	yes	damage
6	141	42.0	0.45	no	yes	OK
Contam. 4-1C	136	38.0	0.47	no	yes	deliberate contam.
region 5	137	41.0	–	no	yes	damage
region 3,5	147	39.3	–	no	yes	damage
region 11-12, and 5	148	39.0	–	no	yes	damage

DISCUSSION AND CONCLUSIONS

Surface Tools

Ellipsometry: The ellipsometer is sensitive to the anodic film thickness and structure as well as to contamination. Since the light beam does not change or contaminate the surface, ellipsometry is truly nondestructive and very useful for detecting all forms of contamination, such as incorrect film thickness from processing errors, damaged films from handling and contamination from the various sources. It has been shown in this report that Δ is the range of 80 to 110° and ψ in the range of 37 to 46° for anodic films less than 1000Å, whereas properly prepared anodic films ($\sim$ 3000 to 4000Å) have a range of $\Delta \sim$ 160 to 180° and $\psi \sim$ 37 to 41°. Therefore, ellipsometric results (as SEM) can easily detect if the film is below 1000Å (in the danger region). However, the high sensitivity to film thickness is a disadvantage with respect to detecting surface contamination because small changes in film thickness might be interpreted as contamination, whereas small changes in film thickness is probably inconsequential with respect to bond strength or durability. The maximum Δ window experiences in this work for a particular alloy and specific anodizing conditions is about 6 degrees ($\sim$ 40Å) and from panel to panel about 12 degrees ($\sim$ 80Å). For example, if a panel had an anodic film of thickness near the top of the Δ window, plus 80Å or organic contamination, the panel would pass the acceptance test. Considering the fact that some organic contamination does not change the wettability, it is conceivable that the 80Å of organic contamination would not be detrimental to bond strength or durability.

The automated intensity ellipsometer, used for mapping panels, is extremely fast since there are no moving parts and the photodetectors have time constants in the nanoseconds. Therefore, the speed of mapping depends upon the speed of the computer data acquisition and the speed of movement of the sensing head with respect to the panel surface. We have shown that an automated nulling ellipsometer would be about twice as sensitive to contamination due to the effect of surface roughness. Although this type of ellipsometer would be much slower than the intensity type, it may well be adequate for most applications. Due to the absolute measurement (phase shift and amplitude ratio of the components of polarized light) the ellipsometer does not suffer from a stability problem. However, due to changes in processing parameters, it is necessary to periodically calibrate with control samples.

Surface Potential Difference (SPD): Although the surface potential difference between the reference electrode and the surface of interest is directly related to the anodic film thickness ($\sim$ -1 mV/10Å) it is much more sensitive to permanent and induced dipoles associated with organic contamination on the surface ($\sim$ +10 mV/10Å). SPD can be very useful for film thickness as well as contamination. The prime difficulty with SPD is that the reference electrode is unstable with time. This problem is equally important for vibrating or oscillating systems (e.g. Fokker contamiation tester and ISO probe) as for our radio active system. This is because each system is measuring the work function difference between two electrodes. The drift of the reference electrode is usually not great over a one day period and therefore measurements comparing different surfaces the same day are very revealing, but variations of the reference electrode makes interpretation difficult unless calibration is very frequent.

Photoelectron Emission (PEE): The PEE is driven by a constant voltage source (45 volt battery in our instrument) much larger than the SPD and therefore is not sensitive to reference electrode drift as with SPD. The emission of electrons from aluminum with UV light of 2500Å is primarily from the substrate metal and depends on the surface roughness and the attenuation of the hydroxide film. Due to the porous nature of the phosphoric acid anodic films, the PEE is almost independent of film thickness. The porous outer film is approximately transparent to emitted electrons as they pass along the open channels. Attenuation does occur with respect to a barrier layer at the bottom of the pores. For example, the PEE for a 5 volt film ($\sim$1500Å) on Al7075-T6, was measured eight times over a period of 10 days. The average value of PEE proved to be 38±6 (pA). This can be compared to 59±7 (pA) for a 25 volt film ($\sim$7500Å) and about 500 (pA) for an FPL etched ($\sim$100Å film) sample. The thicker 25 volt film must either have a less attenuating barrier layer or a thinner barrier layer than the 5 volt film. Due to the thickness and nature of the anodic film, the PEE is near background and is insensitive to mechanical damage or contamination. It has been shown to be very useful for other surface preparations of aluminum and titanium[2].

Contact Angles: The water contact angle measurement has the advantage that it is sensitive only to the outer atomic layers and thus is extremely sensitive to non-polar organic contamination. It has the disadvantage that the surface must be contacted (increasing the chance of contamination) and is not sensitive to polar contamination that can decrease bond strength and durability (see Reference 2).

CONCLUSIONS

Examination of all of the contamination maps has been made to establish if the particular surface tool can or cannot detect the contamination near the minimum levels for this study. Table 4 gives the list of contaminants and a y or N below each surface tool. The y indicates (yes) the surface tool is considered to success-fully detect the contamination and N (no) the surface·tool is not considered to successfully detect the contamination. The ellipso-metric parameter Δ is considered successful for 26 of the 32 con-taminants. Four of the contaminants not detected by Δ were not detected by SPD or contact angle and were probably not present (due to evaporation). Three of the contaminants were not detected by the ellipsometer, but were detected by SPD or contact angle. Seventeen of the contaminants were detected by SPD. SPD was particularly un-successful for processing errors and handling damage. Contact angle measurements detected 15 contaminants, most of which were organic

Table IV. Representative Contamination Due to Various Sources and Surface Tool Utility

		Ellips.	SPD	ΘH_2O
Processing Errors	Anodize time	y	N	N
	Anodize voltage	y	y	N
	Contamination from bath	y	N	
	Delay in H3PO4 before rinse	y	N	N
Handling Damage	Cotton Glove	y	N	N
	Kraft paper	y	N	N
	Kimwipe	Y	N	N
Human Contamination	Fingerprints	y	y	y
	Cough or sneeze	N	N	N
	Cigarette smoke	N	N	N
	Cigarette ashes	N	Y	
	Food remnants	y	y	N
Representative Constitutents of Smog	N Docosane	y	y	y
	16-Bromo-9-hexadecanoic acid	y	y	
	Dotriacontane	y	y	y
	Stearic acid	y	N	y
	Erucic acid	y	y	y
	Brassidic acid	y	y	y
	Decanoic acid	y	N	y
	Benzoic acid	y	y	y
	Amino-Benzoic acid	y	y	y
	1,1,2 diamino dodecane	y	y	y
	1-12-diamino decane	y	y	
	decadiene	N	N	
	decacylene	y	N	N
	1-Eicosene	y	N	y
	1-Hexadeclamine	y	y	y
	Anthracene	N	N	N
	Adamantanol	y	y	N
	2-Adamantanone	y	y	N
	1-Adamantone carbonitrile	N	y	y
	1-Adamantane carbonylic acid	y	N	y
		6N	15N	13N
		26Y	17Y	14Y

and non-polar. Contact angle was unsuccessful for the detection
of process errors, handling damage and human contamination (except
for greasy materials such as fingerprints, lipstick, etc.). The
spaces left blank under the contact angle column were not tested.
It is concluded that the best NDI system would include both ellipsometry and SPD. However, the best single tool is ellipsometry.

ACKNOWLEDGEMENTS

This work was conducted under Contract No. F33615-75-C-5235
for the U.S. Air Force Nondestructive Evaluation Branch, Metals
and Ceramics Division, Air Force Materials Laboratory, WPAFB, Ohio
45437.

REFERENCES

1. G.M. Hidy, "Characterization of Aerosols in California
    (A CHEX)" ARB Contract No. 358, April 1975. (Rockwell Inter-
    national Science Center, Thousand Oaks, Ca. 91360)
2. T. Smith, "Mechanisms of Adhesion Failure between Polymers and
    Metallic Substrates", AFML-TR-74-73, June 1974.
    (Wright Patterson Air Force Base, Dayton, Ohio 45437).
3. T. Smith, J. Appl. Phys. 46, 1553 (1975).
4. T. Smith, Surface Sci. 56, 212 (1976).
5. A. W. Bethune, SAMPE 11, 4 (1975).

DETECTION OF SURFACE CONTAMINATIONS IN METAL BONDING BY SIMPLE METHODS

H. Kollek and W. Brockmann

Institut für angewandte Materialforschung

Lesumer Heerstr. 36, 2820 Bremen 77, Germany

The adhesional properties of metal surfaces have to be tested nondestructively, if optimum strength of metal bonds is to be reached. For use in production lines, special methods are required for this purpose. Easy handling and short time of measurement are obvious virtues. Three methods are reported, which fulfill these considerations. Wetting or measurement of contact angle, and the exo-electron emission are already used for testing the cleanliness of metal surfaces. The third method, the remission photometry, is new in this field. Our preliminary investigations using the remission photometry show that it is a highly sensitive method for detection of surface contaminants.

INTRODUCTION

In metal bonding good and durable adhesion will occur if the surface has been cleaned from grease and an active oxide layer was formed by a pretreatment. So the detection of surface contaminations, grease and inactive oxides, is very important and is mostly done by pretreating and bonding specimens together with the original parts. They are tested then in a destructive way. These tests give no direct information about the bonded part, because only specimens are tested. There is a need for nondestructive testing of surfaces before bonding. Besides bonding specimens and testing them, the finished bond may be checked by supersonic methods.[1] But they give no information on smaller differences in the adhesional strengths and the durability of the bond. If metal bonds fail, often the adhesion was poor. On the other hand, the adhesion depends strongly on the pretreatment of the metal. Therefore, it is evident that metal surfaces be tested after pretreatment and before bonding by nondestructive methods.

For laboratory purposes, methods like SIMS, ESCA, AES, etc. can be used. They are not really nondestructive, because even the evacuation of a surface may change it, for instance by dehydration. Equipment and analysis are rather expensive and complicated. So they can hardly be used for routine tests on a production line.

Today, for routine testing, only two tests are used. One of these is measurement of contact angle, the other one is the determination of the exo-electron emission. Ellipsometry, which is nondestructive, too needs a rather complicated measuring set-up and the interpretation of the results is not always easy. Another optical method, the remission photometry, seems to be more suitable and a few measurements made using this method will be presented.

MEASUREMENT OF CONTACT ANGLE

Testing of the adhesional properties of a metal surface by its wettability has been surveyed very well. Relationships between the contact angle, the surface tension of the wetting liquid, and the surface energy of the metal are well known.[2,3,4]

In its simplest form, one checks whether or not the adhesive spreads; where spreading indicates good adhesion. A few exceptions from this simplified rule are known, but today they seem to have no significance for industrial bonding.[5,6] If the viscosity of the adhesive is too high, a test liquid of known surface tension may be used; but then the test is not really nondestructive. The test liquid may be adsorbed on the metal surface and the adhesional properties may be affected. From the contact angle the surface energy of the surface can be calculated. Splitting the surface energy into a polar and a nonpolar term gives more detailed information about the adhesional properties.[7,8] There is a correlation between the polar term of the surface energy and the adhesional strength of the bond.

MEASUREMENT OF EXO-ELECTRON EMISSION

The principle of this method is known since 1802[9] and it was found to be a nondestructive test for measuring surface properties.[10] Measuring set-ups are available and Fokker and a few other European firms use this system for quality control of metal surfaces to be bonded.[11]

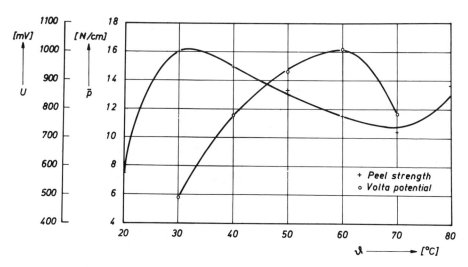

Figure 1. Peel strength (+) and exo-electron emission (o) as a function of etching temperature.

Testing is done by placing a probe on the surface, which touches it only with a small metal ring, and reading the value of exo-electron emission from the measuring instrument. A disadvantage of this technique is that there is no unambiguous relation between the value of the exo-electron emission and the strength of a metal bond. Figure 1 shows, for example, the peel strength of CSA-etched aluminium surface and its exo-electron emission as a function of the etching temperature. The adhesive was a phenolic resin and the etching was done in a solution of 275 g H_2SO_4 and 75 g $Na_2Cr_2O_7.2H_2O$ in 650 ml water. At a value of 800 mV of exo-electron emission, the peel strength may have the value of 15 or 11 N/cm. At a certain value of the emission, the peel strength is at its optimum. But if this value is not reached, adhesion may be better or - in most cases - worse. On the other hand, this method is easy to carry out and is very sensitive. Therefore, it is a good technique for nondestructive testing of the pretreatment of metal surfaces. Optimum values for exo-electron emission for various pretreatments of aluminium alloys and other metals are published by Bijlmer and Brockmann.[12,13,14]

REMISSION PHOTOMETRY

Photometry is a standard method in chemical analysis. For solids there are two techniques for measuring their colour. Measurement of the reflection of light from a surface is used, e.g. on paint surfaces (gloss metering); the reflection of light is highly

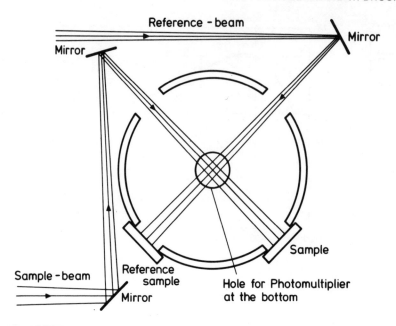

Figure 2.   Principle of integrating sphere.

affected by the surface roughness. The other technique is the remission photometry. Remission spectra are not influenced strongly
by the roughness of the surface. So there is more information
about the chemical compounds on a surface in a remission spectrum.[15]

     An integrating sphere, Figure 2, mounted into a photometer
allows measurements of spectra in the region from 250 to 500 nm.
The influence of roughness is small; and can be neglected in most
practical cases, if the roughness of the surface does not change
very much.

     In photometric measurements, the sample is compared with a
standard to eliminate the non-linear spectral sensitivity of the
photometer. From a spectrum, we get two informations: the wavelength of absorption and the height of absorption. Using Beer's
law, quantitative determinations can be done, in some cases even
for remission measurements. The principle of remission photometry
is often the same as in normal photometry with the light passing
through the sample. First investigations shall be mentioned here
using three examples.

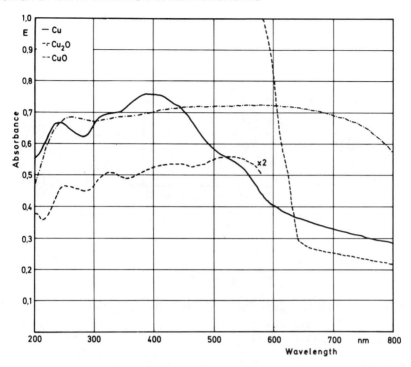

Figure 3.   Remission spectra of copper and its oxides.   (The part
of spectrum marked with x2 was recorded with half sensitivity)

Oxidation of Copper

The remission spectrum of oxidized copper is shown in Figure
3.   The standard was a freshly ground copper surface.   Comparing
this with the spectra of the copper oxides, measured against magnesium oxide as a standard, there are the same maxima in the spectrum
of the oxide layer on copper as in the spectrum of copper(I)-oxide.
So it is possible to detect certain oxides on a metal surface, if
their remission spectra are known and if the spectra of the
various possible oxides differ from each other.

Freshly ground copper oxidizes quickly at room temperature,
and this can be determined by remission photometry.   At a wavelength of 500 nm, the copper alone absorbs less light; whereas the
copper(I)-oxide has its first maximum.   The absorbance of a freshly ground copper surface against magnesium oxide at 500 nm versus
time is given in Figure 4.   It represents the oxidation of copper
at room temperature in the first 15 minutes after grinding.

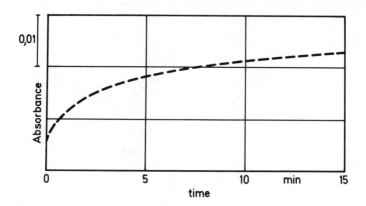

Figure 4.   Oxidation of copper at room temperature.

Detection of Organic Compounds

A lot of organic compounds, like grease, are coloured or
absorb light in the UV-region. In this case, a phenolic resin was
used as a model compound. Its absorption maximum is at about 280
nm. The resin was labelled and could be determined by measuring
its activity too. The spectra given in Figure 5 are from an
anodized (sulphuric acid) and from a CSA-etched aluminium surface.

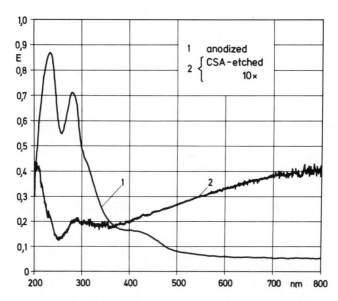

Figure 5.   Remission spectra of Phenolic Resin adsorbed on
Aluminium surfaces. (E-scale for No 2: 0 - 0.1 E)

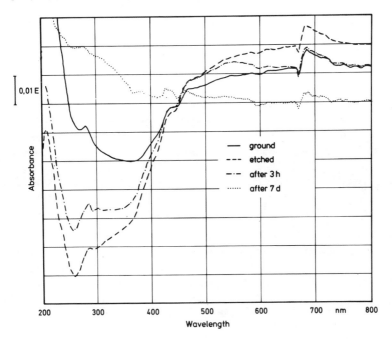

Figure 6.   Corrosion of an etched steel surface, stored at 23°C
in air.

On these surfaces the resin was adsorbed from a benzolic solution.
The anodized surface adsorbs more than 10 times the resin than the
CSA-etched surface.   The peak of the phenolic resin was splitted
into two peaks at 280 nm and 230 nm.   This indicates chemical
reactions of the resin with the anodized surface.   On the CSA-
etched surface no chemical reactions could be found, and the maximum absorption of light is still at a wavelength of 280 nm.   The
amount of resin on this surface was determined to 0.5 $\mu g/cm^2$ by
activity measurements of the labelled product.   This is the lowest
amount of resin which could be determined by this method.

The determination of organic compounds on metal surfaces
depends on the adsorptivity of the compound.   The adsorptivity
becomes high in aromatic compounds or in molecules with hetero-
atoms like oxygen or nitrogen.   In practice, amounts of grease at
a level of about 1 $\mu g/cm^2$ should be detectable.   For a quantitative determination, the amount should be higher.

Corrosion of Steel

A steel surface was ground, etched with dilute hydrochloric
acid, rinsed with water and methanol, and dried in warm air.

Its remission spectrum was recorded with a steel surface as reference specimen, which was only ground under the same conditions.
If both surfaces were absolute equal, the spectrum should be a
straight line. But there will never be absolutely equal surfaces
so the spectra recorded at a 10 times higher sensitivity always
show absorption bands. The spectrum of the ground surface in Figure 6 show a lower absorption of light at 350 nm than in the region
of 500 to 800 nm. This indicates that the surface was less oxidized than the reference specimen, because the absorption of light
at 350 nm is due to oxidation products. Immediately after etching
the absorption of light is lowered again in the 200 to 400 nm
region. This indicates that the oxides have been removed from
the surface. After etching, this surface and the ground reference
specimen were stored under laboratory conditions. After three
hours the etched surface was oxidized and absorbed more light.
This oxidation goes on and seven days later there are more oxidation products on the surface than immediately after grinding. As
for all spectra the same reference specimen was used, which was
stored under the same conditions, one can say that an etched surface oxidizes faster than the reference specimen. If the oxidation of both surfaces is the same, there should be no change in
the spectra or the first spectrum of the ground surface should be
reached again.

CONCLUSIONS

Three methods are discussed which can indicate contaminants
or poor surface treatment of metals in a fast and nondestructive
way. Two of them, wetting and the exo-electron emission, are
already used in testing surfaces to be bonded. The third method,
remission photometry, offers a great deal of potential for diverse
applications. Photometers are available for many purposes. At
least a single beam instrument with an integrating sphere, which
measures only at one wavelength, may be used in a bonding shop.

A few measurements with a microscope photometer showed that
there are local differences in the adhesional properties of metal
surfaces.

Comparing the three methods, it seems that remission photometry may give the best results for testing adhesional properties
of metal surfaces.

REFERENCES

1. A. Matting, "Metallkleben", p. 184, Springer-Verlag, Berlin,
    1969.
2. G. Salomon in "Adhesion and Adhesives", R. Houwink and G.
    Salomon, Editors, 2nd ed. Vol. 1, Elsevier, Amsterdam, 1965.

3.   W. A. Zisman in "Adhesion and Cohesion", P. Weiss, Editor,
     p. 176, Elsevier, Amsterdam, 1962.
4.   P. J. Sell and A. W. Neumann, Angewandte Chemie 78, 321
     (1966).
5.   W. Brockmann and H. Kolleck, in "Proc. of 23rd National SAMPE
     Conf.", p. 1119, SAMPE, Azusa, Cal. 1978.
6.   a. K. L. Mittal, Polymer Eng. Sci. 17, 467 (1977).
     b. K. L. Mittal in "Adhesion Science and Technology", L. H.
        Lee, Editor, Vol. 9A, pp. 129-168, Plenum Press, New York,
        1975.
7.   T. Matsunaga, in "Surface Contamination: Its Genesis, De-
     tection and Control", K. L. Mittal, Editor, Vol. 1, pp. 47-56,
     Plenum Press, New York, 1979.
8.   H. Potente and R. Krueger, Farbe + lack 84, 72 (1978).
9.   Lord Kelvin, Phil. Mag. and J. of Science 46, 82 (1898)
10. J. Kramer, Z. Physik 133, 629 (1952).
11. P.F.A. Bijlmer, This proceedings volume.
12. W. Brockmann, Adhaesion, p. 276 (1973).
13. W. Brockmann, Adheasion, p. 335 and 448 (1969); p. 52 and
     250 (1970).
14. P.F.A. Bijlmer and R. J. Schliekelmann, Fokker Contamination
     Tester, Fokker-VFW-Report, Amsterdam, 1975.
15. W. Brockmann and H. Kollek, defazet p. 166 (1978).

CHARACTERIZATION OF THE SURFACE QUALITY BY MEANS OF SURFACE POTENTIAL DIFFERENCE

P. F. A. Bijlmer

Fokker-VFW B. V. Technological Center
Schiphol - oost.
The Netherlands

The Surface Potential Difference (S.P.D.) measured by means of a dynamic capacitor (Fokker Contamination Tester) is not only dependent on the type of metal, but also on the condition of the surface.

Abraded surfaces of different metals show S.P.D. values in the same sequence as is given in the Electromotive Force Series. The rate of passivation due to oxidation can be measured continuously until the passive state is reached. After chemical or electrochemical treatments the S.P.D. value is indicative of the quality of the surface film formed during and after the treatment. Chromic-sulphuric acid pickling prior to adhesive bonding can be inspected by this method on optimal surface morphology. The effect of hot rinsing, after pickling, on hydration is indicated by a decreasing S.P.D. Demineralized water was found detrimental for the bondability if used as rinsing water at temperatures exceeding 60°C. Mechanical damage of the oxide layer causes a change of the S.P.D.in the positive direction. Cracks in anodic layers were found to increase the S.P.D. in relation to the crack density. The S.P.D. value on anodised material is defined by the potential of formation, the dimensions of the pores, and the rinsing practice. Because the type of alloy is an important factor defining the pore dimensions, copper diffusion in clad material influences the S.P.D. value after anodizing.

Hot water hydration or salt solution treatments to seal the anodic layers are dependent on the porosity as to the sealing effectivity. In this case S.P.D. measurements are useful as an inspection method for sealing quality. Large differences in S.P.D. values can be measured through a primer coating provided this organic material does not influence the S.P.D. on its own.

INTRODUCTION

The S.P.D. method applied to the mechanically or chemically pretreated surface can give information about the electrical state of the surface layer, that is always present at normal environments. A device was constructed based on the vibrating capacitor known as Kelvin's method, modified by Zisman[1,2]. A reference plate is vibrated as near as possible to the surface to be measured. At different work functions of the metals, an alternating electrical field is present in the splice. A compensating device changes the potential of the reference electrode to the same contact potential as the metal to be measured. This compensating voltage is recorded on a display in mV directly. As a reading takes only a few seconds, a large area of a surface can be measured in a short time. The reference plate mounted in a screened probe is made from gold plated material, see Figure 1.

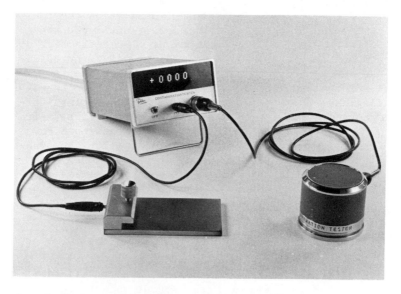

Figure 1. A surface potential difference device, Fokker Contamination Tester. Patented in U.S.

The scope of this paper is to show the results on aluminium alloys prepared for adhesive bonding, painting and corrosion resistance. Correlations with bondability, if any, are given. It must be stressed that a direct correlation with the work function as a physical parameter is not to be expected. The S.P.D. indicates only the condition of the surface after a given treatment.

It is believed that the oxide configuration with its typical absorption ability is responsible for the observed differences in S.P.D. If a certain treatment is varied, all other following steps must be kept as identical as possible. Rinsing and drying methods can interfere with the results. If extensive drying is given to a specimen, the S.P.D. will change until an equilibrium state is obtained.

THE MATERIAL IN THE DELIVERED CONDITION

The condition of the surface is not specified in material specifications. It is, however, known that large differences in surface condition can be found due to differences in manufacturing methods and storage life. The initial condition of the aluminium surface can be extremely important for processing the material in chemical solutions, especially in mild ones. With the S.P.D. method the initial conditions can be monitored. A heavy oxide layer will show a low S.P.D. compared to a material covered with a thin layer, provided no mechanical damage has broken the passivity. The qualitative results on initial surfaces are summarized in this section only to show where differences can be expected.

- The type of alloy defines the condition of the oxide layer. A low concentration of alloying elements favours the passivation of the surface because an undisturbed homogeneous oxide layer can be formed, as is well recognized. The layer is more dense and protective, but grows slower.

- The amount of alloying elements in clad layers due to diffusion is an unpredictable parameter in processing aluminum alloys.

- As most of the work was carried out on clad materials, the copper content, if it varied, caused a lot of scatter.

- The manufacturing procedure was found to give different conditions of the surface. It appeared that American-produced sheet was more active (thin oxide layer) compared to European-made sheet, with S.P.D. values of 800 mV and 200 mV respectively.

- Both sides of a sheet can differ considerably in surface condition. Differences of 200 mV are no exception. This is probably caused by milling and cooling procedures. It is, however, worthwhile to select a certain side and mark it well because differences in pickling rate and surface morphology are possible.

- Thin material was found to be much lower in S.P.D. values compared with thick materials.

- Storage in a non-corrosive environment for years can lower the S.P.D. value considerably. Storage in a chemical laboratory can change the S.P.D. overnight to higher values. It can in fact be used as a method to measure the environmental conditions.

- Mechanical damage naturally changes the S.P.D. value to higher levels.

- Material of the same specification with a small difference in heat treatment condition can differ in S.P.D. for years. A typical example was found to be the aluminum magnesium alloy 5056. One sheet was sensitive to intercrystalline corrosion in the chromic sulphuric acid bath, the other not. They differed considerably in S.P.D. in delivered condition. The sensitive material showed a S.P.D. of 1450 mV, while the sound material indicated 750 mV. It might have to do with the fact that sensitivity to granular corrosion protects the surface against uniform corrosion, while a resistant material is more prone to uniform oxidation.

These facts indicate that the initial material can differ considerably in oxide condition, which might be important to know before starting bondability studies after careful pretreatment.

PRACTICAL APPLICATION OF S.P.D. MEASUREMENTS

The Metal Surface After Abrasion

Abrasion is considered a mechanical method for removing a contaminated oxide layer from the surface. The bondability of an abraded surface on aluminium alloys is however poor, if aging requirements are present.

Besides the removal of oxides, abrasion deforms the surface into an amorphous metal layer with a high dislocation density and increased roughness. The surface is considered highly active in the sense of decreased work function.

The S.P.D. against gold increases after abrasion and decreases again in time due to a combined action of dislocation movement to an equilibrium position of the lowest possible energy level and of oxidation of the surface.

In Table I some measured values on different materials after abrasion are given.

If these S.P.D. values are arranged in order of the standard oxidation potential E° at 25°C (Ref. 3) the following sequence is formed as shown in Table II.

From this summary, it can be seen that the noble materials show a low S.P.D. against gold and the effect of abrasion is low. As a reference electrode it is essential to have a material of a noble metal. A gold plated reference is standard in our investigations although rhodium is a possible alternative because of its higher hardness. A very important feature is that the rate of passivation in time can be measured. The bondability of stainless steel after abrasion decreases in time as is found in practice. The effectiveness of abrasion can be measured and was found to be prone to a high degree of scatter. Observation of abraded surfaces under the microscope reveals different degrees of coverage depending on the type of abrasive paper and the abrasion time.

Table I. S.P.D. After Abrasion in mV

Material	Hours after abrasion					
	1	10	42	114	153	V1–V153
Stainless steel	238	75	78	27	27	211
Low carbon steel	143	48	48	18	28	115
Copper	56	−60	−16	−60	−33	89
Brass	235	231	220	141	145	200
Titanium	435	266	251	181	185	250
2024–T3 clad	1190	982	953	903	870	320
2024–T3	1185	995	961	996	870	315
7075–T6	1165	1030	995	936	906	959

Table II.   E° versus S.P.D.

	E°Volts	S.P.D. after 1 hr. mV	S.P.D. change after 152 hrs. mV
Al	1.66	1190	320
Ti	1.63	435	250
Cr	.74	238	211
Fe	.44	143	115
Cu	−.34	56	89
Au	−1.50	−	−

Figure 2 is given as an example of coverage after cross abrasion on Al 1100.   The S.P.D. was raised from 800 to 1510 mV (Paper used was 320).

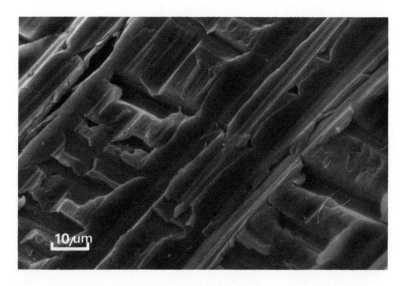

Figure 2.   S.E.M. micrograph of an abraded aluminium surface

S.P.D. After Chemical Cleaning

In general, chemical cleaning is carried out to transform a surface (oxide) layer with unknown and possibly unfavourable properties to a known favourable one. Being a transformation action, it is logical that physical changes will occur which can be measured. The electrical double layer on the surface due to the condition of the transformed oxide layer and its absorptive capacity might be indicative of the quality of the surface in relation to its bondability. The S.P.D. is only a signal that this condition is really reached or not. Only by extensive correlation measurements, it is possible to find out which S.P.D. is a guarantee for the optimal condition of the surface. This optimal condition is not always the same for adhesive bonding or for example spotwelding.

Alkaline cleaning: For aluminium alloys, inhibited alkaline cleaners are preferred prior to deoxidizing treatments. This inhibiting is necessary to prevent the alkaline etching of the aluminum alloy so as to reach a homogeneously cleaned surface. If grease is not homogeneously covering the surface, cleaned parts of the surface might start etching while other parts are still covered. Silicate-free alkaline cleaners are preferentially used in cleaning lines for adhesive bonding. If the inhibiting capacity is optimal, a passive surface will not transform into an active one. Measuring the S.P.D. before and after alkaline cleaning will show the inhibiting capacity of the cleaner. Some interesting facts were found after alkaline cleaning.

- The inhibiting capacity of a silicate containing cleaner at higher temperatures is about the same as at lower temperatures.

- A fresh solution has a low inhibitive capacity due to a low solution rate of the silicates.

- Chromate-inhibited cleaners are sensitive to changes in temperature. At higher temperatures the inhibiting capacity is reduced and etching occurs. High S.P.D. values are the result of this reduced inhibiting capacity.

To measure the behaviour of an alkaline cleaner, it is necessary that the surface is measured before and after cleaning. The S.P.D. before cleaning shall be low, that means an undisturbed oxide layer must be present. If due to handling the oxide layer is mechanically damaged, the S.P.D. is generally high and the effect of alkaline cleaning cannot be recorded. Tape protected coupons are used for this type of measurements. To reach a low contact resistance at spotwelding, the inhibiting capacity of the alkaline cleaner prior to deoxidizing must be constant, otherwise the deoxidizing rate in the normally used weak deoxidizers is not

constant, resulting in different contact resistance values after
treatment. To demonstrate the effect of S.P.D. after alkaline
cleaning on surface activity, highly passive panels were cleaned
in a chromate containing alkaline cleaner with increasing temperatures.

Lapjoint specimens were prepared and tested. Increasing
S.P.D. values resulted in increased lapjoint strength as is shown
in Table III. Each entry is the mean of 7 lapjoint specimens out
of 1 panel.

These results were obtained with Redux 775 on 2024-T3 clad
material which had a very inactive surface, tape protected to
prevent interaction of mechanical damage.

The residual oxides, however, were found to be weak, so peel
tests showed increasing values but were too low value to be of
practical significance. It was therefore concluded that increasing
the activity of the surface by increased removal of the initial
oxides is a necessary but not a sufficient requirement.

Another fact to be considered is the decreasing value at the
highest S.P.D. obtained at the highest temperature of the alkaline
cleaner, indicating an increased weakness of the oxide layer.

S.P.D. After Deoxidizing

Deoxidizing is considered a cleaning action for removing the
initial oxide present on materials as delivered. Solutions of this

Table III.   Lapjoint Values After Chromate Inhibited Alkaline
             Cleaning for 10 Minutes at Different Temperatures.

	series 1		series 2		series 3	
$T^\circ C$	SPD mV	τ kgf/mm^2	SPD mV	τ kfg/mm^2	SPD mV	τ kgf/mm^2
40	−40	1.33	−140	.92	−110	1.34
50	110	1.93	120	1.41	160	1.41
60	660	2.35	520	2.17	590	2.13
70	910	3.02	890	3.05	980	3.10
80	1030	3.34	1070	3.18	1040	3.10
90	1160	2.61	1210	2.57	1230	2.83

kind are mostly composed of nitric (or sulphuric) acid, hydrofluoric
acid and chromates. As they are used at ambient temperatures, the
oxidation action of the chromic acid is low. The surface formed in
this kind of solution is clean and free of residual oxides.

Increasing S.P.D. are found depending on the deoxidizing time,
rinsing procedures, and drying method. In most cases, the S.P.D.
is high (>1000 mV) indicating a high absorptive capacity.

The oxides are too weak to form a bondable surface of high
peel strength value.

The treatment in cold deoxidizers is not considered as a
suitable method for adhesive bonding. In practice they are used
as a pretreatment prior to chromatizing, anodizing or spotwelding.

S.P.D. After Etching

Etching in hot solutions of sulphuric acid and chromic acid
(or sodium dichromate) is considered a cleaning action to remove
the initial oxide and replace it by a strongly adherent one. Due
to the high solution temperature (60°C) not only the etch rate is
increased but also the oxidation capacity of the chromates. A
porous oxide layer is formed if a balanced composition is present
as is shown in Figure 3. This type of surface is considered as
a bondable surface if suitable adhesives are used.

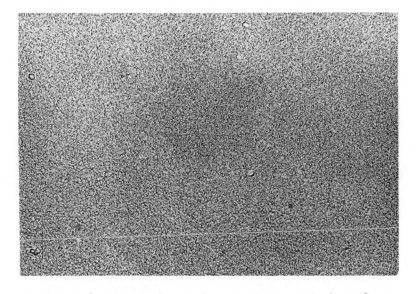

Figure 3. Morphology of a chromate etched surface

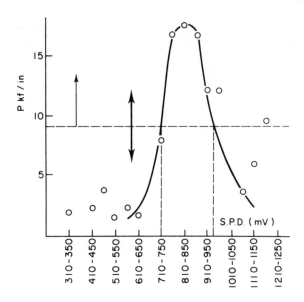

Figure 4. S.P.D. in relation to bondability after pickling in sulfuric-chromic acid solutions under various conditions of temperature, concentration and immersion times.

At the 10th Conference on Adhesion and Adhesives, the importance of the surface morphology after chemical treatment on bondability was shown[4]. Clearly four typical surface configurations are possible after chromate etching. Besides the surface observations, S.P.D. values were also taken for the panels to be adhesive bonded. A frequency table was formed of the readings as given in Table IV. The peel strength values found within a certain range of S.P.D. values were averaged and are shown in the Table under $\bar{p}$ in kgf/m.

In a range of 760 - 1000 mV, high peel strength values are found, the optimal condition at about 850 mV as is shown in Figure 4.

It is interesting to note that at increasing S.P.D. values, increased peel strength values are found up to 850 mV after which the strength is reduced reaching an unacceptable value over 1000 mV. This is comparable with the results after alkaline cleaning where lapjoint values are reduced over the 1000 mV range.

This correlation can only be established if all panels have the same treatments after pickling. The drying method has the most influence on the S.P.D. At drying temperatures over 40°C, the values of the S.P.D. tend to be lower, diminishing the differences. This makes the assumption realistic that the water absorption is

Table IV.

Reference F.p range mV	$\bar{p}$ kgf/m	n	A	B
10 – 50				
60 – 100				
110 – 150				
160 – 200				
210 – 250				
260 – 300				
310 – 350	1,9	2		
360 – 400				
410 – 450	2,4	1		
460 – 500	3,8	2		
510 – 550	1,6	3		
560 – 600	2,2	1		
610 – 650	1,8	3		
660 – 700				
710 – 750	7,9	6	2	
760 – 800	16,7	15		1
810 – 850	17,5	6		
860 – 900	16,8	15		4
910 – 950	12,3	7		4
960 – 1000	12,3	5	3	
1010 – 1050				
1060 – 1100	3,6	2		
1110 – 1150	6,3	1		
1160 – 1200	9,7	3	1	
1210 – 1250				
1260 – 1300				
1310 – 1350				
1360 – 1400				
total		72	6	9

$\bar{p}$ is the average value of the peel strength.
n is the number of measurements in the Fokker Potential range.

Note: Table IV, column A specimens rejected, but high peel
        strength. B accepted but low peel strength.

responsible for the measured differences. If a panel is overdryed
in hot air, it shows lower S.P.D. than what is normally found after
drying at 40°C; the value increases with time after exposure to
ambient conditions for several hours. The overdryed panel surely
absorbs water until an equilibrium is attained. Magnesium aluminium alloy 5052 was found to be a difficult material to etch for
optimum bondability. After chromate etching, the material was investigated by surface morphology, S.P.D. and peel strength. The
pickling time was varied from zero up to 30 minutes while hot and
cold drying were included in the test. The results are given in
Figure 5 for the relation between S.P.D. and pickling time, and in
Figure 6 for the relation between peel strength and pickling time.
The surface morphology is shown in Figure 7. It can be seen that
the surface morphology develops to the optimal configuration within 15 minutes. The underdeveloped suface configurations show a
S.P.D. exceeding 1000 mV combined with a low peel strength values
(less than 9 kgf/inch). With this method all pickled honeycomb
panels for satellites were inspected as a go-no-go check before
application of the adhesive primer. The definite inspection was
carried out by means of the electron microscope on each panel. If

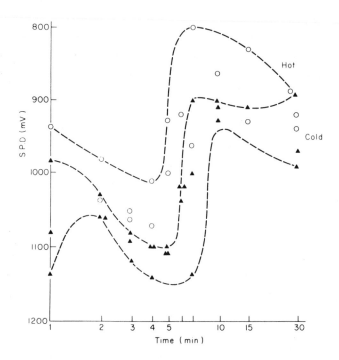

Figure 5. S.P.D. after pickling (5052 material) with standard
chromic-sulfuric acid process, o→ hot drying; ▲→ cold drying as a
function of time. The dotted lines are the upper and lower limits
of the readings, and the middle line is the average line.

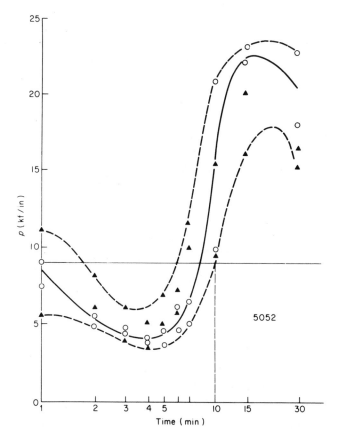

Figure 6. Peel strength after pickling (5052 material) as a function of time. (For details, see legends to Figure 5.)

a S.P.D. over 1000 mV was recorded, the panel was rejected. Under 1000 mV, the panel was accepted and a coupon for E.M. was taken in both cases. Comparison of the S.P.D. values with E.M. morphology pictures showed that, except in a few cases, they all matched.

This proved the usefulness of S.P.D. measurements in quality control of chromate ectched aluminium alloys.

S.P.D. After Ammonium Tartrate Anodizing

Anodizing is a treatment for improving the surface oxide layer by increasing its strength and thickness, to resist mechanical and chemical attack. Normally pore forming formulations are used as anodizing solutions such as sulphuric acid, chromic acid and phosphoric acid resulting in increased pore dimensions respectively.

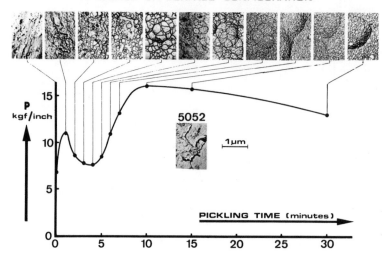

Figure 7.   Surface morphology after pickling (5052 material) and
its relation to bond strength.

As the anodic layers consist of a pore free barrier layer on the
top of which a porous oxide is present, S.P.D. studies were carried
out, in the first place, on barrier layer type coatings formed in
ammonium tartrate.   The rinse intensity and aging time were chosen
as parameters.   Table V summarizes the results of these tests.
From these results the following conclusions are drawn.

-   The S.P.D. is a function of the rinse conditions.   The
    higher the rinse intensity, the higher the S.P.D. values.

-   The S.P.D. increases after exposure to ambient conditions,

Table V.   Effect of Rinsing and Aging on S.P.D.

Rinsing		S.P.D. after aging mV				
t min.	T°C	0 h	2 h	1 d	11d	21d
1	20	−1550	−1500	−1220	−740	−550
10	40	−300	−540	−1170	−630	−455
20	60	−110	−210	−500	−600	−510

reducing the differences due to the variation in voltage
of formation.

A possible explanation is the formation of electrical polar-
ised dipoles in the barrier layer, because the oxide is grown under
a high electrical field. After some time, due to thermal influences, the dipoles turn into a more random distribution.

If hot rinsing is carried out at 40°C and 60°C, the S.P.D.
is increased. Typically after several hours the S.P.D. is decreasing again after which the normal trend in time is followed.
In Figure 8, this is shown for a 150 Volt ammonium tartrate barrier
layer which had been rinsed for 1 minute at 20°C, 10 minutes at
40°C and 20 minutes at 60°C in demineralized water. From the
graph, it can be seen that effects other than the temperature are
responsible for changes in S.P.D. measured after drying and
storage. If the anodizing time is varied, another effect is found.
In the first few minutes of anodizing, extreme differences in
S.P.D. are measured. One can see a large scatter in the recorded

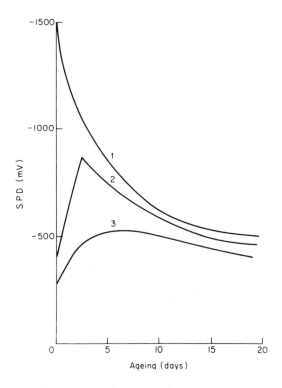

Figure 8. S.P.D. after pore free anodizing in ammonium tartrate
of aluminum 2024-T5. 1. rinsing 20°C for 1 min. 2. rinsing 40°C
for 10 min. 3. rinsing 60°C for 20 min.

values. The same occurs if thin anodic layers are used as a base
for adhesive bonding. A large scatter in values occurs until a
certain oxide layer thickness is reached (For chromic acid, this
limit is around 22 mg/dm^2). Once this value is reached, change
in bondability is less severe but still can be recorded. In the
case of ammonium tartrate, peel strength values are of a low level,
because the oxide is weak and interface fracture occurs, if structural adhesives are used. For that reason, fluctuations in bondability due to different S.P.D. values are not easily detected.
The fluctuation in S.P.D. was found in all practical anodizing
systems as will be shown in next paragraphs. Figure 9 shows the
influence of anodizing time on S.P.D.

Sulphuric Acid Anodizing

A more practical anodizing system was chosen for examination
of bondability. In the first place, sulphuric acid was of interest
because it gives fine porous structure. As it became clear that
the voltage of formation influences the S.P.D. considerably without direct relation to bondability, the voltage was kept constant
during the tests. The parameter to be investigated was the anodizing time. Again, a large scatter was found but it seemed a
strictly organized scatter and more or less reproducible. For that
reason, the graphs were drawn by connecting each individual point.
In Figure 10, the influence of the anodizing time on S.P.D. values
is given. A similar graph was found as in the case of ammonium
tartrate anodizing; with a phenolic adhesive hardly any influence

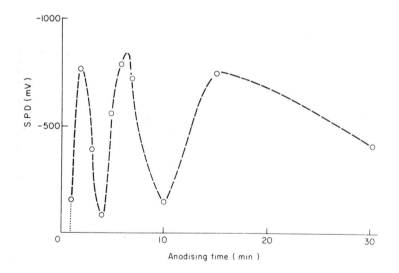

Figure 9. Pore free anodising in NH$_4$ tartrate of aluminum 2024-T3
alclad at 150 Volt with various treatment times.

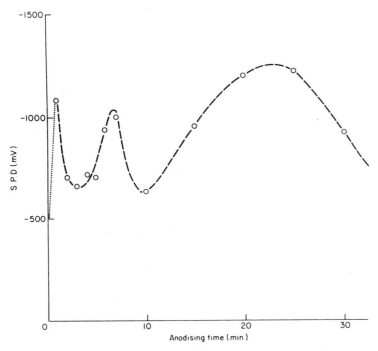

Figure 10.   S.P.D. values after sulphuric acid anodizing of aluminum 2024-T3 alclad at various times.

could be detected.   Either the adhesive was cohesively fractured
or the anodic layer was fractured (brittle).   From one test in
which an epoxy adhesive was used (EC 2216) without adhesive primer,
a large scatter was found as shown in Figure 11.

Connecting the individual values, a similar graph was obtained
as is found in Figure 10.   If Figure 10 is turned upside down, the
similarity is striking.   This transformation can be carried out
because the S.P.D. is a relative value.   This might indicate that
the bondability is reduced in the negative direction of the S.P.D.
value of the anodic layer, provided no fracture in the oxide layer
occurs as is the case with more rigid adhesives.   (Redux 775
practically always shows anodic layer fracture on sulphuric acid
anodized material testing).   As the S.P.D. is only a signal from
the oxide layer in connection with its condition (what that condition might be) and is not directly related to adhesive forces, the
S.P.D. values can be false due to excessive drying without apparent
influence on bond strength. Either there is no influence or the mechanical test method is not able to detect it. One of the most dis-
turbing actions is rinsing because it can "wash out" the indications

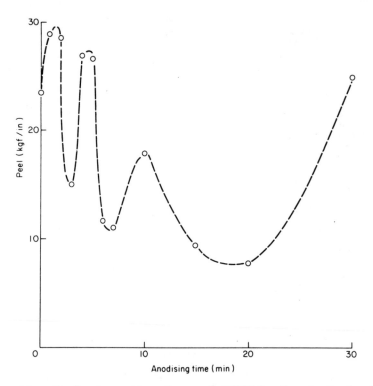

Figure 11. Peel strength values of EC2216 adhesively bonded panels after sulfuric acid anodizing of 2024-T5 clad at various anodizing times.

with a leveling effect on existing differences. It was found that the pore dimensions are related to this "wash out" effect. Layers with large pores are very sensitive to rinsing, increasing the S.P.D. values after drying. This might be a very important fact because if the bath contamination influences the pore dimensions, which is practically always the case, the S.P.D. method can be used as an inspection method for this matter.

For sulphuric acid anodizing this was not investigated systematically; only a trend was observed in a product unit, where increased S.P.D. values were measured after ageing of the solution.

Chromic Acid Anodizing

Chromic acid anodizing is a widely used method to improve bondability and corrosion resistance of adhesive-bonded joints. They differ from former described methods in all aspects. The pore dimensions are larger, the layer is 1-3 μm (NH_4 tartrate 0.2 μm, and sulphuric acid 14 μm) if production processes are used.

No detectable chromic acid is found in the layer; whereas in the
case of sulphuric acid, large amounts of sulphate groups are
present and can not be rinsed out. Because sulphuric acid is a
contaminant in praxis, so it was used in the investigation. The
sulphuric acid is dragged in from former processes such as chromate
etching. It was found that the pore dimensions, as measured by
means of the electron microscope, are influenced by the amount of
sulphuric acid. The sulphuric acid is built in the anodic layer
and cannot be rinsed out as was shown by means of X-ray analysis.
The S.P.D. however, is strongly influenced by the rinsing action
as is shown in Figure 12.

It was found that the uncontaminated solution gave larger
pores compared to 50 mg/l H_2SO_4 contaminated solutions. The S.P.D.
after 5 second rinsing increases proportional to the pore dimensions from −880 to −190 mV. After 30 minute rinsing, only extreme
contamination can be detected. A limit of 200 ppm sulphuric acid
is found in most specifications. After adhesive bonding with a
phenolic adhesive, differences in peeling behaviour were found, as
to the mode of fracture. Panels pretreated in the most contamin-

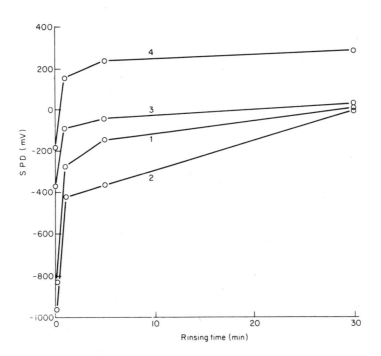

Figure 12. The influence of rinsing time on the S.P.D. after CrO_3
anodizing in contaminated solutions. 1. 0 ppm H_2SO_4; 2. 50 ppm
H_2SO_4; 3. 500 ppm H_2SO_4; 4. 5000 ppm H_2SO_4.

Table VI.   Peel Strength Values After Anodizing and Redux Bonding

	rinsing time min.	1/12	1	5	30
Contamination ppm H_2SO_4					
0		45 /\/\/	45-	50-	45-
50		40 /\/	45 /\/	40-	45-
500		35 /\/	40 /\/	40 ∧	45-
5000		0x	0x	0x	0x

x   anodic layer fracture

–   no slip stick

/\/\/   slip stick

ated solution (5 g/1 H_2SO_4) showed fracture completely in the anodic layer. The Redux peel strength decreased in mean value at higher contamination levels, but the average is strongly influenced by an oscillating behaviour during peeling ("slip stick" in the table indicated as /\/\/). It was found that this "slip stick" behaviour decreased at more extended rinsing times.

The influence of sulphuric acid contamination on bondability was also measured with a cold curing adhesive (EC 2216). First of all, no anodic layer fracture was observed with the weakest oxide layer, as Redux did. This type of adhesive seldom induces this type of failure during testing. Typically (Table VII) the solution contamination favoured bondability by formation of larger pores.

The increased contamination leads to increased peel strength values and more scatter. There is an indication that a too pure

Table VII.   Peel Strength Values After Anodizing and Cold Cured
             Adhesive Bonding

Contamination ppm H_2SO_4	Peel Strength kgf/inch
0	15 – 20
50	25 – 30
500	30 – 40
5000	20 – 40

solution will not give the optimal results. A small amount of
contamination is needed to improve bondability. This typical
effect was already found in chromate etching. The influence of
anodizing time on S.P.D. was found to be identical to what occurs
with other processes, however, less pronounced, due to a higher
sensitivity to the "wash out" effect. In Figure 13, the influence
of anodizing time on S.P.D. is shown. Again a lot of scatter is
found. Not very successful correlations with bondability were
found up until now. It seems that for larger pores, S.P.D. indications are less significant and the mechanical tests are more
complicated due to the fact that a higher absorption of adhesives
takes place and more cohesive failures are found. Not all structural adhesives react in the same way in mechanical testing on
quality differences of the oxide layer. Different modes of failure
can occur. Redux 775 can show different cohesive strength values
due to different ratios of the liquid and powder. Differences in
curing can change the mode of fracture. Undercuring can give

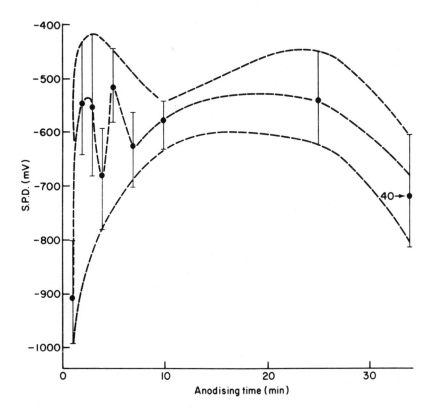

Figure 13. The influence of anodizing time on S.P.D. of aluminum
2024-T3 in chromic acid.

better peel behavior than over-curing which is more critical for
interface failure. Clad layer deformation, anodic layer fracture,
and slip stick effects can mask differences in surface condition.
FM 123-5 with primer BR 127 is even more complicated because its
peel strength value depends strongly on primer thickness. Only
in a narrow range of effective primer thickness, the peel strength
is indicative of surface conditions. Some data, however, suggest
that extensive further investigation might pay off.

In another tests, it was found that panels bonded with
FM 123-5 correlated with the S.P.D. values as shown in Figure 13;
the more negative the S.P.D. values, the lower the peel strength.

It was proved at a later stage that the method of application of the voltage of formation is important for the quality of
the top layer. To reach a voltage of 40 Volts normally takes 10
minutes. A one minute anodise after reaching 40 Volts means an
immersion time of 11 minutes. In laboratory tests, the voltage
of 40 V is applied in half a minute by hand. The end result is not
the same, however. Too fast an application will burn the surface
and spoil it. A better parameter for thin anodic layers is the
energy of formation Vxixt. This value integrated over the whole
duration of the treatment is directly related to coating weight.
Because the voltage of formation has a high influence on S.P.D.
values without a proportional influence on bondability it is
necessary to correlate S.P.D. values to peel strength at constant
voltage of formation. Some main effects are shown in Table VIII.

Table VIII. Main Effects of CrO_3 Parameters on S.P.D. Each Value
is the Mean of 54 Observations

T °C	clad 2024 S.P.D. mV	bare 2024 S.P.D. mV
30	−630	−200
40	−600	−220
50	−550	−100
t min		
5	−570	−260
25	−600	−170
40	−600	−90
V volts		
10	−20	−300
25	−680	−150
40	−1040	−80

This means that at   higher temperatures, larger pores are
present; this effect is more pronounced for bare materials than
for clad material. The anodizing time does not influence the
pore dimensions on clad, but larger pores are found on bare at
extended anodizing times. The voltage of formation shows different effects on clad and non clad material. Clad 2024 increases
in thickness if the voltage of formation is increased, with less
absorptive capacity (high S.P.D. values). Bare 2024 is more
anodically dissolved at higher voltages but its absorptive capacity
is greatly improved. Bondability tests did not give the expected
results for reasons as given earlies. The bare material showed
anodic layer fracture if adhesive-bonded with Redux. Results are
summarized in Table IX.

This result shows that panels A have a weak oxide layer which
is practically stripped off from the barrier layer. These are
found at high voltage of formation and high solution temperatures
and long immersion times. The cold cured adhesive results show
that panels pretreated under the same conditions show somewhat
higher values. This again makes it acceptable to conclude that a
less negative value indicates better absorptive capacity.

S.P.D. After Phosphoric Acid Anodizing

Phosphoric acid anodizing was used because it gives large
pores in the sequence of anodizing processes. From a systematic
investigation on 2024-T3 clad, the results can be summarized as
shown in Table X.

For the clad material all indications of the S.P.D. are in
correlation with the peel strength values.

-   Increasing the process time results in higher S.P.D. and
    peel strength values.

Table IX. Chromic Acid Anodizing on Bare Materials. Each Entry
           is the Mean of 39 Observations.

	Redux bonded panels		EC 2216 bonded panels	
	A	B	A	B
S.P.D. (mV)	−140	−210	−80	−230
Peel Strength (kg/inch)	4	40		

A Anodized at high voltage, high temperature, long immersion times.
B Anodized at low voltage, low temperature, short immersion times.

Table X.   S.P.D. and Bondability After H_3PO_4 Anodizing.   (Each
Entry is the Mean of 54 Observations) Redux and FM
123-5.

	Material 2024 Clad			
	Redux bonded		FM 123-5 **bonded**	
Process time	$\bar{V}$	$\bar{P}$	$\bar{V}$	$\bar{P}$
1 min	−90	9,4	−120	31,8
5 min	+50	19,5	0	53,1
25 min	−10	27,3	+10	90,3
Anodizing voltage				
10 Volt	+110	28,6	+110	63,1
25 Volt	−10	17,0	−50	55,5
40 Volt	−150	10,5	−180	56,5
Concentration				
100 g/l	−90	15,4	−90	42,6
200 g/l	−10	18,8	−70	58,2
300 g/l	+50	21,9	+50	74,5

∇ $\bar{V}$ → S.P.D. in mV
$\bar{P}$ → Peel strength, kg/inch

- Increasing the voltage of formation results in lower S.P.D.
  and peel strength values.

- Increasing the H_3PO_4 concentration results in increased
  S.P.D. and peel strength values.

THE S.P.D. METHOD TO DETECT CONTAMINATION WHICH INTERFERES WITH BONDABILITY

As the surface is considered to be contaminated with an oxide
layer in which all kinds of constituents can be absorbed, it must
be defined what contamination means, if bondability is concerned.
In practice, no adhesion is ever measured, only the bond strength
can be established (joint performance). This joint performance
may depend on adhesion but it is likely that it is more dependent
on stress distribution which is defined by the type of adhesive,
geometical factors, and the strength of the oxide layer. Looking
at the oxide layer, contamination can be due to the process in the
sense that the wrong oxide configuration is formed resulting in

interfacial failure at low strength values. This has been discussed earlier. If the optimal surface configuration is reached,
the contamination of the metal surface favours the bondability.
If this optimal oxide is transformed to an unfavourable one, the
bondability is reduced. This degradation of the oxide layer can
occur during post-treatments such as rinsing and drying or may be
due to storage and handling. A practical spoiling action is found
in hot rinsing if high purity water is used after etching. In the
first place, hydration takes place and the oxide gets weaker. In
extreme cases if hot water is used, the oxide is cracking in the
top layer; this results in a sharp decrease of bond strength values.
The S.P.D. value is decreasing to a low value with a minimum around
60°C. Rinsing at higher temperatures results in an increase of the
S.P.D., perhaps due to cracking of the oxide layer. This should increase the activity of the surface and it is likely that it does so
but no bondability is reached due to the cracked and flaky oxide
layer. Mechanical handling can damage the top of the oxide layer.
Wipe cleaning is detrimental to an optimal oxide configuration.
Touching the surface even with white clean gloves should be avoided.
This mechanical damage can be measured shortly after the condition
is changed. After a few days or even hours the original value is
reached provided nothing has been added to the surface (solvents,
water, etc.). It might be that the equilibrium state is disturbed
due to the mechanical action. If the same equilibrium state can
be resotred, the original S.P.D. returns after a while. Contamination due to strange constituents absorbed in the oxide layer can
only be detected if they are polar or if non polar material dis-
places polar substances. Absorption tests with palmitic acid dissolved in benzene showed that the change in S.P.D. was due to
benzene and not due to the palmetic acid. Surface impedance showed
increasing values with no appreciable change in S.P.D. Absorption
tests are still in progress but the results indicate that the evaporative rate can be measured. It is however not sure that the
evaporative rate correlates in some way to bondability in the sense
of bond strength.

CONCLUSION

     The S.P.D. is not directly related to bondability. It only
indicates a certain condition of the oxide layer, which is responsible for the bondability of the surface. Changes in S.P.D. can
indicate transformation of the oxide layer due to removal or absorption of water and solution compounds. Alkaline cleaning can
be inspected for its inhibiting capacity. Deoxidizers with fluor-
ides can remove scale and heavy oxides easily, but they do not
transform it to a useful laver where upon structural adhesives will
find a sound base. The S.P.D. values found after cold deoxidizers
are higher than 1000 m^{T} indicating a poor surface. Pickling in
chromate solutions can be controlled with suitable panels. The

type 3 and type 4 surface configurations can be distinguished
provided after-treatments are given carefully. Hot rinsing after
pickling spoils the surface and decreases the S.P.D. values up to
60°C if pure water is used for rinsing. At temperatures higher
than 60°C, an increase in S.P.D. value indicates disturbance of
the oxide layer with increased weakness, resulting in reduced bond
strength values. On anodic coatings, the S.P.D. value is indicative of the porosity of the layer. Nonporous layers after ammonium tartrate anodizing show values of -2000 mV. Coatings with
fine pores obtained with sulfuric acid anodizing give values of
-1500 mV. Layers with medium sized pores formed by chromic acid
anodizing have S.P.D. values of about -400 mV, whereas phosphoric
acid anodized surfaces with larger pores show values around zero.
The bondability increases in the direction of increasing pore
size as indicated by increasing S.P.D. values.

REFERENCES

1. Lord Kelvin, Phil. Mag. 46, 82 (1898).
2. W. A. Zisman, Sci. Instrum. 3, 369 (1932).
3. H. H. Uhlig, "Corrosion and Corrosion Control," p. 28, John
    Wiley, New York, 1963.
4. P. F. A. Bijlmer, J. Adhesion 5, 319 (1973).

ACKNOWLEDGEMENTS

I thank all my colleagues who worked on this project, in
particular Bob Exalto and Arie Kwakernaak.

APPLICATIONS OF ELLIPSOMETRY FOR MONITORING SURFACE CONTAMINATION

AND DEGREE OF SURFACE CLEANLINESS

W.E.J.Neal

The University of Aston in Birmingham,
Gosta Green,
Birmingham, B4 7ET, United Kingdom.

Ellipsometry is an optical technique which is eminently suitable for surface examination and can add to knowledge of surfaces in its own right or complement information obtained by using other analytical techniques.   The paper outlines the ways in which ellipsometry has been used to characterise surfaces by determining optical constants prior to investigations of the growth of surface contamination and corrosion layers.   The various states of surfaces which can affect the optical constants such as oxide layers, gas adsorption, roughness and anisotropy are discussed.

Surface characterisation can be made in terms of the instrument parameters ψ and Δ both in the case of film free surfaces and also (in some instances)on composite surfaces.

Specific examples are given of oil contamination on gold in a vacuum system and oxide layer growth on (1) aluminium film, (2) "as received" and etched aluminium sheet and (3) niobium and tantalum.   Other examples are briefly mentioned and the references give access to applications in various disciplines.

INTRODUCTION

The increased interest in the properties of surfaces which
has taken place in the last ten to fifteen years in particular
has resulted in the development of instruments such as ESCA (XPS),
LEED and SEM.   In parallel with the application of these surface
analytical techniques there has also been a revived interest in
the use of more classical optical methods for surface examination.
The availability of computers for the otherwise tedious calculations involved has assisted in these developments.   Ellipsometry
is an optical technique which is eminently suitable for surface
examination and can make an independent contribution to the
knowledge of surfaces or complement information obtained from
other analytical techniques.   Although the basic theory was
developed by Drude[1,2,3,4] over 90 years ago the technique was not
used to any great extent until 1933 when Tronstad[5] investigated
the chemically produced passivity of iron and steel mirrors and
the optical constants of mercury which he used as a standard.
Essentially, the technique involves the determination of the
optical properties of a reflecting surface or composite film
covered surface.   This is achieved by measuring the ellipticity,
after reflection from a surface, of light which is initially plane
polarized.   Drude observed that the ellipticity of light reflected from the freshly cleaved surface of rock salt was small but
increased quite quickly when exposed to air.   Similarly Lord
Rayleigh noticed that ellipticity produced by reflection from a
"water" surface was eliminated when the surface grease film was
removed.   Early instruments were often normal spectrometers
which had been modified.   The word ellipsometry was first introduced in 1945 by Rothen[6] to describe the technique.

Surfaces play an important role in a variety of material
based engineering, chemical and semi-conductor industries and
consequently surface technology is very much an interdisciplinary
activity extending even into the life sciences.   A surface can
be affected by the environment or by treatments such as mechanical
or chemical polishing.   A surface may have a chemisorbed or physisorbed film superimposed or it may be damaged, rough, anisotropic,
inhomogeneous contaminated or corroded.   The state of a surface
can depend on its previous history so that it is often necessary
to characterise a surface as a preliminary to further experimental
tests.   Corrosion processes involve the formation of a film
between a 'clean' surface and the environment.   For many metals
such films prevent further corrosion because of reduced chemical
activity of the metal oxide and are usually considered as inert
or passive.   The physical and mechanical properties of oxide
films can be affected by the chemical nature of the surrounding
environment, for example see Westwood et al[7].   According to
Grosskreutz[8] adsorbed water on naturally occurring aluminium

oxide can reduce the elastic modulus by up to 50%. Optical
techniques enable a non destructive examination of a surface
during the growth of corrosion films, there being little or no
influence on the reaction because of the low interaction energies
involved.   This paper will deal with the characterisation of
surfaces by ellipsometry and the examination of subsequent changes
resulting from exposure to an environment from some form of
surface treatment.

PRINCIPLES OF THE METHOD AND THE BASIC INSTRUMENT

Film Free Surfaces

In the consideration of reflection of light from a surface
there are two principal directions to which reference is normally
made. They are parallel and perpendicular to the plane of
incidence and are often called azimuthal directions. Plane
polarized light (not polarized in either azimuthal direction)
incident on a metal surface, in general, gives rise to reflected
light which is elliptically polarized.   This is illustrated in
Figure 1 where reflection at the interface between the two media
introduces a phase difference between the parallel (p) and
perpendicular (s) components of the incident light and changes

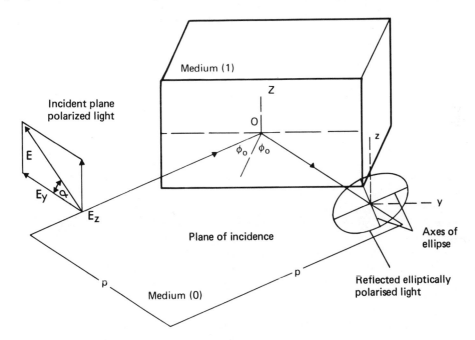

Figure 1. Reflection of plane polarized light at the interface
between media (0) and (1).

the amplitude ratio. For two isotropic non absorbing homogeneous media the reflection coefficients for light travelling from medium 0 to 1 are given by:

$$r_{01(p)} = (n_o \cos\phi_1 - n_1 \cos\phi_o)(n_o \cos\phi_1 + n_1 \cos\phi_o)^{-1} \qquad (1)$$

$$r_{01(s)} = (n_o \cos\phi_o - n_1 \cos\phi_1)(n_o \cos\phi_o + n_1 \cos\phi_1)^{-1} \qquad (2)$$

where ϕ_o and ϕ_1 are the angles of incidence and refraction respectively; n_o and n_1 are the refractive indices of the two media. For media which are absorbing n_o and n_1 would be replaced by the complex quntities $N_o = n_o - ik_o$ and $N_1 = n_1 - ik_1$ respectively.

The basic instrument equation is:

$$r_{(p)} \Big/ r_{(s)} = \tan\psi \, \exp(i\Delta) \qquad (3)$$

where $\tan\psi$ is the modulus of the amplitude ratio for the two components and Δ which is the phase difference between them can be determined from the orientation of the ellipse. One arrangement of the components of a basic instrument is illustrated in Figure 2 and the angles ψ and Δ are obtained from the instrument settings at extinction. Discriptions of the basic instrument and methods of use are described by Winterbottom[9], Archer[10] and various authors[11].

When a surface is examined by ellipsometry the instrument angles ψ and Δ relate to the state of the surface and permit the calculat -ion of the optical constants n and k through equations given by Ditchburn[12]:

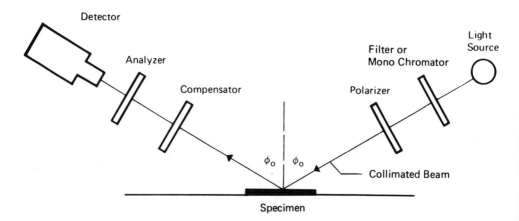

Figure 2. Basic instrument components.

$$n^2-k^2 = \frac{\sin^2\phi_o \tan^2\phi_o (\cos^2 2\psi -\sin^2 2\psi \sin^2\Delta)}{(1+\sin 2\psi \cos \Delta)^2} + \sin^2\phi_o \qquad (4)$$

$$2nk = \frac{\sin^2\phi_o \tan^2\phi_o \sin 4\psi \sin\Delta}{(1+\sin 2\psi \cos\Delta)^2} \qquad (5)$$

Film Covered Surfaces

Figure 3 represents the multiple reflections which take place in a surface film.

For the purpose of this exercise surface films can be divided into two groups (a) non absorbing and (b) absorbing. For a non absorbing film the reflection coefficient for light with electric vectors parallel and perpendicular to the plane of incidence can be written as:

$$r_p = \left[r_{01(p)} + r_{12(p)} e^{xpD}\right]\left[1 + r_{01(p)} r_{12(p)} e^{xpD}\right]^{-1} \qquad (6)$$

with a similar expression for r_s.

$$D = 4\pi n_1 (\cos\phi_1) \, t/\lambda \qquad (7)$$

where t is the film thickness and λ the radiation wavelength.   The

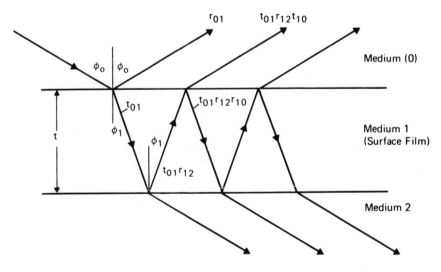

Figure 3. Multiple reflections in a surface film.

state of polarization will be changed from that of a clean surface
as instanced by different extinction settings of ψ_F and Δ_F for
the instrument.   If the values of ψ_F and Δ_F are used in the
Ditchburn equations 4 and 5 the constants obtained will relate to
a composite surface and are referred to as pseudo constants. If
the optical constants n_2 and k_2 of the underlying material are
known, both the refractive index n_1 and the thickness t of the
contamination film can be determined.   For an absorbing film D
is given by:

$$D = 4\pi i \; n_1 (\cos\phi_1) t / \lambda \qquad\qquad (8)$$

For such a film there are essentially three unknown quantities n_1,
k_1 and t to be evaluated with only two instrument parameters
available. McCrackin and Colson[13] suggest four methods of
increasing the number of instrument readings but not all methods
are applicable in every situation. One method is to vary the angle
of incidence to give differing values of ψ and Δ. This approach
has been adopted by several authors and the method is known as
multiple angle ellipsometry. In practice the ψ and Δ sensitivities vary with the angle of incidence so too wide a variation of
ϕ_0 from the optimum (in the region of 70^0) is not advisable.

Film Growth

Equation 6 in conjunction with equations 7 and 8 enable film
thicknesses on a surface to be calculated if values of ψ and Δ
for the film free surface usually identified as $\bar\psi$ and $\bar\Delta$ are known.
In the early stages of film growth when the film is less than
about 10nm thick the thickness can be determined by using an
approximation first suggested by Drude. The exponential terms
of equation 6 can be expanded in a power series of t/λ and to a
first approximation one obtains for a surface film in air or
vacuum.

$$\bar\Delta - \Delta = At \qquad\qquad (9)$$

$$\bar\psi - \psi = Bt \qquad\qquad (10)$$

where A and B are constants, $\bar\Delta - \Delta$ and $\bar\psi - \psi$ are changes in Δ
and ψ from the film free surface due to film growth. A and B can
be evaluated if the optical constants of the film are known. If
the constants are not known or if t is greater than 10 nm the
exact equations must be used. Film growth on a surface can be
represented on an argand diagram as shown in Figure 4 where the
point S represents a film free surface,or if a plot is made of ψ
against Δ for the growth of a non absorbing film the values of ψ
and Δ will be repeated for thick films and a closed loop is
formed. For absorbing films values will not be repeated and a

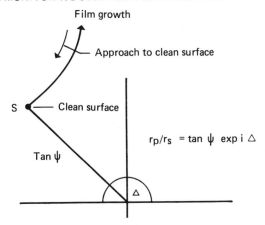

Figure 4.   Representation of film growth on a surface.

spiral will be formed. In most cases, changes in ψ are much
smaller than changes in Δ during early stages of growth.

ELLIPSOMETRIC EXAMINATION OF SURFACES

Surface Contamination and Gas Adsorption

The definition of a 'clean surface' is relative and can only
relate to the limits of detection of foreign material on a surface.
One may wish to define a clean film free surface as that of a
non reactive material which could be achieved by evaporation under
ultra high vacuum conditions. Most surfaces even in an ultra high
vacuum will however eventually have a layer of adsorbed gas superimposed. It has already been mentioned that a surface can be
characterised by the ellipsometer parameters ψ and Δ. At this
point it is necessary to assume a smooth homogeneous isotropic
surface since, as will be indicated later, roughness, structure
and anisotropy can affect the values of ψ and Δ obtained.
Figure 4 shows that departure from a clean surface decreases the
value of Δ.   Conversely cleaner surfaces tend towards higher
values of Δ and in this context it is possible to define a clean
surface in terms of the highest achievable value of Δ after
repeated measurements.   A similar treatment could be applied to
ψ changes but these changes are often within the limits of error
in measurement for thin surface films. Once a clean surface has
been characterised, ellipsometry can be used to measure the
presence of contamination or adsorbed gas. Problems do arise
in the interpretation of measured quantities in terms of percentage coverage when there is less than a monolayer present. Partial
coverage is essentially a two dimensional system whereas film

growth as has been discussed earlier is three dimensional. The matter is considered by Archer[14] and Meyer and Bootsma[15]. Such considerations lead to a modification of equations 9 and 10. Archer gave Δ and ψ as:

$$\Delta = \overline{\Delta} - \Sigma \alpha_i \sigma \qquad (11)$$

$$\psi = \overline{\psi} + \Sigma \beta_i \sigma_i \qquad (12)$$

by assuming that a monolayer could be represented by a distribution of scattering centres characterized by three scattering indices σ_1, σ_2 and σ_3 which are proportional to surface layer coverage. α_i and β_i are constants for given values of ϕ_o, radiation wavelength and optical constants of surface and contamination or adsorbed layer.

Recently, Fiebelman[16] has questioned the use of this classical approach particularly for waves with the electric vector perpendicular to the plane of incidence and where there is submonolayer or monolayer coverage. He derived formulae for surface optical properties assuming rapid spatial variation for s waves and showed that the contribution to surface reflectivity is a measure of the induced surface charge. More recent measurements of Passler and Stiles[17] indicate that the classical model is satisfactory for both p and s waves in the energy range 0.5 to 4.5 eV.

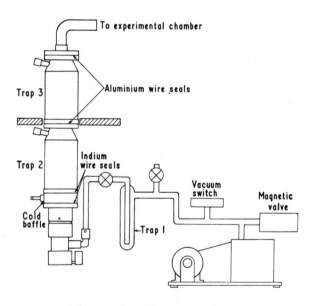

Figure 5. Vacuum system.

Fane and Neal[18] used ellipsometry to investigate any traces
of oil contamination on surfaces in a relatively inexpensive UHV
system (1×10^{-8}Pa.) evacuated by conventional diffusion pumps.
Gold films were selected as non reactive test surfaces and two
methods for the preparation of the gold base were employed.
(i) Evaporation in a standard coating unit with subsequent
transfer to the experimental UHV system (ii) Evaporation in situ
in the UHV system.   Figure 5 shows the arrangement of the
vacuum system.   The gold films were mounted on a thimble, in the
vacuum chamber, and its temperature could be varied from 77K
to 600K.

Figure 6 gives a sequence of operations and the consequent
thicknesses of surface oil contamination determined ellipsomet-
trically.   A clean surface could always be reproduced (maximum
Δ value) by heating the thimble to 100 °C.

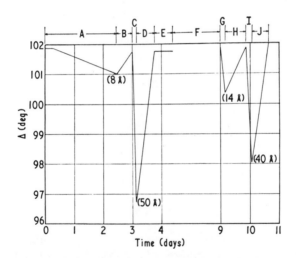

Figure 6.   Contamination on evaporated gold film.
A. Thimble, traps 2&3 20 °C;p=5×10^{-6}Pa;
B. Chamber 100 °C.
C. Thimble 77K;
D. System 100 °C;
E. Traps 2&3 77K; System 100 °C;p=4×10^{-7}Pa;
F. System 100 °C; traps 2&3 77K;
G. Trap 3 75 °C; thimble 77K;
H. Chamber 100 °C;
I. Trap 3 100 °C; thimble 77K;
J. System 100 °C;

For a specimen held at 77K in an untrapped system the rate of growth of an oil film was 2 nm per hour compared with a fully trapped system where no contamination could be detected over extended periods of time. An oil layer of 1 nm gives a pseudo constant for gold significantly different from that of a clean surface as is shown in Table I.

Table 1.   Pseudo Optical Constants of Gold with Oil (n_1 = 1.58, k_1 = 0) Contamination at a Wavelength of 546.1 nm. ϕ_o = 64.5°.

Oil thickness (nm)		n_2	k_2	$\delta\psi°$	$\delta\Delta°$
Clean surface	0	0.340	2.37	0	0
	0.5	0.332	2.33	+0.02	−0.52
	1.0	0.326	2.30	+0.04	−1.03
	2.0	0.315	2.25	+0.08	−2.04
	5.0	0.286	2.12	+0.21	−4.85
	10.0	0.249	1.93	+0.44	−9.37

Table I also shows changes in ψ and Δ as the oil layer is increased. $\delta\Delta/\delta t$ is approximately 1° per nm for thin layers and in most instruments it is possible to measure angles to within 0.01°. With the same angle of incidence $\delta\Delta/\delta t$ for oil on silver would be 1.2° per nm and for mercury (an absorbing layer) on chromium 0.58° per nm. Greater sensitivity can be achieved by optimising the angle of incidence and in the case of oil on silver for an angle of incidence of 72° the sensitivity would be increased to 1.3° per nm.

Changes in optical constants from those of a clean surface also result from adsorbed gases. Archer and Gobeli[19] observed changes in Δ with time when cleaved silicon was exposed to oxygen and at a point when there was a marked decrease in the rate of change of Δ, they assumed a monolayer coverage. They estimated changes of ±0.02 monolayers. Quentel et al[20] measured the adsorption of xenon on graphite ellipsometrically and found that Δ varied linearly with the coverage ratio (θ) and with the number of adsorbed layers. They were able to detect ±0.01 monolayer coverages and their results agreed with volumetric, LEED and Auger measurements. Changes in Δ of 5×10^{-3} degrees to 10^{-1} degrees have been reported by Abdel-Hamid[21] when argon pressure over polycrystalline gold was increased from 10^{-7} Pa to 10^{-3} Pa. Some reports suggest that changes can take place in the base material as a result of adsorption. For example, Carrol and Melmed[22] combined ellipsometry and LEED to investigate the exposure of a (011) tungsten surface to oxygen. A change in the

rate of decrease in ψ corresponded to a peak intensity of a LEED
spot when plotted against oxygen exposure. The angle Δ changed
in direction at an exposure slightly less than that corresponding
to the LEED peak intensity.   They suggested a rearrangement of
tungsten and oxygen atoms before the completion of a 50% monolayer
coverage. For monolayer adsorption there was no evidence of
readjustment as determined from the LEED patterns.

Corrosion Layers

The preparation of a surface prior to corrosion investigations
can itself introduce variables in the starting point parameters.
In the derivation of the equations relating to surface films it
was assumed that films were parallel sided and homogeneous. Most
commercially produced materials are unlikely to conform to such
assumptions and attempts to prepare an ideal reflecting surface
could produce a surface which is damaged and of variable structure.
This problem was discussed by Hayfield in 1976[23].   The majority
of oxide and hydroxide layers on metals are non absorbing (k = 0)
in the visible region of the spectrum.   A plot of ψ against Δ for
the progressive addition of such layers results in a closed loop
as has already been mentioned.   For a reactive metal such as
aluminium it is difficult to measure the optical constants n_2 and
k_2 in the absence of a film and unless great care is taken,
pseudo constants will be measured. Studies on both aqueous
corrosion and gaseous oxidation have resulted in various methods
of characterising a clean surface. Aqueous studies use electrochemical methods in the study of corrosion layers and surfaces
are inevitably covered by a layer produced by exposure to the
environment from the start.   The effect of an oxide layer on a
sample of pure aluminium would be to reduce the value of Δ by a
few degrees from that of a clean surface.   Table II illustrates
the point for an aluminium film prepared in U.H.V.

Table II.   Growth of Aluminium Oxide on an Aluminium Film.
            (λ = 549 nm)

Oxide Thickness (nm)	$\delta\Delta^{\circ}$	$\delta\psi^{\circ}$	Surface constants n_2	k_2
0 *	0	0	1.45	6.12
0.5	-0.7	0.00	1.39	5.99
1.0	-1.38	0.02	1.33	5.87
5.0	-6.98	0.11	0.98	5.01

* Represents a clean surface immediately after deposition.

It is seen that changes in ψ are small for thin oxide films.
n and k values (other than for zero oxide thickness) are pseudo
constants.   By contrast the optical constants for commercial
aluminium sheet were found by the author to be $n_2 = 0.675$, $k_2 =$
4.73.   These correspond to pseudo constants for metal with 20-30
nm of oxide grown on the surface at high temperatures during sheet
preparation.   It would be possible to characterise the underlying
base metal by using multiple angle ellipsometry to determine n_2
and k_2 for the base and n_1 and t for the existing film although
such a procedure presents difficulties.

Kucirek and Melmed[24] have investigated the way in which an
initial contamination layer can influence the computed optical
constants and thickness of additional layers. They computed the
errors for various materials resulting from taking the pseudo
constants to characterise a surface to use as a starting point
for subsequent film growth.   Their results show considerable
differences in the errors depending on the base material. In all
they made in the region of 500 computations using substrates of
iron, gallium arsenide, germanium, niobium, silicon, tantalum,
titanium and tungsten together with seven theoretical substances
as substrates. They concluded in general that there was a
tendency to an increasing percentage error, in taking a composite
surface pseudo constants to characterise a surface, as the thickness of the initial contaminant film increased.   The percentage
error decreased with increasing thickness of the subsequently
grown film. They recommend that calculations should be undertaken
for the specific materials involved since exceptions to general
trends did occur.   In extreme cases quoted they found a 9.24%
error by using pseudo cosntants to characterise a surface of
silicon when growing a film on silicon but only 0.21% in the case
of iron. Table III gives computed changes in instrument angles
ψ and Δ when aluminium oxide is grown on commercial aluminium
sheet with pseudo constants $n_2 = 0.675$ and $k_2 = 4.73$ used as a
starting point in order to characterise the surface.

Table III.   Oxide Growth on Aluminium Sheet.

Oxide thickness (nm)	$\Delta\psi^o$	$\delta\Delta^o$
0	0	0
2	+0.04	−2.82
6	+0.13	−8.28
10	+0.23	−13.48

By comparing Tables II and III it can be seen that the Δ sensitivity for early stages of oxide growth are similar whether pure

aluminium or commercial sheet are taken as starting points for
further corrosion tests.

If the contamination layer is thin then its thickness can be
determined if the optical constants of the film are known since
equations 9 and 10 apply. If there is gross contamination and
neither the base material constants or film constants are known
one or both have to be determined independently.

Evaporated Films

Early determinations on optical constants of many pure metals
as a means of characterising a surface were made by evaporating
films of the materials in vacuum chambers held at pressures of
the order of 10^{-4} Pa.   Values of the optical constants determined at such pressures are suspect due to the rapid growth of
oxide surface layers and also to the possibility of "volume"
oxidation during evaporation unless high evaporation rates can be
maintained. Much of the earlier work has been repeated since the
introduction of ultra high vacuum techniques. The consequences
of both surface and volume oxidation on optical constants of
surfaces have been reported by Fane and Neal[25] as part of a study
on optical and electrical properties of thin films. Aluminium
films were evaporated in a system with a base pressure of
10^{-8} Pa at a rate of 10 nm/min and the optical constants were
determined in less than 10 minutes after deposition. A clean
surface was defined in terms of the highest value of the angle Δ
which could be achieved.   This was found to be 148.56° at a
wavelength of 549 nm. and an angle of incidence of 63°51'. The
corresponding value of ψ was 41.55° leading to a complex refractive index of 1.45 - i 6.12.   The estimated oxide thickness within
10 minutes of deposition was less than 0.5 nm. Exposure to the
atmosphere reduced the value of Δ by 7.5° and increased ψ by
0.12° corresponding to a pseudo constant of 0.95 - i 4.98 and an
oxide layer in the region of 5.5 nm. A film deliberately
prepared at a pressure of 10^{-4} Pa. had a refractive index of
0.72 - i 5.65 as measured in the vacuum system and a decrease in
Δ of only 1.1° when exposed to the atmosphere. It was concluded
that a surface layer could not produce a combination of n = 0.72
and k = 5.65.   A value of n = 0.72 could be achieved with a
surface oxide film in the region of 9.5 nm, but other evidence
suggested that such a thickness would be unlikely. The evidence
suggested that volume oxidation during preparation at the higher
pressure was mainly responsible for the large differences in
optical constants.

Halford et al[26] made a comprehensive study of the optical
properties of ultra-pure aluminium films and they were able to

show that the value of 'n' for vacuum deposited films of aluminium could be consistently reproduced by controlling both the pressure and deposition rate of the surface region, i.e. the optical skin depth region of films. They demonstrated this by evaluating the optical constants for variations in the terminal ratio of pressure /deposition rate.

Allen[27] using the combined techniques of Auger electron spectrometry and ellipsometry measured the elemental chemical composition and optical properties of clean and contaminated aluminium under vacuum. His measurements showed a linear relation -ship between changes in amplitude of Auger peaks associated with both oxygen and aluminium oxide and the oxide thickness measured optically. He gave 0.19 nm as the estimated attenuation length of aluminium Auger electrons (67eV) in the oxide.

Argon ion bombardment

If an examination is to be made of the gaseous corrosion of commercial samples then a clean surface can be achieved by remov- ing the initial corrosion layer by argon ion bombardment in an ultra high vacuum followed by exposure to the required environment. This is a technique which has been adopted in the examination of surfaces by ESCA (XPS) in order to carry out a chemical profile analysis of surface layers. In the course of bombardment it is possible to cause surface damage which can modify the optical properties (depending on the energy of the ion beam) and therefore make optical characterisation very difficult. Optical examination does however permit changes from the cleaned base material to be measured. Investigations are currently being undertaken in the author's laboratory to compare the results of ESCA, EM and ellipsometric studies of oxide and hydroxide growth on aluminium and aluminium alloys under varying environmental conditions for both the early stages of growth and after prolonged exposure following ion etching. The optical examination of surfaces is providing additional information and is proving to be a useful complementary technique.

Table IV shows the measured changes in ψ and Δ obtained by Rehal[28] during experiments on aluminium oxide growth on commercial aluminium sheet which had first been heated at 350 °C for 15 minutes and then etched with argon ions (accelerating voltage 4kV, beam current 20μA) to obtain a clean surface. It was possible to reproduce values of Δ to within 0.1° after repeated etching cycles during corrosion tests on the same sample. Values of ψ were not reproducible due to possible damage to the surface (see also reference 29).

Table IV. Oxide Growth on Etched Commercial Aluminium Sheet.

ψ	Δ	Exposure (Langmuirs)	Oxide Thickness nm (Ellipsometry)
43.18	143.56	0	0
43.18	143.35	2	0.2
43.18	143.03	10	0.5
43.20	142.65	2×10^2	0.9
43.50	142.22	10^3	1.3

Table V gives values of ψ and Δ for various samples of aluminium which have been used by the author to characterise surfaces prior to subsequent corrosion tests.

Table V. Ellipsometric Characterisation of Aluminium Surfaces.

Sample	ψ	Δ
Commercial sheet (with oxide layer)	39.9	123.40
Etched commercial sheet	43.0	143.56
Evaporated film on etched sheet	44.18	144.10
Sheet (preheated to 350°C) and etched	43.0	143.47
Evaporated film on glass substrate	41.54	148.53

Characterisation by other methods.

Other methods have been adopted by various investigators in order to obtain a clean surface. Yolken and Kruger[29] during investigations on the protective layers on iron and steel used hydrogen gas as a reducing agent to react with the layers. Fane et al[30] bombarded stainless steel with electrons in ultra high vacuum to remove the oxide coating and investigated subsequent film growth at different gas pressures. An oxide free surface was characterised by the Δ parameter which could be reproduced to within 0.14° after cyclic changes involving oxidation and electron bombardment. This departure in Δ corresponded to less than an average surface coverage of oxide of 0.1 nm. The stable oxide protective layer measured to be $\sim$ 3.4 nm was at the bottom of the range reported by Farr and Zubillaga[31]. The method can be adopted without the complication of gas admission and the subsequent problem of removal of hydrogen from the vacuum system.

Hayfield[23] drew attention to the problem of surface characterlisation as between optical and other techniques for surface examination. High resolution electron microscopy, X-ray microanalysis and electron spectroscopy analysis reveal inhomogeneities

on a micro-scale.   Ellipsometry is limited by the poor resolution
in the plane of the surface.   Chang and Boulin[32] using the
combined techniques of ellipsometry and Auger spectroscopy
revealed a difference between oxide thicknesses on silicon and
aluminium measured by the two methods.   They suggested that the
difference was a consequence of the finite extent of the oxide/
base material interface resulting in the measurement of different
physical parameters in the two techniques.   In other words a
difference in characterisation between the two methods.

In the case of a metal where the rate of growth of an oxide
layer on a freshly exposed surface is slow it is possible to use
another approach to evaluate the optical properties of a film free
surface.   The time is taken from the instant of exposure of the
surface to the particular environment.   Values of ψ and Δ are
then plotted against time during a prolonged exposure.   Extrapolation to zero time gives values of ψ and Δ corresponding to a film
free surface to a first approximation.   Figure 7 shows the changes
in Δ due to growth of oxide on niobium and tantalum after removal
from the U.H.V. chamber in which they were prepared.   The growth
follows the law given by Vermilyea[33]

$$t = A_1 + B_1 \log T$$

where t is oxide thickness, T is the exposure time and A_1 and B_1
are constants.

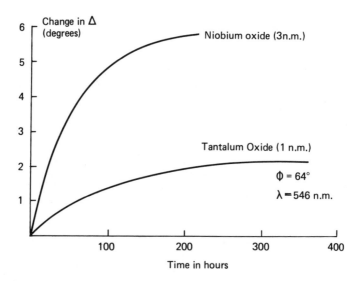

Figure 7.   Oxide growth on niobium and tantalum films.

Other factors which influence surface characterisation.

The ideal surface for optical examination is one which is perfectly smooth, specularly reflecting isotropic and homogeneous. The optical constants of an actual surface may depend on its previous history.   There have been limited studies on the relationships between structure variation and optical constants. Several workers have reported on changes in the angle ψ due to chemisorption and annealing.   Vedam and So[34] have attributed 'anomalous' changes in ψ to possible changes in surface structure. Mechanically polished silicon samples had variations in ψ of as much as 2^O from sample to sample and up to 5° higher than values expected for a strain free, film free silicon surface. On the other hand, values of ψ for chemically polished samples agreed to within $\pm 0.03^O$.   In order to establish any correlation between surface defects and optical properties it is necessary to use some other technique in conjunction with ellipsometry. Photo-emission and Auger spectroscopy have been used with ellipsometry by Smith[35] to monitor aluminium samples during fatigue cycling and large changes in both ψ and Δ were attributed to the formation of a sub-microscopic network of surface cracks. Grimes et al[36] using X-ray line broadening found that high values of the constant n for aluminium films were associated with high defect concentrat- ions. Meyer[37] using photo-emission and electron loss measurements related optical changes to changes in surface state distributions for semi-conductor surfaces, following chemical adsorption.

Surface roughness is another factor which can also cause ψ and Δ to differ from clean surface values. In general chemical or electrochemical polishing have the advantage of providing the most reproducible surfaces for "commercially" produced specimens. For vacuum studies a test surface can be prepared by argon ion bombardment. Even so real surfaces are not atomically smooth. Fenstermaker and McCrackin[38] investigated the errors introduced by neglecting roughness.   They considered roughness up to 50 nm on materials with 'n' values ranging from 0.18 to 4.05 and 'k' values from 0 to 4.54.   Unlike most other materials glass was found to have apparent values of 'n' close to its true value even for roughnesses of the order of 30 nm. The apparent value of k/n did change significantly with roughness. To yield the same values of ψ and Δ as a rough transparent surface an equivalent smooth surface must be assumed to be absorbing. Morton and Chan[39] have even suggested that for many metals surface roughness and not an oxide layer was responsible for the variation of ψ and Δ from that of a truly smooth bulk material.   Deleterious effects of surface roughness can be reduced by using angles of incidence less than 70°. It must however be borne in mind that the instru- ment sensitivity can be decreased if angles of incidence depart too far from the optimum as has already been mentioned in dealing with multi angle ellipsometry.

Anisotropy is another factor which can influence the ellipsometric characterisation of a surface. Anisotropy can be detected by rotating a specimen about an axis perpendicular to the surface. Variations in optical properties with direction will mean variations in the instrument settings for extinction, i.e. variations in ψ and Δ at a fixed angle of incidence as the specimen is rotated. This was observed by Cathcart and Petersen[40] in studies on the epitaxial growth of Cu_2O films on single crystal copper. When they assumed optical constants were independent of direction and calculated oxide thicknesses for different specimen orientations they found periodic variations in thickness with angle of rotation. To date there have been very few reports of studies on optical anisotropy in surfaces and its effect on surface characterisation but I am convinced that there will be considerable efforts in this direction in the next few years.

CONCLUDING REMARKS

In this paper I have tried to show that ellipsometry has a role to play in surface examination and in surface character-isation. A 'clean' surface can be defined in terms of the instrument parameters ψ and Δ, althouth care must be exercised in interpretation, when values of optical constants deviate from those of a clean surface. In addition to gas adsorption, contamination, oxidation and corrosion, roughness, structural defects and anisotropy can influence the optical constants of surfaces. It has been shown that the instrument can be used under certain circumstances to monitor changes taking place on a film covered surface as long as the starting point is character-ised by the pseudo constants of the composite surface. It has the advantage of being non destructive. The instrument averages over the area of surface covered by the light beam and consequently has poor resolution in the plane of the surface. If used in conjunction with other surface analytical techniques, useful information can be obtained concerning the state of a surface and extent of surface coverage.

ACKNOWLEDGEMENTS

I am grateful to the Institute of Physics (UK) for permission to reproduce Figures 5 and 6.

REFERENCES

1. P.Drude, Ann.Phys.Leipzig, 32, 584, (1887).
2. P.Drude, Ann.Phys.Leipzig, 34, 489, (1888).
3. P.Drude, Ann.Phys.Leipzig, 36, 532, (1889).
4. P.Drude, Ann.Phys.Leipzig, 39, 481, (1890).
5. L.Tronstad, Trans.Faraday Soc. 29, 502, (1933).
6. A.Rothen, Rev.Sci.Inst.16, 26, (1945).
7. A.R.C.Westwood, R.K.Wiswanadham, and J.A.S.Green, Thin Solid Films, 39, 69, (1976).
8. J.C.Grosskreutz, J.Electrochem.Soc. 114, 882, (1967).
9. A.B.Winterbottom, Nature (London) 140, 364, (1937).
10. R.J.Archer, J.Opt.Soc.Am. 52, 970, (1962).
11. E.Passaglia, Editor, "Ellipsometry in the Measurement of Surfaces and Thin Films" National Bureau of Standards, Washington, D.C., 1964.
12. R.W.Ditchburn, J.Opt.Soc.Am. 45, 743, (1955).
13. F.L.McCracken and J.P.Colson, see Ref.11, page 61.
14. R.J.Archer, see Ref.11, page 255.
15. F.Meyer and G.A.Bootsma, Surf.Sci. 16, 221, (1969).
16. P.J.Fiebelman, Phys.Rev.B, 14, 762, (1976).
17. M.A.Passler and P.J.Stiles, J.Vac.Sci.Technol, 15, 611, (1978).
18. R.W.Fane and W.E.J.Neal, Proc.of 4th Int.Vac.Congress, p.510, (1968).
19. R.J.Archer and G.W.Gobeli, J.Phys.Chem.Soc.26, 343, (1965).
20. G.Quentel, J.M.Rickard and R.Kern, Surf.Sci. 50, 343, (1975).
21. S.M.M.Abdel-Hamid, PhD Thesis, University of Aston (1976).
22. J.J.Carrol and A.J.Melmed, Surf.Sci. 16, 251, (1969).
23. P.C.S.Hayfield, Surf.Sci. 56, 488, (1976).
24. J.Kuricek and A.J.Melmed, J.Opt.Soc.Am. 65, 611, (1975).
25. R.W.Fane and W.E.J.Neal, J.Opt.Soc.Am. 60, 790, (1970).
26. J.H.Halford, F.K.Chin and J.E.Norman, J.Opt.Soc.Am, 63, 786, (1973).
27. T.H.Allen, J.Vac.Sci.Technol, 13, 112, (1976).
28. A.Rehal, (1977), unpublished data.
29. H.T.Yolken and J.Kruger, J.Opt.Soc.Am. 55, 842, (1965).
30. R.W.Fane, W.E.J.Neal and R.V.Latham, J.Appl.Phys, 44, 740, (1973).
31. J.P.G.Farr and C.Zubillaga, Soc.Chem.Ind.Monograph No.28, p.436 (1968).
32. C.C.Chang and D.M.Boulin, Surf.Sci. 69, 385, (1977).
33. D.A.Vermilyea, Acta Metall. 6, 166, (1958).
34. K.Vedam and S.S.So, Surf.Sci. 29, 379, (1972).
35. T.Smith, Surf.Sci. 45, 117, (1974).
36. N.W.Grimes, J.M.Pearson, R.W.Fane and W.E.J.Neal, Phil.Mag. 21, 177, (1970).
37. F.Meyer, Phys.Ref. B.9, 3622, (1974).
38. C.A.Fenstermaker and F.L.McCrackin, Surf.Sci,16, 85, (1969).
39. J.P.Morton and E.C.Chan, J.Appl.Phys.45, 5008, (1974).
40. J.V.Cathcart and C.F.Petersen, see Ref.11, p.201.

IDENTIFICATION AND ELIMINATION OF ORGANIC CONTAMINANTS ON THE

SURFACE OF PLZT CERAMIC WAFERS

J. A. Wagner

Sandia Laboratories

Albuquerque, New Mexico   87185

PLZT (lead lanthanum zirconate-titanate) hot-pressed
ferroelectric ceramics are being used as electrooptic
wafers in thermal/flash protective goggles. Surface
contamination of the wafers has caused nonuniform appearance of the goggle lens in the transmitting (open) and
protective (closed) states. Experiments were devised to
identify the contaminant and to find a method of effectively removing it from the ceramic surface without
changing the electrical properties of the PLZT. Infrared spectroscopy, secondary ion mass spectroscopy, Auger
electron spectroscopy and scanning electron microscopy
were used to determine the type and extent of contamination and the effectiveness of cleaning methods. The contaminant was identified as the residue from an adhesive
used in processing the wafers. Various cleaning methods
were compared: washing and soaking wafers in a 50% by
weight volume potassium carbonate/deionized water solution;
solvent cleaning with various combinations of tetrahydrofuran, methylene chloride, methylethylketone; oxygen
plasma; and ultraviolet radiation. Results indicated
that although solvents and potassium carbonate solutions
removed the gross contamination, there was still a sufficient amount present to cause "mottling." In addition,
some adverse effects on the ceramic surface were discovered
in the case K_2CO_3 cleaning. Auger scans proved ultraviolet
to be the most effective method, with no contamination remaining after a two hour exposure. Scanning electron
micrographs of exposed wafers showed no apparent surface
damage. A model explaining the behavior of the contaminated surface with respect to various solvents was developed.

INTRODUCTION

Hot-pressed transparent ferroelectric ceramic PLZT (lead lanthanum zirconate-titanate), is being used for lenses in thermal/flash protective goggles. During electrical testing of the lenses, "mottling" appeared. This adversely affected the efficiency of both the transparent and opaque phases of the goggles by allowing light to leak in during the transparent phase. Nonuniform electric fields are responsible for the mottling effect.

Polar molecules residual on the surface of the ceramic were thought to be a possible cause of the electrical phenomena. Since the ceramic wafers go through a series of production steps and are subjected to a variety of organic compounds, an organic contaminant was postulated as a possible source of the polar molecules.

After the wafers were ground and polished by an optical laboratory, they were delivered to a production company for electroding and bonding to form a bonded lens assembly. Upon receipt of the wafers at the production company there was a visually noticeable thin film contamination on the surface of many of the units. Experiments were devised to identify the contaminant, to show that it was responsible for the mottling problem, and to find a method of effectively removing it from the ceramic surface without changing the electrical properties of the PLZT.

Infrared spectroscopy, secondary ion mass spectroscopy, Auger electron spectroscopy and scanning electron microscopy were used to determine the type and extent of the contamination and the effectiveness of the various cleaning methods.

Numerous methods of cleaning the wafers were compared; washing and soaking in deionized water/K_2CO_3; solvent cleaning with various combinations of tetrahydrofuran, methylene chloride, methylethylketone; oxygen plasma; and ultraviolet radiation.

EXPERIMENTAL

Samples of the thin film contaminant were obtained from the production company by scraping the visible residue from a contaminated lens. Samples of the suspected source material, an optical adhesive used in the grinding and polishing process, were acquired from the optical laboratory responsible for these production steps. The samples were taken from an acetone soak used to remove gross adhesive from the ground lens. A Perkin Elmer Model 21 Double Beam Infrared spectrophotometer was used to measure infrared spectra of both film residues. All solvents and chemicals used were reagent grade. The ceramic wafers were random samples taken from preproduction and production groups.

Wafers were subjected to cleaning in an activated RF generated oxygen plasma in a five inch diameter reaction chamber. The
activated gas equipment was made by International Plasma Corporation of Hayward, California, with a control console, Model #PM
101, and a power convertor, Model #PM 301. Conditions used were
0.5 Torr pressure and 50 watts power for 30 minutes.

Ultraviolet cleaning was performed in a stainless steel
chamber with interior reflecting surfaces of Alzak[1] and a mercury
vapor lamp rated at 3 watts/m[2] at 20 mm from the lamp. The air
inside the chamber was room air which was sealed in the chamber
during the cleaning process.

Surface analyses both before and after the various cleaning
techniques were done by secondary ion mass spectroscopy, Auger
electron spectroscopy and scanning electron microscopy.

RESULTS AND DISCUSSION

The bonded lens assembly for the goggles includes the electroded wafer held between cross polarizers. Mottling on the lens
surface appears with the application of voltage to the electrodes.
The electrodes are formed by plating from solution, gold over
nickel, and are placed in shallow (1.5 mils) grooves in the
ceramic wafer. During the phase when voltage is applied to the
electrodes, regions of high $\bar{E}$ cause color effects and regions of
abnormally low $\bar{E}$ result in gray shades in an otherwise optically
clear lens. When the electrodes are shorted, there is a residual
space charge effect in these areas causing light spots in an
otherwise relatively opaque lens. (Figure 1 and 2.) This is
probably caused by adsorption of the charge by polar molecules
causing non-uniform distribution across the surface of the wafer.

Other phenomena were noted when the mottled wafers were subjected to various attempts at solvent cleaning.[2] When the wafer
was rinsed with isopropyl alcohol, the mottling occured over the
surface. Methyl alcohol rinses showed some improvement but mottling was still evident. Methylethylketone produced better results but there was still some mottling present. Experiments
with humidity showed that water vapor increased the problem.
Baking of mottled wafers at 65º-80ºC resulted in uniform transmission for a short time but the mottling returned after 24 hours
storage on a clean bench. Wafers which showed none of the mottling problems were not affected by any of the above treatments.
These results again indicated the presence of polar molecules
on the surface of the ceramic.

The most probable cause of the contamination was the optical
adhesive being used by the laboratory which grinds and polishes
the wafers. Infrared spectra from both these samples are shown
(Figures 3 and 4) proving that they are identical compounds.

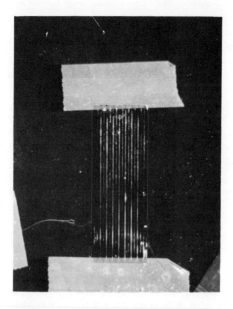

Figure 1.   Contaminated, electroded wafer in "open" or "transmit-
            ting" phase showing "mottled" areas.

Figure 2.   Contaminated, electroded wafer in "closed" or "pro-
            tective" phase showing light spots from residual
            space charge.

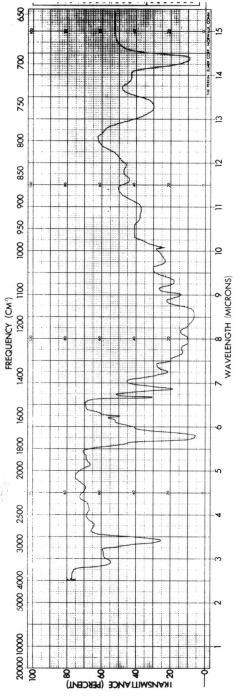

Figure 3: Infrared spectrum of contaminant film obtained from company processing PLZT wafers.

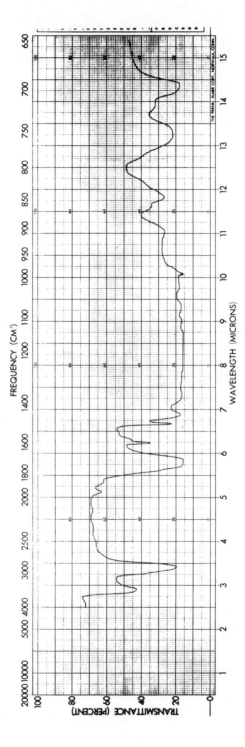

Figure 4.   Infrared spectrum of contaminant film obtained from "cleaned" wafers.

   Another experiment was carried out to support the finding
that the residues from the optical adhesive were causing the
mottling problem. An uncleaned ground lens was obtained from the
supplier and subjected to the same kind of acetone cleaning used
by the supplier. A very small amount of this solution was placed
on an unmottled electroded PLZT wafer to determine if "mottling"
could be induced. When the voltage was applied to the inter-
digital electrode array, widespread mottling appeared on the wafer.

   Information on the particular adhesive was not available
from the optical laboratory because of a proprietary application
which they had developed. For positive identification so that
cleaning procedures could be defined, samples of the film were
also analyzed by mass spectrometry. The main constituents were
styrene, the dimethyl ester of phthalic acid, and an unsaturated
hydrocarbon of 4-6 carbons. Based upon these analyses and a
knowledge of possible adhesives used, the material was tentative-
ly identified as Summers Lens Bond M-62 with catalyst, manufactured by Summers Laboratory, Fort Washington, PA, a styrene modified polyester cured with methylethylketone peroxide. The
following structures are present in the adhesive:

     styrene   dimethylphthlate   MEK peroxide

Polymerization of these compounds would leave many polar sites
including the carbonyl groups containing electronegative oxygen
with two pairs of unbonded electrons. Water, isopropanol and
methanol would readily form hydrogen bonds with such a surface.
Methylethylketone having a carbonyl structure itself would not
absorb easily in this environment. This could explain the detrimental effect of the first three solvents on the contaminated
wafer. Hydrogen bonds with a bonding energy of 5 Kcal/mole[3] are
considered to be an especially strong kind of dipole-dipole attraction; so, at ambient temperatures they would remain intact.
The heating of the wafers to 150-200°F would, however, introduce
enough thermal energy to cause the hydrogen bonds to break free-
ing the solvents and lessening the mottling problem.

   Several approaches to cleaning the wafers were examined.
Since the gross adhesive is removed from the wafers with acetone

but residues remain, a more powerful solvent was tried. Soaking, boiling and rinsing in tetrahydrofuran, methylethylketone was done in various combinations. The solvent treatments were followed by thorough rinsing with deionized water and a blow-dry with nitrogen. In all cases, Auger analysis showed remaining contamination on the surface (Figures 5 and 6). The presence of a strong carbon peak was used as an indication of organic material as the ceramic contains no carbon.

Potassium carbonate/deionized water (50%) solutions are often used to remove greases, oils, etc., from glasswork. A potassium carbonate solution of the same strength was used as a soaking bath for contaminated wafers. The wafers were soaked for periods of time from 2 hours to 24 hours at 65°C in both pyrex and teflon beakers, after which they were thoroughly rinsed with deionized water. It was found that a side-reaction between the solution, the wafer and the pyrex beaker was taking place which generated amorphous silica containing slivers and etched the surface of the wafer as seen in Figure 7. Another undesirable result was the deposition of white K_2CO_3 on the surface of the wafer soaked for four hours in a teflon beaker (Figure 8). There was a greater removal of the residual organic from the surface after the K_2CO_3 treatment but the possible etching and deposition of crystals detracts from the use of this method. Auger scans of wafers processed by soaking in K_2CO_3/DI H_2O (50%) and rinsing with deionized water confirmed the presence of both carbon and potassium on the surface (Figure 9).

The results of subjecting the contaminated wafers to oxygen plasma indicated that complete cleaning could be accomplished if proper conditions were set. The contaminated wafers were exposed to activated oxygen plasma for a period of 30 minutes at 50 watts and 0.5 Torr pressure after which secondary ion mass spectrometry showed a sizeable decrease in carbon concentrations on the surface. Auger analysis confirmed these data. Exposing the wafers for longer periods of time would have completed the elimination of the organic contaminant. Since production schedules had to be met by the production company and because they had no access to equipment for plasma cleaning without large expenditures, it was decided to attempt ultraviolet cleaning instead.

The UV cleaning procedure works consistently only if the samples to be cleaned are properly precleaned prior to UV irradiation to remove gross contamination. Since acetone is used to remove the gross residues from the wafer surface during processing the precleaning step is already completed. Ultraviolet light of two wave lengths is generated by the mercury discharge tubes, 1849 Å which generates ozone, and 2537 Å which is absorbed by most hydrocarbons and by ozone. When both wave lengths are present ozone is continuously formed and destroyed producing atomic oxygen, a very strong oxidizing agent.

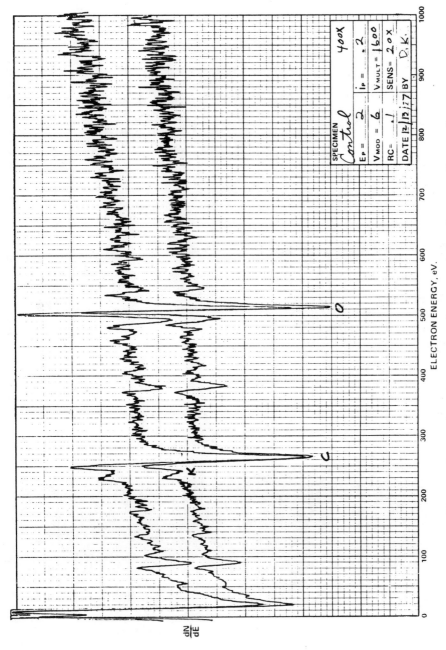

Figure 5. Auger spectrum of contaminated control sample.

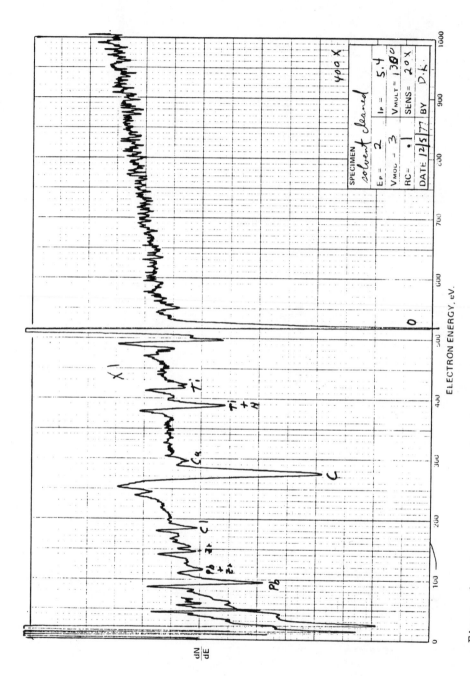

Figure 6.   Auger spectrum of solvent cleaned wafer showing residual contamination.

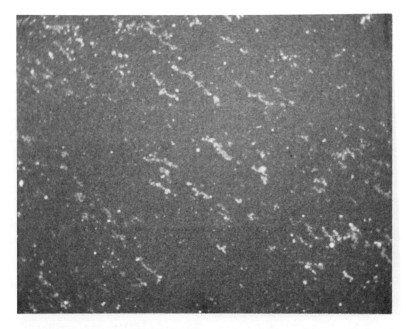

Figure 7. Scanning electron micrograph of ceramic surface after soaking in K_2CO_3. (x300)

Figure 8. Scanning electron micrograph showing K_2CO_3 crystals formed on surface of ceramic. (x1000)

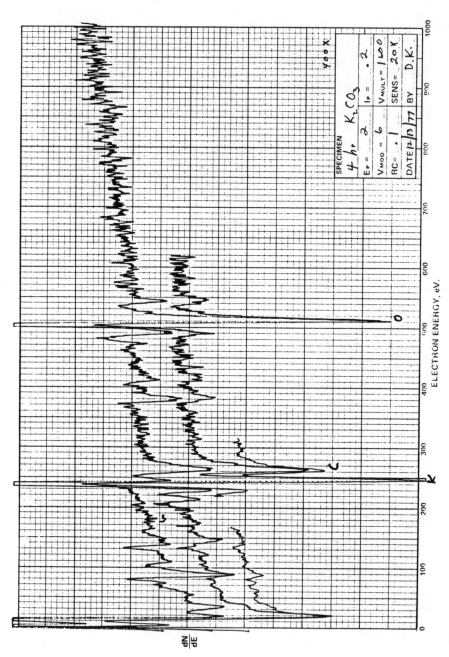

Figure 9.   Auger spectrum after soaking ceramic in K_2CO_3 solution.   Note residual carbon and strong potassium peaks.

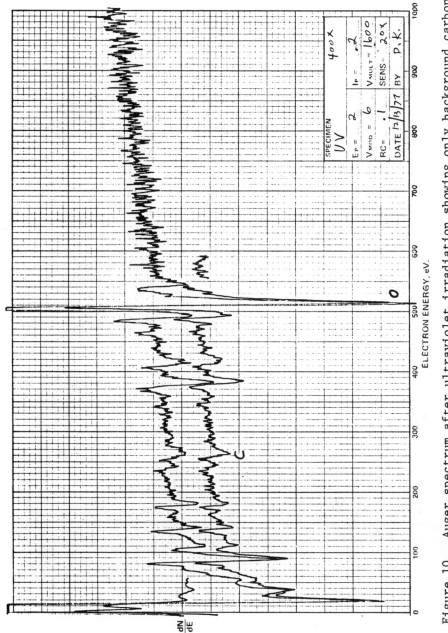

Figure 10. Auger spectrum after ultraviolet irradiation showing only background carbon traces remaining.

A number of samples of ceramic wafers which were precleaned
by the optical laboratory in acetone baths were used in the UV
experiment. The samples were placed about 20 mm from the source
and irradiated for 2 hours. In all cases, Auger spectroscopy
performed immediately after removal from the UV chamber showed
only background carbon peaks (Figure 10). The Auger spectra of a
clean ceramic surface which had been exposed to air even a very
short time invariably showed a trace of carbon as background.
The amount was negligible and wafers which showed only this trace
are considered to be "clean."[6] Because of the active surface of
the ceramic, contamination below this background level would
probably not be possible to achieve. Scanning electron micrographs of UV cleaned wafers showed no adverse effects on the surface.

CONCLUSIONS

This study confirmed the problems associated with even a minute amount of polar organic contamination on the surface of PLZT
interdigitally electroded ceramic wafers. Positive identification of the contaminant showed that insufficient removal of the
optical adhesive used for grinding the lens was a source of the
"mottling" problem. Results based on the behavior of the contaminated wafers with regard to various solvents precluded solvents as the final cleaning method.

Potassium carbonate solutions cleaned better than solvents
but had some undesirable side effects such as possible etching of
the ceramic surface and deposition of potassium containing crystals on the surface.

Oxygen plasma was indicated as a viable method but cost and time
considerations made plasma cleaning unsuitable.

Ultraviolet light proved to be the most effective method
for removing the residual organics from the wafers. More careful precleaning with acetone to remove gross contamination should
precede the UV irradiation for greatest efficiency. There were
no detectable changes in the properties of the PLZT ceramic wafers as a result of cleaning by ultraviolet irradiation.

ACKNOWLEDGEMENTS

The author wishes to thank D. K. Kramer (5825) for the Auger electron spectroscopy, R. L. Courtney (1472) and D. H. Huskisson (5822) for scanning electron microscopy, and R. R. Sowell (5834) for fruitful discussions on ultraviolet irradiation. This work was supported by the U. S. Department of Energy.

REFERENCES

1. Alzak is an aluminum alloy processed to provide high reflectivity. This process is licensed to several manufacturers by the Aluminum Co. of America, Pittsburgh, PA.
2. Memorandum issued October 21, 1977, by Honeywell, Inc. SRC Division, Minneapolis, MN.
3. R. T. Morrison, and R. N. Boyd, "Organic Chemistry," p. 307, Allyn and Bacon, Inc., Boston, 1959.
4. J. R. Vig, in "Surface Contamination: Its Genesis, Detection and Control," K. L. Mittal, Editor, Vol. 1, pp. 235-254, Plenum Press, 1979.
5. J. G. Calvert and J. N. Pitts, Jr., "Photochemistry," pp. 205-209, John Wiley and Sons, New York, 1966.
6. D. Kramer, private communication, 1978.

A NEW MONITORING TECHNIQUE FOR SURFACE CONTAMINATION —— THE TEST SURFACE METHOD

I.Anzai and T.Kikuchi

Faculty of Medicine, University of Tokyo, Hongo, Bunkyo-ku, Tokyo 113, Japan and Research Center for Nuclear Science and Technology, University of Tokyo, Yayoi, Bunkyo-ku, Tokyo 113, Japan

The test surface method, a new monitoring technique for surface contamination, was developed. In the proposed technique, a piece of test surface, which is made of un-woven fabric attached on an adhesive tape, is first applied to the surface to be monitored. After a certain period of time, the test surface is stripped off and the radioactivity deposited on the test surface is counted with an appropriate radiation detector. The merits of the method as compared with the smear technique are as follows: (1) The test surface method measures the total accumulation of the fixed and the loose contamination originated in each working period. (2) The measured values do not depend upon the kind of surface to be monitored, and are free from the skill of the performers. (3) Test pieces can be applied to the surface of a com-plicated shape, and also to the sole of the worker's footwear, which enables personnel monitoring of the surface contamination. (4) The routine procedure is only to strip the old test piece for radiation counting and to replace with a clean test piece.

INTRODUCTION

The measurement of surface contamination is one of the im-portant tools for checking the presence of potential risk of internal contamination.[1] The smear method is the most widely used method of surface monitoring. Several different techniques for determining the level of loose contamination have been developed in addition to the smear method, such as the gummed tape technique and the "Smair" method.[2,3] The major reason why the smear method

785

has most popularly been put into practical use lies in its sim-
plicity. In the Union Carbide Y-12 Plant in Oak Ridge where an
extensive number of samples have to be measured, a special method
was developed in which smearing was directly performed with the
electronic computer card in order to carry out the procedure from
counting to data processing automatically.[4] Basic and practical
problems have also been studied concerning the liquid scintillation
counting of the smear samples of soft-beta contaminants.[5,6,7]

The authors have experimentally examined the dependence of
smear efficiency on the manner of smearing, the kinds of radio-
nuclide, the surface materials, and the time lapse after the
occurrence of radioactive contamination by using $^3H, ^{35}S, ^{60}Co, ^{90}Sr$-
^{90}Y and ^{137}Cs as contaminants of glass, stainless steel, and lino-
leum. In spite of many advantages of the smear method as a
practical monitoring means, there are some drawbacks to it. For
example (1) the physical condition of the contaminated surface may
affect the smear efficiency, (2) smearing cannot always be practiced
under the same condition when the surface has a complicated shape,
(3) smear efficiency may also depend on the manner of smearing,
which differs with each performer, (4) the fixed contamination
cannot be determined by this method, and (5) it is not commonly
possible to compare objectively the smear data measured by differ-

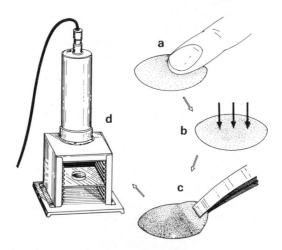

Figure 1. Procedure of *the test surface method* for monitoring
surface contamination. (a) A piece of test surface is attached to
the laboratory table, floor, door handle, drawer knob, elevator
button, etc. in workplace to be monitored. (b) Contamination by
contact with worker's hands, footwear, etc.; and by gravitational
deposition of airborne radiosubstances. (c) The test piece is
stripped after a specified monitoring period. (d) The activity
accumulated on the piece of test surface is counted.

ent operators in different radioisotope facilities because the
data are influenced by many factors mentioned above.

The present study was aimed at developing a simple technique
for surface monitoring that overcomes some of the imperfections
of the smear method.

A NEW TECHNIQUE ── THE TEST SURFACE METHOD

A new technique devised by the authors, which is illustrated
in Figure 1, can most appropriately be called *the test surface
method*. In this method, a piece of test surface is first applied
to the surface of the workplace to be monitored. After a certain
period of time, the test piece is stripped off and the activity
accumulated on its outer surface during the working period is
counted with an appropriate radiation detection system.

The best method for measuring the fixed surface contamination
is, in principle, to remove a piece of surface material for counting. In the test surface method, a removable test surface is introduced instead of whittling the surface material itself.

A practicable test surface must be (1) highly efficient in
the cumulative collection of the contaminants, (2) sufficiently
durable to withstand the worker's trampling, and (3) easy to stick
on and strip off. The authors made pieces of test surface from
materials such as paper tape, vinyl tape, and woven and unwoven
fabric tape; and examined their properties with regard to the
requirements mentioned above. Unwoven fabric attached on an adhesive tape was found to be most satisfactory. Figure 2 is the
photograph of a piece of test surface developed by the authors.

Figure 2. A piece of test surface made of unwoven fabric attached
on an adhesive tape.

Many pieces of test surface have been used at the actual work-places of the radioisotope handling facilities of the Faculty of Medicine, University of Tokyo. Each piece of test surface had a diameter of 23 mm so that it could be placed in a usual sample dish for counting. Liquid scintillation counting was found to be possible for the soft-beta-emitting contaminants.

Figure 3 shows a device for stripping the test piece from the surface with ease.

Figure 4 indicates a piece of test surface sandwiched between the polyester pasteboard and the polyethylene cover that were developed in consideration of the disposability.

ADVANTAGES OF THE TEST SURFACE METHOD

Distinctive advantages of the test surface method are as follows :

(1) The test surface method can be regarded as a monitoring technique of integration type that enables us to obtain information about the total accumulation of radioactive contaminants in each operation period.

(2) The data obtained by the test surface method do not depend on the kind or condition of the surface to be monitored, and are not subject to operators' errors. This distinguishing feature is important since it allows unbiased comparisons of data measure in different workplaces.

(3) A piece of test surface can even be applied to complicated surfaces such as door handles, drawer knobs, elevator buttons, etc.

(4) A piece of test surface can also be applied to the sole of a worker's footwear to obtain important information on the

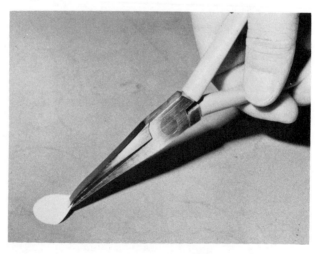

Figure 3. A device for stripping the test surface.

surface contamination in large work area. The test surface thus applied can well be regarded as "walking smear" that, in a sense, enables personnel monitoring of surface contamination.

(5) The test surface method is a simple procedure because all that one must do is to strip off the old piece of test surface for counting and to stick on a new one.

DISCUSSION

There may be a different viewpoint concerning the fact that the test surface method does not separately measure the fixed and the loose contamination. The level of loose contamination is of primary importance for hazard evaluation. But, from the basic standpoint of radiation protection, it is quite important to minimize the total level of contamination and to use monitoring information to assess the adequacy of radiation control practices. The test surface method and the smear technique, when used together, will mutually supplement each other.

The piece of test surface developed in the present study was found to be satisfactory for practical use in ordinary radioisotope facilities. It may, however, be useful in other applications to make the piece of test surface from a material that has physico-chemical properties similar to those of the surface to be monitored. This would probably yield a better measure of the level of contamination on the surface under consideration.

The authors have used the test surface method discussed in this report together with the smear technique in the routine monitoring of the workplaces of the Research Center for Nuclear Science and Technology, University of Tokyo, for 18 months since May, 1974.

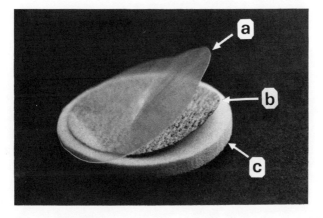

Figure 4. A piece of used test surface (b) sandwiched between a polyester pasteboard (c) and a polyethylene cover (a) which were developed in consideration of easy disposability.

They used 1870 pieces of test surface and 2280 smear samples. The measurement of radioactivity was performed monthly using liquid scintillation counting. The coefficient of correlation between the number of contaminated smear samples and that of contaminated pieces of test surface in each month (0.82) suggests the practicality of the test surface method as a routine means of radiation control.

REFERENCES

1. International Commission on Radiological Protection, "General Principles of Monitoring for Radiation Protection of Workers", Publication 12, p.8, Pergamon Press, Oxford, 1968.
2. International Atomic Energy Agency, "Monitoring of Radioactive Contamination on Surfaces", Technical Report Series No.120, IAEA, Vienna, 1970.
3. G.W.Royster Jr. and B.R.Fish, "Techniques for Assessing Removable Surface Contamination", ORNL-TM-1045 (1965).
4. M.Sanders, Health Physics 10, 341 (1964).
5. C.H.Wang and D.E.Jones, Biochem. Biophys. Res. Commun. 1, 203 (1959).
6. J.R.Prince and C.H.Wang, "A Method for Evaluating Surface Contamination of Soft Beta Emitters", AT(45-1)-1367 (1964).
7. J.D.Eakins and W.P.Hutchinson, "The Estimation of the Level of Tritium Contamination on Metal Surface by Smearing", AERE-R5988 (1969).

REPLICATION TECHNIQUE FOR EXAMINING DEFECTS IN THE INTERFACE

OF A METAL-TO-GLASS CERAMIC BOND

R. K. Spears*

General Electric Company
Neutron Devices Department
St. Petersburg, Florida 33733

Epoxy replicas were made of the interface of a molybdenum and glass-ceramic assembly and examined by scanning electron microscopy.

Replications of this interface were produced by first removing the molybdenum from four assemblies using a nitric acid-based etchant. The glass-ceramic was etched from the epoxy in a hydrogen fluoride-based acid etchant. The resulting replicas resembled the texture of the molybdenum surface with the interface defects shown in detail as projections. This process revealed some unusual interface problems which appeared to be associated with the evolution of gas from the molybdenum piece parts.

INTRODUCTION

At the General Electric Neutron Devices Department (GEND), a vacuum tube subassembly (Figure 1) consists of an insulator made of glass-ceramic bonded between two concentric molybdenum cylinders.[1] To uniformly distribute mechanical stress, as well as to decrease the gas permeation path, it is necessary that the interfaces be as free of defects as possible. Large defects (0.5 mm) can be detected by ultrasonic inspection. However, it is known that these interfaces usually contain many voids smaller than the

*Work supported by the U. S. Department of Energy under Contract No. DE-AC04-76DP00656

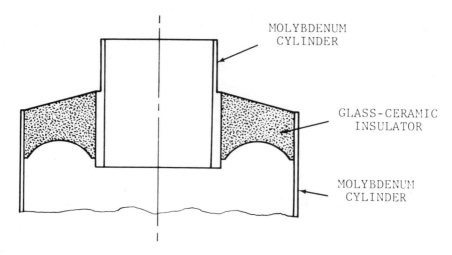

Figure 1.   Subassembly

detection limit of the ultrasonic scanner. The source of these
defects is commonly attributed to contamination or glass preform
defects. Due to the complexity of the subassembly molding process, the precise source of flaws is difficult to ascertain.

An approach to determining the cause of defects consists of
a scanning electron microscopic (SEM) examination of the glass
interface after the molybdenum is etched from the insulator. SEM
photographs of a typical insulator treated in this manner are
shown in Figure 2. Although the shape of the defect at the interface can be measured, the extent of the defect or its internal
features cannot be characterized. It is important that these features be determined since, with this knowledge, the source of the
voids could possibly be found. This could be accomplished by
characterization of the features of defects whose sources were
known and then comparing them to defects whose sources were unknown.
Examining the insulator-molybdenum interfaces by etching the molybdenum from the insulator was initially performed by W. Andrews,
GEND, with the aid of personnel in the GEND Chemistry Laboratory.

It was proposed that one way of obtaining details of these
features would be to replicate the surface of the glass-ceramic
with an elastic polymeric material, to strip the glass-ceramic
from the surface and to examine the surface of the replica using
SEM.[*] One material used at GEND on previous occasions for this
purpose was RTV 602[†]—an unfilled silicon rubber. When this was
_____
[*]The author was first made aware of this technique by SEM personnel
at GE Research & Development Laboratory, Schenectady, N.Y., in 1967.
[†]Trademark, General Electric Co.

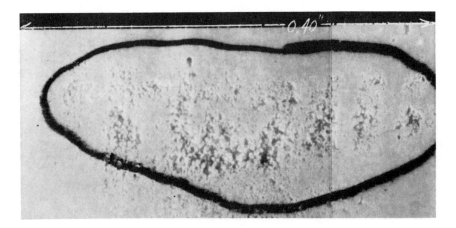

Figure 2-A. SEM photograph of a defective region in the insulator at the insulator-molybdenum interface. (This is a photograph of the insulator after the molybdenum had been etched away.)

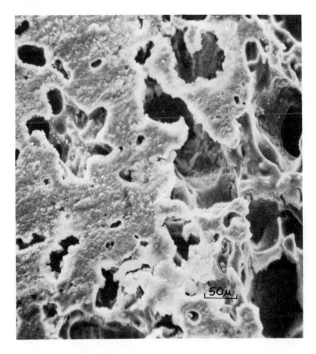

Figure 2-B. Highly magnified SEM photograph of a region shown in Figure 2-A. (Surface detail can be observed but details concerning the structure of the voids are not observable.)

tried on the insulator, the rubber tore at the rubber insulator
interface due to the sharpness of the glass and the fact that the
void increased in size within the insulator capturing the RTV.
It was then proposed that the insulator be etched from the replica
using a hydrofluoric acid-based etchant. Since the acid would
attack the silicone rubber, a rigid unfilled epoxy resin, which
is not attacked by the etchant, was used as the replicating medium.
The result revealed what is thought to be the true structure of
the defects at the interface, and it is hoped that this technique
will prove beneficial for future studies of glass-ceramic-to-metal
interface defects.

EXPERIMENTAL PROCEDURE

     Four subassemblies were examined in this study. The first
was a unit rejected for ultrasonic flaws (tube No. 811411). The
second was a complete unit that leaked gas at the inner sleeve
(tube No. 732115). The other two units were subassemblies used
for preliminary evaluations on the effect of polishing the inner
sleeve on interface defects (subassemblies No. 451-1 and 451-2).

     The technique used in fabricating epoxy replicas evolved over
a period of time. The process ultimately used was as follows:

1.   Excess molybdenum was removed from the subassembly by cutting
     the inner and outer sleeve along a plane perpendicular to the
     tube axis at points shown in Figure 3.

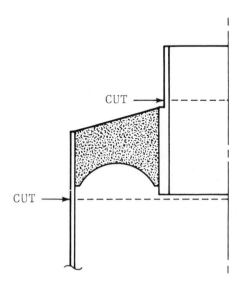

Figure 3. Subassembly molybdenum cut-off locations to eliminate
excessive molybdenum from unit to achieve faster removal.

2. The insulator portion was then placed in an etchant consisting
    of the following ingredients: HNO_3, 80 ml; H_2SO_4, 20 ml; HCl,
    2 ml; H_2O, 150 ml. The part was introduced while the mixture
    was still hot ($\sim63°C$) from the exothermic mixing reaction. It
    took approximately 40 to 50 minutes to etch the molybdenum
    from the insulator.

3. The insulator was ultrasonically cleaned from 5 to 10 minutes
    in alcohol and acetone. It was then blown dry and placed in
    a heated vacuum oven (~1 torr at 135 to 150°C) overnight.

4. The insulator was removed from the oven and the interfaces
    to be examined were brush coated with a deaired epoxy resin
    consisting of 100 parts of Epon 828* and 12 parts of DEA
    (diethanolamine). It was then deaired (1 torr minimum) and
    placed in an aluminum cup. The remaining resin was poured
    around the insulator to a level shown in Figure 4 and pres-
    sure (100 psi) cured at 71°C for 16 hours.

5. After polymerization of the epoxy, the aluminum foil con-
    tainer was stripped from the assembly and the glass-ceramic
    was etched from the epoxy by immersing in an etchant con-
    sisting of hydrofluoric acid.†

---

*Trademark, Shell Oil Co.

†It is believed that a 1-to-1 H_2SO_4-water buffer solution
should be added to the HF.

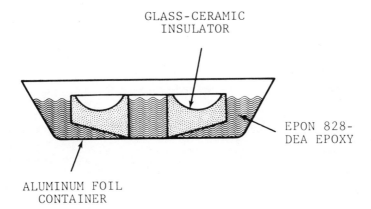

GLASS-CERAMIC
INSULATOR

EPON 828-
DEA EPOXY

ALUMINUM FOIL
CONTAINER

Figure 4. Method of encapsulating insulator in epoxy resin.
(Insulator is placed in an aluminum cup and resin is poured
around it. The top surface of the insulator is left uncovered
so that HF etchant can remove the glass-ceramic.)

6. The epoxy replica was ultrasonically cleaned to remove resi-
    due and then specific sections having defects of interest
    were cut out, carbon coated and examined by SEM.

RESULTS

Figure 5 is a series of photographs taken of a replica of the
outer sleeve—insulator interface of a subassembly rejected for
ultrasonic flaws. This was the initial series of replicas examined. The replication procedure for this part was not exactly as
described in the Experimental Procedure section of this report in
that the insulator was not thoroughly cleaned and vacuum dried
prior to epoxy encapsulation. It is believed that this is the
reason some depressions are observed. (See arrow on Figure 5.)
There is no reason to expect molybdenum to have indentations as
large as those shown by this replica. Examinations of replicas
made using the solvent cleaning and vacuum drying procedure do
not show artifacts of this nature. An examination of the inner
sleeve of this unit (Figure 6) showed similar artifacts.

Figures 5 and 6 show defects that are quite rugged in their
exterior features. The large projection in Figure 5 is approximately 0.3 mm in diameter which is below the resolution of the
production ultrasonic scanner(which will measure defects approximately 0.5 mm in diameter and larger). Most other defects on
Figures 5 and 6 are smaller than 0.5 mm in diameter.

The replica from the production tube that leaked at the inner
sleeve exhibited an unusual surface condition. Figures 7, 8 and
9 are inner interface replicas while Figures 10 and 11 are outer
surface replicas. The inner sleeve machining grooves are quite
evident in Figures 7, 8 and 9. In the bottom of these grooves are
evenly distributed projections as large as 0.04 mm in diameter.
At this time the exact cause of these has not been confirmed;
they are presently believed to be attributed to outgassing of CO
from the molybdenum although confirmation has not been established.
Similar defects are observed at the outer interface of this tube
as is seen in Figures 10 and 11. This type of projection is
smooth as might be anticipated if attributed to gas evolution.

Results of examinations made on interfaces of other assemblies
for this type of defect vary. Some show no sign of these defects
on either the inner or outer sleeve while others show this type
of defect on the inner sleeve only. On the outer interface of
the production "leaker," the bubbles are much smaller than on the
inner interface as shown by comparing Figures 7, 8 and 9 with
Figures 10 and 11. Figure 11 also shows a large irregular defect
which measures 0.75 mm (0.060 in.) in diameter. The reason the
leak occurred at the inner sleeve was not specifically resolved

because no clear leak path was observed on the replica. This
unit was from a fully assembled tube; a leak path consisting of
an interconnection of the voids seen in Figures 9, 10 and 11 could
have occurred during one of the many thermal cycles the subassembly
is subjected to during tube build.

Figures 12, 13, 14 and 15 were taken of the two laboratory
subassemblies. Figures 12 and 13 are defects which are smooth
in texture rather than rugged. Figure 12 is a defect from unit
No. 451-2 which appeared as a chain of voids on the glass insulator. Figure 14 is a rugged projection similar to those observed
on previous surface examinations. The ultrasonic scans of both
these units show flaws at the outer sleeve interface. Ultrasonic
scans of the inner sleeves of unit No. 451-1 were free of defects
while No. 451-2 showed a single flaw. As mentioned previously,
the inner sleeves of these units were polished (600 grit for
No. 451-1 and 15 μ for No. 451-2) and etched prior to sealing.
Figure 15 is the resin replica of No. 451-1 showing no defects.
Note that the surface texture duplicates the 600 grit polished
and etched molybdenum surface.

SUMMARY

These examinations are representative of the initial effort
that has been directed toward the characterization of interface
defects between the insulator and the molybdenum sleeves. Exact
correlation between ultrasonic flaws and defects was not made
since that was not the prime mission at the time the examinations
were made.

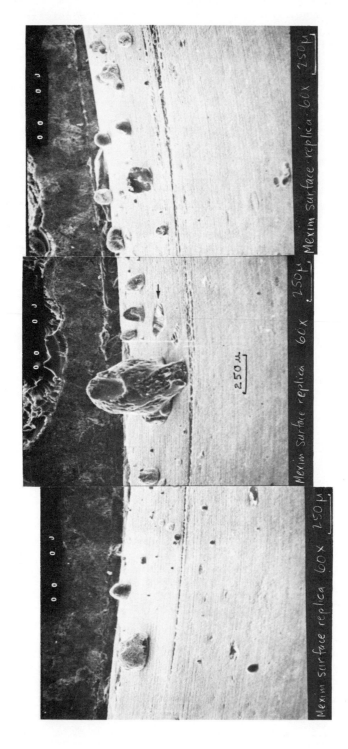

Figure 5. A series of three photographs taken of a replica of the defects at the outer interface of Tube 811411. (This was the first insulator examined in this manner and the insulator may not have been adequately cleaned; this could have resulted in the depressions shown by the arrow.)

Figure 6. SEM photograph of the outer interface of Tube 811411. (Depressions similar to those seen in Figure 5 are shown.)

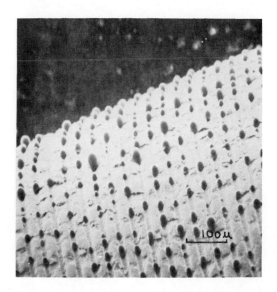

Figure 7. SEM photograph of a resin replica of the inner sleeve of production Tube 732115 showing protrusions from the base of machining grooves. (These protrusions are believed to be gas pockets which result from CO outgassing from the molybdenum.)

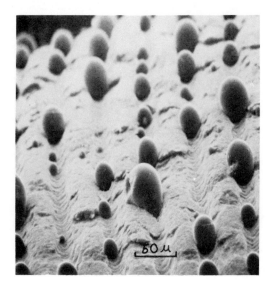

Figure 8.   A magnified view of the inner sleeve of unit No. 732115.

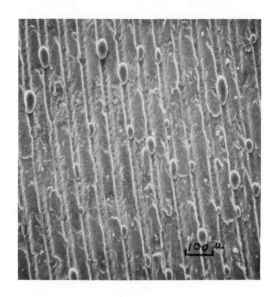

Figure 9.   SEM photograph of a replica of the inner sleeve of
unit No. 732115.   (View is perpendicular to the surface.)

Figure 10. SEM photograph of a resin replica taken of the outer interface of tube No. 732115. (Projections similar to those seen on the inner sleeve are shown.)

Figure 11. SEM photograph of a resin replica of a large defect observed on the outer interface of tube No. 732115.

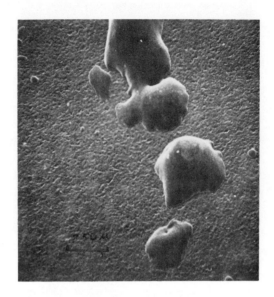

Figure 12. SEM photograph of resin replica of the outer interface of laboratory-sealed subassembly No. 451-2. (Projections
are in line as if defects were caused by a scratch.)

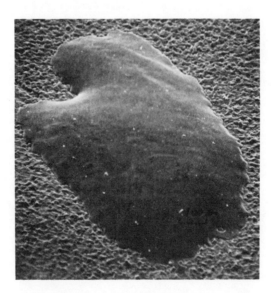

Figure 13. SEM photograph of a resin replica of the outer interface of Unit 451-1 showing a smooth textured defect.

Figure 14.   SEM photograph of a resin replica made of the outer interface of laboratory-produced subassembly No. 451-1.   (The small projections observed about the large projection are believed to be caused by contamination and not outgassing of the molybdenum because other replicas of this surface are quite clean.)

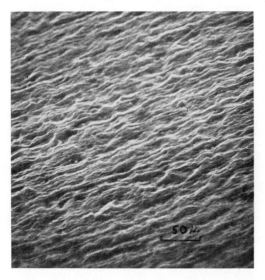

Figure 15.   SEM photograph of a resin replica of the inner sleeve of Unit 451-1.   (This replica reveals the structure of the molybdenum quite well.   No gross interface defects are observed.)

ACKNOWLEDGMENTS

The author wishes to acknowledge the following GEND personnel
for their contributions to this study: R. Antepenko and A. Sidney
of the Chemistry Laboratory removed the molybdenum from the insulator. R. Hansen of the Materials Engineering Laboratory, under
the direction of K. Creed, performed the resin encapsulation.
D. Hubble of the Metallurgy and Ceramics Laboratory etched the
glass-ceramic from the resin. J. Spinks of the Metallurgy and
Ceramics Laboratory performed the SEM examination. W. Andrews,
with the aid of the Chemistry Laboratory, initially examined
insulator interfaces by etching the molybdenum from the assembly.

REFERENCE

1. R. J. Eagan, "Development of a Low Permeability Glass Ceramic
    to Seal to Molybdenum," Sandia Labs, Albuquerque, N.M.,
    Sandia Report No. SAND74-0140, March 1975.

MICROFLUORESCENCE TECHNIQUE FOR DETECTING AND IDENTIFYING

ORGANIC CONTAMINATION ON A VARIETY OF SURFACES

H.A. Froot

IBM Data Systems Division, East Fishkill

Hopewell Junction, N.Y. 12533

A new microfluorescence method has been developed for the rapid, non-destructive detection
and identification of organic contaminants on a
variety of surfaces. The equipment consists of
an automated, computer controlled, modified incident light microscope. The various components,
operation, and examples of results are described.

INTRODUCTION AND BACKGROUND

In the broad spectrum of contamination control, a major concern has been the presence on various surfaces of extraneous micro-
meter and sub-micrometer particulate matter. This is especially
true in the aerospace and semiconductor manufacturing industries
where extremely low concentrations of these particulates have had
a disproportionately high impact on yield, reliability, and performance. In order to control these particles, it is necessary
that the detection method be rapid and not affect the surface, and
that any identification technique be both non-destructive and
capable of analyzing the particle on the surface in question. The
usual detection techniques such as normal incidence and oblique
light microscopy are generally limited to simple surfaces with a
minimum of topographical features, while analytical tools such as
the microprobe, which are satisfactory for analyzing inorganic
particles, are incapable of analyzing organic particles especially
if they are complex compounds or mixtures.

This paper describes a system, based on the phenomenon of
luminescence, that can rapidly and non-destructively detect and
identify organic particulates without the necessity of removing
them from, or altering, the surfaces in question. It can also
identify particles embedded in transparent layers and detect the
presence of thin organic films. The system also has the advantage
of being relatively inexpensive when compared to other analytical
tools such as the plasma chromatograph mass spectrometer and Auger.
Its price runs about $60,000.

THEORY

Because of its extreme sensitivity, luminescence has been used
for many years on a broad range of problems. Gilbault, (Reference
1) is an excellent source for its application, theory and additional
reading. The theory, very briefly, makes use of the fact that every
molecule possesses a series of closely spaced energy levels and can
be excited from a lower to a higher energy level by the absorption
of a discrete quantum of energy equal to the difference between the
two energy levels.

During the time the molecule is in the excited state, energy
in excess of that for the lowest vibrational state for that level
is rapidly dissipated. Because of this loss of energy before
emission can occur, the energy emitted as the molecule returns to
the ground state is of lower energy (longer wavelength) than the
energy that was absorbed. If this energy is in the visible range,
the phenomenon is called luminescence.

In an organic molecule, the electrons of interest move in poly-
centric orbitals called molecular orbitals where the energy associated with each orbital is dependent on the spacing between nucleii as
well as their type. This results in a unique energy configuration
for every compound. In addition, the intensity of a given absorption or emission is dependent on the probability of a given transition occurring, which in turn is dependent on the overall energy
configuration. Therefore, the probability that two compounds will
have exactly the same emission intensity versus wavelength "fingerprint" is small.

In practice, since the emission peak tends to be broad, luminescence is not a useful property for qualitative analysis when the
unknown compound may be any one of the entire universe of organic
compounds. However, in most contamination control situations, the
population of solid organic contaminants that might be present is
limited. These are usually residues from previous operations,
particles from the wear and breakdown of processing equipment or
externally generated particles and fibers from packing, garments

and/or handling.

Identification of these types of contaminants is made by matching the luminescent spectrum of the unknown against a library of the spectra of possible contaminants. In cases where these comparisons are not unique, insight into the process plus knowledge of other parameters such as shape will usually be sufficient for positive identification.

APPARATUS

The system is shown schematically in Figure 1 and consists of a modified Leitz MPV-II microspectrophotometer. The major optical components are described below.

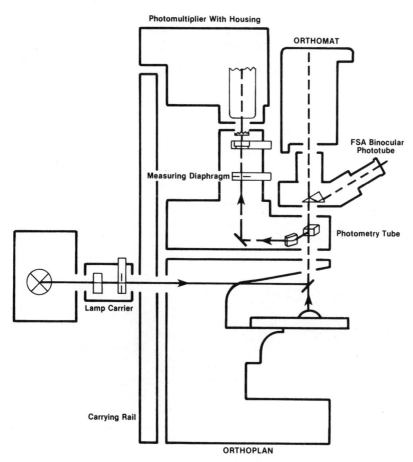

Figure 1. Diagrammatic representation of the optical system. Courtesy of E. Leitz, Inc.

Excitation Source

Since for any given material, the luminescent intensity is dependent upon the intensity and spectral distribution of the exciting
light source, high intensity, broad band UV source such as a Xenon
lamp is preferred to a Mercury lamp. The short arc XB075 was
selected because of its high luminous density. The visible components of the light source are removed by a combination of filters
consisting of a 2mm KG-1 heat absorption filter, a 4mm BG-38 red
suppression filter, and a 2mm UG-1 UV exciting filter. This combination will restrict the excitation to the 300 to 400 nm band.

Vertical Illuminator (Leitz Ploempak 2)

The illuminator, Figure 2, utilizes a dichroic beam splitting
mirror which reflects wavelengths shorter than 400 nm through the
objective onto the specimen. The longer wavelength luminescent
light from the specimen is transmitted through the mirror. A suppression filter removes any residual exciting light.

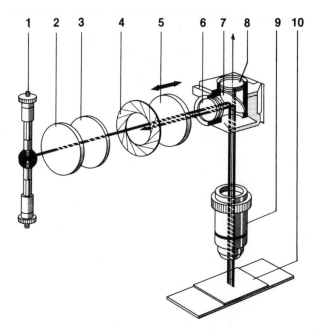

Figure 2. Diagrammatic representation of the Ploempak 2.
Courtesy of E. Leitz, Inc. (1) light source; (2) KG-1 filter;
(3) BG-38 filter; (4) field diaphragm; (5) adjustable lens for
focussing the field diaphragm; (6) UG-1 filter; (7) Dichroic
beam-splitting mirror; (8) suppression filter; (9) objective;
(10) specimen.

Objective

Since the intensity of the luminescence is directly proportional to the fourth power of the numerical aperture of the objective. A high numerical aperture, moderately low magnification objective that contains no autofluorescing components is preferred for survey work. The Leitz APO25X.0.65 objective satisfies these requirements.

Monochromator

The luminescent light is scanned by a Schoeffel GM-100 grating monochromator ($\Delta\lambda/2$ = 4.5 nm) optically coupled to the microscope and photomultiplier tube. It is computer driven and its position sensed by an absolute optical encoder.

Photomultiplier

The photomultiplier is an extended S-20 type, EMI 9658R maintained at -50 deg C $\pm$ 1/2 deg C by a temperature controlled housing.

Automation

Instrument control, data collection, storage and reduction is accomplished by an IBM 5100 computer with an IBM 5106 auxiliary tape driver unit and an IBM 5103 printer. An IBM 7406 device coupler is used as a software controlled interface.

Self-promoting APL programs allow operator selection of the wavelength range to be scanned as well as the wavelength interval between readings. Wavelength versus intensity data is automatically stored on the auxiliary tape. Additional programs allow the plotting of raw or scaled data, comparison of the data with that of known materials, and normalization of the data for non-linearity in the response of the system.

RESULTS

In the normal mode of operation, the surface in question is visually scanned until a particle is found. The computer is then programmed to drive the monochromator and to record the wavelength and the associated PMT signal. Figure 3 is the emission intensity vs wavelength plot of a particle detected on a semiconductor wafer while Figures 4 and 5 are the plots of the data for two standards that the unknown might match.

The compare program is then used to scale the data for the unknown and each standard and determine the absolute intensity

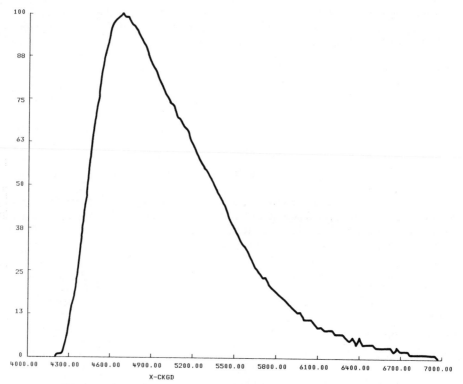

Figure 3.   Intensity vs wavelength plot of unknown.

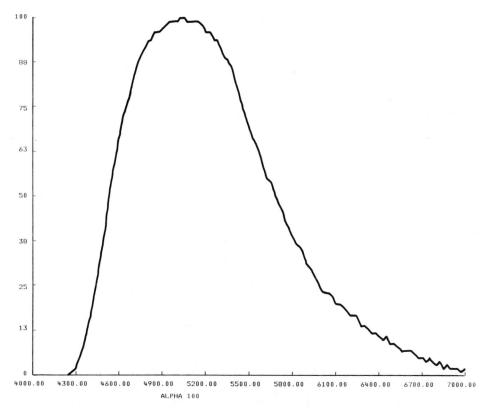

Figure 4.   Intensity vs wavelength plot of α-100 flux.

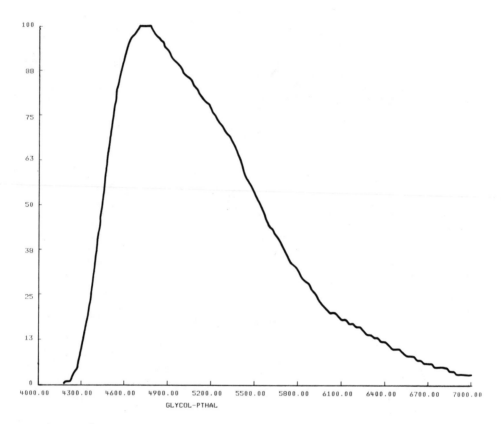

Figure 5.   Intensity vs wavelength plot of Glycol phthalate.

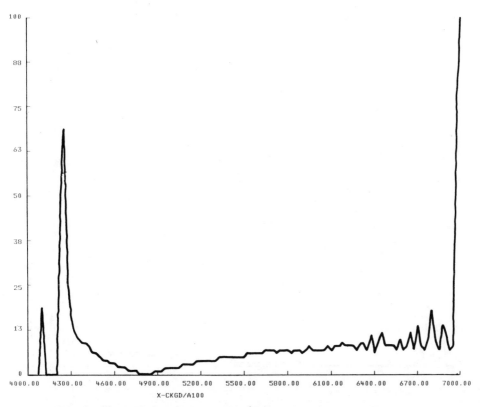

Figure 6. Plot of $I_{unknown}/I_{\alpha-100\ flux}$ vs wavelength.

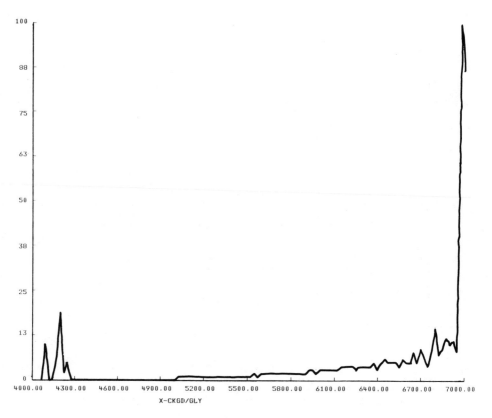

Figure 7. Plot of $I_{unknown}/I_{Glycol\ phthalate}$ vs wavelength.

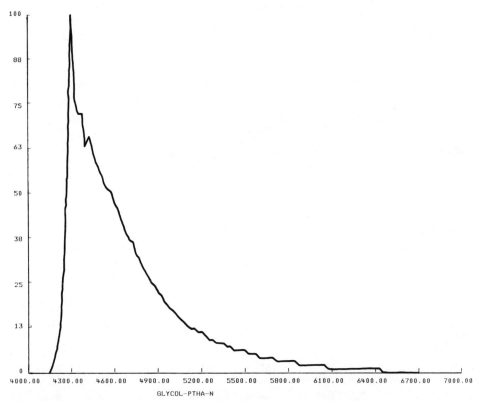

GLYCOL-PTHA-N

Figure 8. Normalized plot of intensity vs wavelength for Glycol phthalate.

difference between the unknown and each standard at every wavelength. The plots of these results are shown in Figures 6 and 7. Obviously, the best match for the unknown was Glycol Phthalate.

Normally, the above method is sufficient for identifying a material when all the data is taken on one system. However, if comparisons are to be made between different systems, the data must be corrected for the non-linear response of the systems. This is done by developing a high precision normalization set of data of intensity vs wavelength for a light source of known uniform intensity. To normalize the data for any sample, a program has been written to perform a wavelength by wavelength division of the scaled sample intensity by the scaled normalization data. Figure 8 is an example of the result of this normalization.

DISCUSSION

For industrial application, simple visual versions are used to monitor the manufacturing line at various critical points in the process. Since these are relatively inexpensive and simple to use, a large number of these can be installed.

Whenever an unacceptable level of contamination is detected, samples are taken off line to the fully automated analytical tool for identification so that proper corrective action can be taken.

The equipment is also capable of detecting the presence of thin organic films on various surfaces. It is particularly sensitive on either textured surfaces or those containing some form of pattern. For this application, the field diaphragm is stopped down and the area inside the disc compared to the remainder of the field. Since only the area within the disc is being illuminated by U.V. light, if any organic film is present within this area, it will appear brighter than the background. On patterned surfaces, the pattern will become visible because the film will act as a visible light source illuminating the pattern. Using this method, films have been detected that were as little as 0.3 nm thick as determined by ellipsometry.

Although the use of this technique has just begun to be explored, it has already been proven to be a powerful tool in the control of organic particulate contamination and can be added to the long list of applications of fluorescence.

ACKNOWLEDGMENTS

The author wishes to gratefully acknowledge the work of R. Hill, H. Hendrick, M. Cangolesi, and K. Pockett in automating the equipment.

REFERENCES

1. G.C. Guilbault, "Practical Fluorescence", Marcel Dekker, Inc., New York, 1973

ANALYSIS OF ORGANIC SURFACE CONTAMINATION

BY PLASMA CHROMATOGRAPHY/MASS SPECTROSCOPY

T. W. Carr and C. D. Needham

IBM East Fishkill Facility

Route 52, Hopewell Junction, NY 12533

Plasma chromatography is an analytical technique uniquely suited for the analysis of trace organic con- taminants absorbed on semiconductor substrates. Its advantages include very high sensitivity, atmospheric pressure operation, and compound separation based on molecular properties. Combining the plasma chromato- graph with a quadrupole mass spectrometer allows ionic mass identification of the observed mobility peaks, and greatly expands the capability of the instrument for qualitative analysis. This paper describes the principles of plasma chromatographic analysis, and reports some applications to problems which occur in semiconductor device manufacturing.

INTRODUCTION

One of the more difficult analyses of modern chemistry has been the identification of trace organic contaminants absorbed on inorganic substrates. Such analyses are important in the manufacture of semiconductor devices, where problems of yield and reliability involving adhesion, corrosion, and embrittlement are frequently encountered. The combining of the techniques of plasma chromatography and mass spectrometry has been found to offer a unique way of identifying such trace contaminants.(1-6)

The plasma chromatograph (or ion mobility spectrometer) consists of an ion-molecule reactor coupled to an ion-drift tube. The trace vapors desorbed from the sample surface are transported by the carrier gas (usually nitrogen or zero-grade air) into the ion reaction zone of the chromatograph, where a ^{63}Ni beta emitter (about 11 mC) generates a series of reactive ions (both positive and negative) from trace impurities (H_2O, NH_3, CO_2) in the carrier gas. The reactive ions, such as NH_4^+, NO^+, $(H_2O)_n H^+$, in turn react with the trace organic vapors to form organic ions. These ion-molecule reactions are of low energy, and usually involve hydride abstraction, charge transfer, or proton donation, resulting in ions of mass M-1, M, or M+1, respectively. Since, generally, no fragmentation occurs, a particularly simple spectrum results.

Under an applied electric field of about 214 volts/cm, the ions are accelerated through the drift tube, typically 10 cm in length. An electric grid separates the reaction zone from the drift region, and is normally biased so as to prevent transmission of the ions. Periodically the grid bias is removed for a short time (0.2 ms) allowing a burst of ions to enter the drift region. An electrometer detector records the ion current as a function of time; from this the drift time of each ion is measured. Under normal operating conditions, most ions will have drift times between 10 and 40 ms. Because of the rapid "elution time" of the chromatograph and the very low current, a multi-channel signal averager is used to collect the mobility spectrum. The technique of repetitive scanning is used to build up the signal to noise ratio. The drift tube is continually swept by a counter-flow of gas (usually nitrogen), and the entire tube is kept at a temperature near 200 C to prevent condensation of material within the tube.

The drift time of an ion is a function of the tube length, the applied electric field, the temperature and pressure of the drift gas, and the ion mobility. The ion mobility in turn depends upon the mass of the ion and its collisional cross section with the drift gas, as well as the molecular weight of the drift gas. Knowing the above experimental parameters, we can calculate the mobility of an ion from its observed drift time.

There are three major advantages of the plasma chromatograph, one of which is its very high sensitivity. As little as one picogram of material may be detected for many organic substances. The atmospheric pressure of the plasma tube and sample inlet system offer a second advantage, since no loss of volatile sample occurs, as with such techniques as ESCA (electron spectroscopy for chemical analysis) and SIMS (secondary ion mass spectrometry), where the sample must be under a high vacuum.(7,8) The third advantage is that the information obtained is molecular rather than atomic in nature; for organic analysis, this is essential.

Since there exists no standard compilations of ion mobility data covering an extensive range of organic compounds, the user must create his own library of standard ion mobility spectra. To aid in the interpretation of the mobility spectra, it is useful to have another piece of molecular information, such as molecular mass, for which extensive tabulations already exist. The plasma chromatograph operates at rather low resolution so that mobility alone is usually insufficient to identify unambiguously an observed ionic species. But with the additional ionic mass data, assignments can be made with much greater certainty.

A quadrupole mass filter may conveniently be coupled to the plasma drift tube so that some of the ions may enter into the mass analyzer.(9-12) A pinhole aperture following the electrometer allows some of the ions and drift gas to enter the focusing lens assembly, where the ions are injected into the quadrupole assembly, and the neutral drift gas is pumped away in the vacuum system. The size of the aperture is determined by the speed of the pumping system and the desired operating pressure of the mass spectrometer. The transmission of ions through the aperture is low enough that ion detection at the detector must be done by pulse counting.

The combined plasma chromatograph/mass spectrometer may be operated in any of four operating modes:

1) PC Mode: The electrometer detector is used to obtain mobility data, and the quadrupole mass filter is not used.

2) Total Ion Mode: The mass filter is set to transmit all ions regardless of mass, so that it acts as a non-discriminating ion detector. The mobility spectrum is recorded as above, but the signal is considerably weaker and the drift times are slightly longer.

3) Tuned Ion Mode: The mass filter is set to transmit ions of a single mass number only. Since the detector sees only one ionic species, the mobility spectrum consists of only that peak corresponding to the ion. In this way a mass may be assigned to each peak in the mobility spectrum.

4) Mass Spectrometer Mode: The gating grid is kept open rather than pulsed, so that ions are continually arriving at the quadrupole mass filter. The mass setting of the quadrupoles is continually scanned over a desired mass range, so that the mass spectrum of the ions is recorded without regard to their mobilities.

EXPERIMENTAL

The coupled plasma chromatograph/mass spectrometer used for this work was the model MMS-160 manufactured by PCP, Inc., West

Palm Beach, Florida. A schematic diagram of the system is shown
in Figure 1. The plasma tube has a 5 cm reaction zone and a 10 cm
drift region in the normal configuration, with the first grid as
the gate. It is also possible to use the second grid as the gate,
which gives a 10 cm reaction zone and a 5 cm drift zone. The mass
filter is a specially modified Extranuclear Laboratories Spectr-El
quadrupole mass spectrometer.

The readout system consists of a Nicolet Model 535 Signal
Averager with 4K memory, a readout scope, a Hewlett-Packard Model
7035 X-Y plotter, a Nicolet Model NIC-283A magnetic tape coupler,
and a Kennedy Model 9700 tape drive.

The sample inlet oven was manufactured by PCP, Inc., and is
controlled by a Theall model temperature programmer. This con-
troller allows operation of the oven from room temperature to
230 C, either isothermally or with a temperature ramp of 0 to 10
C/min.

The operating parameters used in this study are listed in
Table I.

It may be shown that the reduced mobility K_0 of an ion may
be calculated by(13)

$$K_o = (\frac{d}{tE}) (\frac{P}{760}) (\frac{273}{T}) \qquad (1)$$

where d is the drift tube length in cm, t is the drift time in ms,
E is the electric field in V/cm, P is the barometric pressure in
torr, and T is the tube temperature. The drift time of an ion is

Table I. Operating Conditions

Accelerating Field	$\pm$ 214 v/cm
Drift Length	10cm
Tube Temperature	200 C
Tube Pressure	Barometric
Carrier Gas	N_2, 100 ml/min
Drift Gas	N_2, 500 ml/min
Dwell Time	20 μs/channel, 1024 channels
Delay Time[1]	5.12 ms
Gate Width	0.20 ms
Digitizer Resolution	9 bits
Number of Sweeps	512

[1] Time between start of gate opening and start of data
   collection.

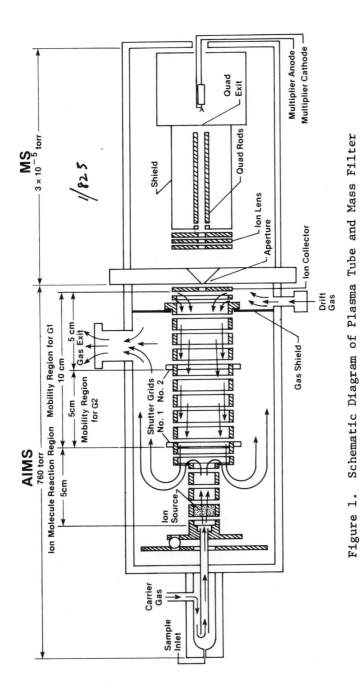

Figure 1. Schematic Diagram of Plasma Tube and Mass Filter

calculated from the channel number (x) of the peak position in the
signal averager:

$$t \text{ (ms)} = .02 \text{ (x + 256)} \qquad (2)$$

(using the tabulated values for dwell time and delay). It is
somewhat more convenient to calculate reduced mobility by using
the first reactive ion (NH_4^+) peak as an internal calibration; the
reduced mobility of this peak is 3.00. Then if x and y denote the
channel numbers of the unknown ion and the NH_4^+ ion respectively,
it may be shown that the reduced mobility of the unknown ion is:

$$K_o = (.06 \text{ y} + 15)/(.02 \text{ x} +5) \qquad (3)$$

RESULTS AND DISCUSSION

Figure 2 shows the positive ion mobility and mass spectra of
nitrogen carrier gas. Carroll and co-workers (14) have identified
these positive reactant ions as NH_4^+, NO^+, and $(H_2O)_2H^+$. These
reactive ions form product ions with the neutral sample molecules
by several types of ion-molecule reactions.

The NH_4^+ and $H(H_2O)_2^+$ ions generally undergo proton transfer
(or gas phase Bronsted acid-base reactions) depending upon the

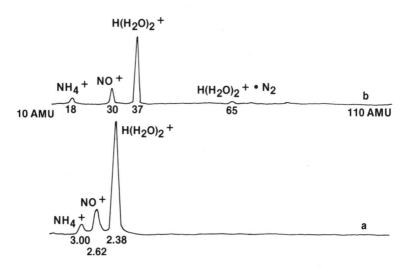

(Reduced Mobility, Ko, cm^2 v^{-1} sec^{-1}

Figure 2. Positive Ions From Nitrogen Carrier Gas.

(a) Mobility Spectrum, Drift Times S-25 MS.
(b) Mass Spectrum, Range 10-110 AMU.

basicity of the sample molecules:

$$H(H_2O)_2^+ + M \longrightarrow MH^+ + 2H_2O \tag{4}$$

The NO^+ reactive ion may undergo several types of reactions; a charge transfer reaction

$$NO^+ + M \longrightarrow NO + M^+ \tag{5}$$

A hydride abstraction reaction:

$$NO^+ + M \longrightarrow HNO + (M-H)^+ \tag{6}$$

or an addition reaction:

$$NO^+ + M \longrightarrow MNO^+ \tag{7}$$

In the charge transfer reaction, the ionization potential of the sample molecule must be lower than 9.5 eV, the ionization potential of NO. Hydride abstraction depends upon the gas phase acidity of the sample molecule. The addition reaction is often electrophilic in nature but can also be an associate ionization.

The assignment of ionic mass to each mobility spectrum peak is illustrated in Figure 3. A silicon disc with a nascent SiO_2 surface was stored for several days in a molded polypropylene container. Before the disc was stored, it was flamed with a propane torch to remove all volatile absorbed material from its surface. The plasma chromatograph analysis shows that several compounds had outgassed from the plastic container and were subsequently absorbed on the surface of the disc.

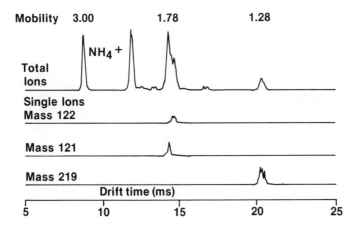

Figure 3. Mass Identification of Peaks in Mobility Spectrum.

The top trace is the mobility spectrum obtained in the total
ion mode, where the mass spectrometer is set so as to detect all
ions regardless of mass. The lower traces are obtained in the
single ion mode of detection, with the mass spectrometer set to
masses 122, 121, and 219, respectively. The coincidence of each
single ion peak with a peak in the total ion spectrum confirms
the mass identification of that particular mobility peak. Thus the
peak of mobility 1.29 is seen to represent an ion of mass 219. This
ion has been identified as resulting from 2,6-di-t-butyl-p-cresol
(molecular weight 220) by the mechanism of hydride abstraction.
This identification was verified by running the mobility spectrum
of the pure compound, a commonly used volatile anti-oxidant. With
the PC/MS combination, we can identify a mobility and mass with
each component; these two quantities, along with an estimated
volatility, make identifications reasonably certain in most cases.
Two compounds of the same mobility will usually differ in mass;
and conversely, isomers will frequently differ in mobility.

It is observed that the peak of mobility 1.78 is actually a
doublet, and in fact two distinct masses, 121 and 122, are associated with this peak. We occasionally find multiple mass assignments, differing by 2 amu, for the same mobility peak. This is a
valuable clue to the presence of elements such as Cl or Br in the
compound. But since no common elements have naturally occurring
isotopes differing by 1 amu, this mass assignment cannot be explained by isotope effects. Three possible explanations are:

1)   A compound of mass 121 has undergone both proton
      transfer and charge transfer.

2)   A compound of mass 122 has undergone both hydride
      abstraction and charge transfer.

3)   Two different compounds are present.

We have not yet identified this component, and so do not know
which of the above explanations is correct. In similar experiments
with other containers we have observed only the mass 122; this would
tend to support the third explanation, but it is also likely that
the type of ion-molecule reaction depends upon the concentration of
the molecule, so that we still cannot readily choose among the above
explanations.

Figure 4 shows a mobility spectrum obtained from a disc stored
in a polypropylene container molded from a different source of
polymer. The experimental details are the same as in the previous
example. As the temperature of the sample is raised, several compounds desorb from the surface and are carried into the plasma tube
for analysis. The most volatile component is 2,6-di-t-butyl-p-cresol
as was seen in the previous example.

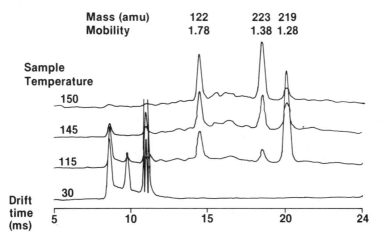

Figure 4. Mobility Spectrum from Silicon Disc Stored in
Polypropylene Container.

A second, less volatile component is also seen, with mobility
1.38. This is the plasticizer diethyl phthalate, molecular weight
222, which undergoes proton transfer to form the positive ion of
mass 223. It is evident that the species of lower mass does not
have the higher mobility (or shorter drift time) as we would expect.
This observation shows the importance of ionic size as well as ionic
mass in determining an ion's mobility. The apparently greater collisional cross section of the ion of 2,6-di-t-butyl-p-cresol more
than compensates for its lower mass, and consequently it has a
lower mobility than does diethyl phthalate. Estimates of molecular
weight based on observed mobility alone can be misleading, since
the ion size effect must also be taken into account. The addition
of a quadrupole mass analyzer to the mobility spectrometer is
therefore essential to interpreting the mobility data.

A third major absorbed component is observed to have a
mobility of 1.78 and an ionic mass of 122 amu; its identity has
not been established.

In another example, a similar silicon disc was precleaned and
stored in a container constructed of polypropylene, polycarbonate,
and butyl rubber. The mobility spectra of the absorbed components
are shown in Figure 5. In addition to the 2,6-di-t-butyl-p-cresol
seen in the previous example, two new components of mobilities 1.58
and 1.54, corresponding to ionic masses 159 and 163, are observed
to desorb from the surface at a temperature near 175 C. It may be
seen that the peak intensities reverse at higher temperature. The
origin of the peaks has been shown to be nicotine, which undergoes
thermal dehydrogenation to form 2,2'-nicotyrine.

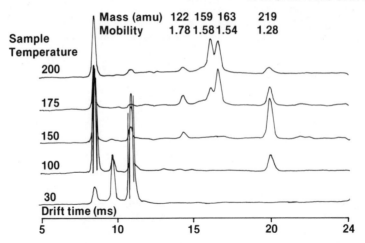

Figure 5. Mobility Spectrum of Silicon Disc Stored in Container of Polypropylene, Polycarbonate, and Butyl Rubber.

$$\text{MW} = 162 \quad\quad\quad\quad \text{MW} = 158 \quad + 2H_2$$

Both nicotine and 2,2'-nicotyrine undergo the proton transfer reactions, because of the basicity of the nitrogen group. The presence of the twin peaks makes this an easily identifiable contaminant; the presence of nicotine on a test surface may be taken as a general indicator of the degree of cleanliness in the working environment.

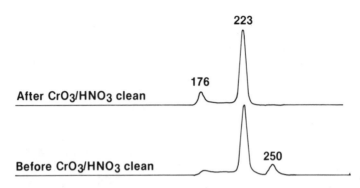

Figure 6. Mobility Spectrum from Silicon Disc Doped with DEP and with Subsequent Acid Cleaning (Upper Trace).

Oxidizing acid solutions are frequently used in the laboratory to remove organic residues from glass surfaces. To test the effectiveness of a chromic oxide/nitric acid cleaning solution, we exposed a clean SiO_2 surface on silicon to the saturated vapors of diethyl phthalate. The exposure was for 3 hours at room temperature. The disc was then immersed in a CrO_3/HNO_3 bath and was subsequently analyzed by plasma chromatography. A control disc which had not been acid treated was also analyzed, and the results are shown in Figure 6. It should be noted that the complete disappearance of the reactive ion peaks indicates an overloading of the plasma tube. The reactive ions are being consumed by proton transfer reactions with the diethyl phthalate faster than they are being produced and under these conditions, the peak area is no longer proportional to the sample concentration. We cannot say from the data how much sample was removed by the acid treatment; nevertheless, the saturation of the instrument response for the cleaned disc indicates that the acid treatment still leaves a large amount of diethyl phthalate. The peak with mobility 1.62 and mass 176 amu corresponds to the $(M-OC_2H_5)^+$ ion, and the non-zero signal level between this peak and the main peak indicates that ionization (with loss of an ethoxy group) is occurring within the drift region of the tube.

CONCLUSION

These examples illustrate several contaminant analyses for which the combined techniques of plasma chromatography and mass spectroscopy are uniquely suited. In the manufacture of semiconductor microcircuit devices, high product yield requires strict control of cleanliness at each process step. Many process yield failures may be traced to the presence of unwanted organic contaminants in amounts so low as to be undetectable by conventional analytical techniques.

The coupling of the mass filter to the plasma chromatograph should result in a rapid growth in understanding the nature of the ion-molecule reactions, as well as provide a more useful analytical tool. Among the expected future developments we might expect to see establishment of mobility-structure correlations and a semi-empirical approach towards spectra interpretation. The technique is presently only semi-quantitative, but a more basic approach might allow for more quantitative data. Finally we should direct some effort towards understanding some of the "non-linear" effects which result from competing reactions, and which lead to non-additivity of multi-component samples.

ACKNOWLEDGEMENTS

We are indebted to Doris Braddock of this laboratory, who performed the analyses reported here, and to Jack Ramsey, whose support and encouragement made this work possible.

REFERENCES

1. F. W. Karasek, Anal. Chem., 46, 710A (1974).
2. R. A. Keller and M. M. Metro, Sep. Purif. Methods, 3, 207 (1974).
3. T. W. Carr, J. Chromatog. Sci., 15, 85 (1977).
4. S. A. Benezra, J. Chromatog. Sci., 14, 122 (1976).
5. F. W. Karasek, Res/Dev., 21, 34 (1973).
6. G. W. Griffin, I. Dzidic, D. I. Carrol, R. N. Stillwell, and E. C. Horning, Anal. Chem., 45, 1204 (1973).
7. D. A. Shirley, Editor, "Electron Spectroscopy," North Holland, Amsterdam, 1972.
8. C. A. Evans, Anal. Chem., 47, (4), 818A (1975).
9. F. W. Karasek, S. H. Kim and H. H. Hill, Anal. Chem., 48, 1133 (1976).
10. T. W. Carr, Anal. Chem., 49, 828 (1977).
11. T. W. Carr, Thin Solid Films, 45, 115 (1977).
12. F. W. Karasek, H. H. Hill, and S. H. Kim, J. Chromatog., 117, 327 (1976).
13. H. E. Rivercomb and E. A. Mason, Anal. Chem., 47, 970 (1975).
14. D. I. Carrol, I. Dzidic, R. N. Stillwell and E. C. Horning, Anal. Chem., 47, 1956 (1975).

DESCRIPTION AND OPERATION OF TWO INSTRUMENTS FOR CONTINUOUSLY DETECTING AIRBORNE CONTAMINANT VAPORS

Robert E. Cuthrell

Sandia Laboratories

Albuquerque, New Mexico 87185

Two nonspecific detectors are described for monitoring contaminants in the environment: (1) a tin oxide detector, which is sensitive to any oxidizable vapor, and (2) a gold-gold adhesion detector, which is sensitive to the higher molecular weight condensable contaminants. The principles of operation, construction, typical response curves, and approximate sensitivity are given for these two instruments used individually and in combination.

INTRODUCTION

Although particulate contaminant levels are well controlled and monitored in many industrial environments, the presence of organic vapor contaminants is usually ignored. These may inhibit thermocompression bonding of electrical lead wires, decrease the adhesion of metallizations, poison electroplating baths, and increase the resistance of electrical switches through frictional polymer formation.

In this paper, the principles of operation and use of a tin oxide vapor detector and of a gold-gold adhesion detector[1] for monitoring airborne organic contaminant levels are described. The former instrument is sensitive to any oxidizable vapor, whereas the latter detects primarily the higher molecular weight condensable gases. When used together, it is possible to distinguish between these two types of contamination. Both detectors sense nonspecific relative amounts of gaseous environmental pollutants and have been used to continuously monitor clean rooms, clean benches, and

831

other production facilities where sensitive operations and assembly are performed.

EXPERIMENTAL

Tin Oxide Vapor Detector

The vapor detector, which is shown schematically in Figure 1,
consists of two platinum wire electrodes separated by a thin coating of heated tin oxide, the electrical conductance of which is
very sensitive to (1) oxidizable organic gases, (2) oxygen, (3)
ozone, and (4) hydrogen. The electrical conductance is increased
on exposure to organics or hydrogen and is decreased on exposure
to oxygen or ozone. Very fast changes in conductance (by as much
as three orders of magnitude in less than 1 sec.) have been noted
on exposure of the vapor detector to organics or hydrogen. The
sensitivity of this detector is about 10 ppm. [2]

One or both of the platinum wires may be wound to form a
helical heating element such that the detector may be operated
over a wide temperature range. Very rapid response is an advantage of operating at elevated temperatures (300-400° C). Very
rapid recovery after exposure to organics is an advantage of operating in an oxygen environment (air). An additional advantage
of operating in air is that automatic regeneration occurs without
the requirement for special treatments or bottled gases. Both
free standing models (in which the tin oxide coating was supported
by the platinum helix) and supported models (in which the tin
oxide coating and the platinum wires were supported by ceramic
substrates) have been constructed.[3] An advantage of the free
standing model is that the size and mass may be minimized (about
2 mm in diameter or smaller). The supported model is more easily
tailored for resistances which are compatible with a wheatstone
bridge or some other nulled readout (about 30 KΩ compared to about
10 MΩ for the free standing model). Although highly sophisticated
electronics may be designed for the detector, it works quite well
using a filament transformer for the heater coil and an ohmmeter
for the detection circuit. For this paper, a fixed potential was
applied across the tin oxide detector and an equal opposing potential was applied across a matching resistor thus producing a null.
As the conductance of the detector changed on exposure to contaminant vapors the potential deviation from the null was ampli-
fied and recorded as a function of time.

Gold-Gold Adhesion Detector

A previously described gold-gold adhesion detector[1] was constructed based on the sensitivity of metallic adhesion to adsorbed

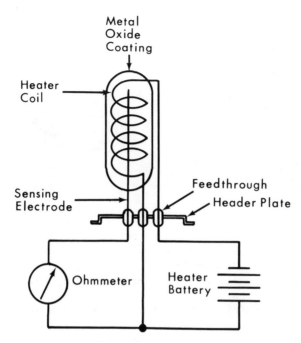

Figure 1.   Schematic representation of the tin oxide vapor detector.   The ohmmeter may be replaced by nulled electronics.   The metal oxide coating (SnO_2, Fisher Scientific Co.) was formed by painting the heater coil and the sensing electrode with an aqueous slurry of tin oxide particles and oven drying the coating at 200° C.

contaminants[1,4,5] and on the temperature and partial pressure dependence of adsorption.   The fully automated instrument (Figures 2 and 3) consists of two gold rods, the rounded ends of which are bonded and debonded repetitively (every 4 sec.) by the action of a push-pull solenoid.   A linear variable differential transformer (LVDT) load cell is used to measure compressive and tensile loads. The gold contacts are partially encircled by a continuously operated ultraviolet (UV) lamp which partially oxidizes gaseous hydrocarbon contaminants (by the UV-Ozone process[6]) thus causing stronger adsorption on the gold surfaces.   Platinum wire resistance heaters (not shown in Figure 2) are used to thermally desorb contaminants from the gold surfaces.   These heaters are turned on during a cleaning cycle and turned off during a recontamination cycle as indicated in Figure 3.   The rate of increase in adhesion during the cleaning cycle is inversely proportional (and the rate of decrease in adhesion during the recontamination cycle is directly proportional) to the amount of gaseous organic contaminants in the detector environment.

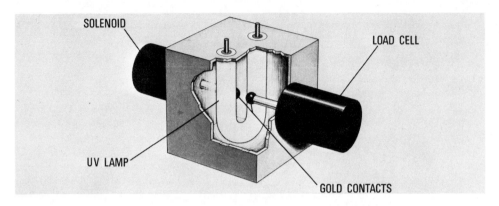

Figure 2. Cut-away drawing of the sensing head portion of the
gold-gold adhesion detector. In early models a high power ultraviolet (UV) lamp provided UV-Ozone cleaning of the gold in addition
to heat for thermal desorption of contaminants from the gold surfaces. Alternate clean-up and recontamination periods were obtained by turning the lamp on and off. In later models a low power
ultraviolet lamp was left on continuously and resistance wire heater windings (not shown) were alternately energized and deenergized
for the cleanup and recontamination cycles.

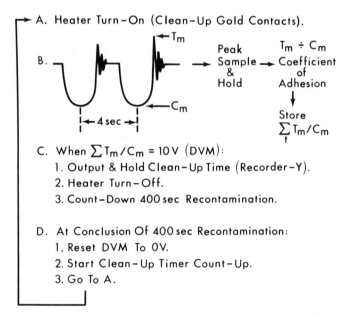

Figure 3. Sequence of automatic repetitive events for data collection and processing using the gold-gold adhesion detector. C_m and
T_m are the maximum bond forming and bond breaking forces, respectively. DVM indicates a digital voltmeter.

In automating the data acquisition and processing functions (Figure 3) of the gold-gold adhesion detector it was found convenient to (1) measure the maximum compressive (C_m) and tensile (T_m) loads applied in making and breaking the gold-gold bonds during the clean-up cycle (heaters on) using peak sample and hold circuitry, (2) compute the coefficient of adhesion (T_m/C_m) using a dividing amplifier, and (3) store successive values of T_m/C_m (obtained every 4 sec.) as incremental potentials on a low leakage capacitor until a predetermined potential ($\Sigma T_m/C_m$ = 10V) was displayed by a digital voltmeter (DVM). The time required for this process to occur was proportional to the amount of contamination in the test environment (long clean-up periods were associated with high contaminant levels and short periods with low levels[1]). As indicated in Figure 3, these clean-up times were fed (as potentials) to a strip chart recorder and held until a new value was obtained after automatically repeating the cycle. Each clean-up cycle was followed by a 400 sec. recontamination period (heaters off).

RESULTS AND DISCUSSION

Tin Oxide Vapor Detector

As indicated in Table I, the tin oxide vapor detector is sensitive to a wide variety of contaminant gases and is insensitive only to those which are completely oxidized. These results imply that the conductance of the tin oxide depends on the valence of the tin and that oxidation-reduction reactions on or near the surface are involved in the detection of organic contaminants.

The typical response of the tin oxide vapor detector to either temporary or sustained exposure to oxidizable airborne contaminants is shown in Figure 4 for natural gas and for carbon monoxide. The peaks represent contaminant levels greater than those present in the laboratory air in which the instrument was zeroed. Thus the detector may be used as a differential sensing device to show changes in contamination from some ambient condition. Body odors and perfumes were easily detected, which suggests that the detector may be used as an intrusion alarm in addition to sensing contaminants in clean benches, clean rooms, and other manufacturing facilities.

The detector was shown to respond to changes in the work shift and to the entry of a security guard in a clean room at a manufacturing facility.[7] Seasonal as well as diurnal changes in the contaminant content of laboratory air were detected. The environment was cleaner at night and in winter than during the day and in summer (presumably a result of reduced contaminant vapor pressures at reduced temperatures). As expected, the detector also responds to polluting activities in a given environment such as soldering, solvent rinsing, machining with cutting fluids, and smoking.

Table I.   Detection Tests Using the Tin Oxide Vapor Detector.

CONTAMINANT	DETECTED	NOT DETECTED
1. Smoke	X	
2. Hydrogen	X	
3. Methyl Alcohol	X	
4. Acetone	X	
5. Butane	X	
6. Natural Gas	X	
7. Trichloroethylene	X	
8. Oxygen	X	
9. Ozone	X	
10. Methylene Chloride	X	
11. Toluene	X	
12. Methyl Ethyl Ketone	X	
13. Hexane	X	
14. Butyl Alcohol	X	
15. Ethylene Glycol Monobutyl Ether	X	
16. Xylene	X	
17. Monochlorobenzene	X	
18. Ethylene Glycol	X	
19. Ammonium Hydroxide	X	
20. Acetic Acid	X	
21. Dimethylsulfoxide	X	
22. Carbon Disulfide	X	
23. Nitromethane	X	
24. Formamide	X	
25. Cyclohexane	X	
26. Chloroform	X	
27. n-Octane	X	
28. Dichloro-difluoromethane		X
29. Carbon Dioxide		X
30. Carbon Tetrachloride		X

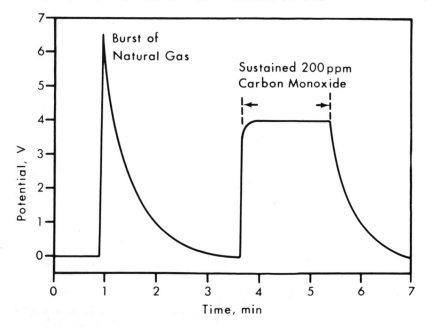

Figure 4. Typical curves for the detection of transient (natural gas) and sustained (carbon monoxide) additions to the laboratory air (arbitrary zero) using the tin oxide vapor detector.

Other uses which have been suggested for the detector are (1) very inexpensive hydrogen leak detection, (2) mine safety monitoring, (3) smoke and fire detection,[8] (4) monitoring automobile emissions, (5) fuel-air ratio detection in combustion, and (6) natural gas leak detection.

Gold-Gold Adhesion Detector

The sensitivity of the gold-gold adhesion detector increases with the molecular weight of the condensable contaminant gas. The detector is insensitive to the normal constituents of clean air (N_2, O_2, CO_2, CO, H_2O, etc.).[1,6,9,10] Figure 5 shows the effect on the coefficient of adhesion of a 5 minute thermal desorption of contaminants at 250°C and the recontamination thereafter on cooling in ultrahigh vacuum, in an electronics laboratory, and in machine shop environments. The initial value of the coefficient of adhesion is directly proportional to the cleanliness, and the decrease in adhesion during the recontamination period is directly proportional to the amount of condensable contamination in the test environment.

When used in the automatic mode (Figure 3), there appears to be no upper limit to the sensitivity while the lower limit is thought

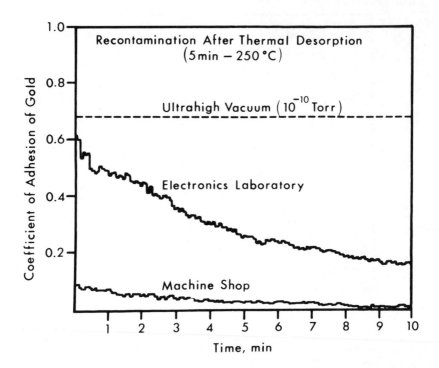

Figure 5. The coefficient of adhesion of gold-to-gold is greater
in cleaner environments immediately after a 5 min - 250°C thermal
desorption period (initial points on the curves). During recontamination periods of 10 min. the gold contacts cool to room temperature and the coefficient of adhesion decreases as a result of the
readsorption of contaminants on the gold surfaces. In prior experiments it was found that a 0.67 coefficient of adhesion was maintained indefinitely in ultrahigh vacuum environments.[10]

to be a single adsorbed molecular layer of pollutant.[1,4] It was
found that clean room atmospheres deteriorate with time in years
since construction, unless a technique which removes gaseous condensable contaminants (such as flow through activated charcoal) is
used in addition to the usual particle filtration.

Combined Operation

    Figure 6 shows curves which were obtained simultaneously using
the tin oxide vapor detector (upper) and the gold-gold adhesion detector (lower). Since the tin oxide vapor detector was zeroed in
ultraclean air (a bottled gas obtained by mixing oxygen and nitrogen from the respective liquids), the initial 1.5 hour portions of
the curves are representative of the response to contaminants in

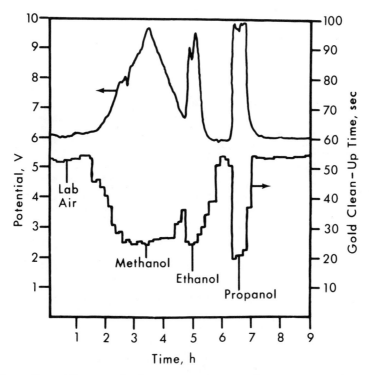

Figure 6. The effects of the contaminants laboratory air, methanol, ethanol, and propanol (with respect to a clean air reference) are shown for the tin oxide vapor detector (upper curve) and for the gold-gold adhesion detector (lower curve).

the laboratory air. The three peaks (upper curve) due to methanol, ethanol, and propanol have the expected shapes and were caused by contaminants purposely added to the laboratory air. The recovery to the initial contaminant level was a result of the removal of these contaminants through an exhaust hood. The more gradual rise and fall of the first peak (upper curve) was a result of a greater separation between the detector and an evaporating drop of methanol. An interesting and important result is shown in the lower curve. The adhesion of gold-to-gold was increased (clean-up times were decreased) coincident with the sensing of methanol, ethanol, and propanol by the tin oxide vapor detector. This result is consistent with prior data which showed that the adhesion of oxide free metals was high when the surface contaminants had relatively low molecular weights.[1,6,9,10] Figure 6 also indicates that the laboratory air contains contaminants of molecular weight greater than that of propanol (60.09 g/mole). It is suggested that tests with progressively higher homologs in the series of normal alcohols may indicate

the approximate molecular weights of those contaminants in the
laboratory air which degrade the adhesion of gold-to-gold. (Figures
5 and 6 indicate that a cross-over from negative to positive peaks
would be expected for an alcohol of molecular weight greater than
those of the contaminants in laboratory air.)

CONCLUSIONS

The construction and operation of a tin oxide vapor detector
and a gold-gold adhesion detector for monitoring airborne contaminant vapors have been described. The vapor detector senses any
oxidizable gaseous contaminant while the adhesion detector senses
primarily higher molecular weight condensable contaminants. Particular utility may be derived from the simultaneous use of the two
detectors. The test results may be interpreted as molecular weight
limits for those contaminants present which degrade the thermocompression bonding of gold to gold.

Although both instruments have been used in various laboratory
and industrial environments (clean rooms, clean benches, assembly
rooms, etc.[7]), the tin oxide vapor detector has potential uses in
many other applications (hydrogen leak detection, smoke detection,[8]
intrusion alarm,[7] etc.).

ACKNOWLEDGEMENTS

The author wishes to acknowledge the contributions of D. M.
Mattox, who initially suggested the application of gold bonding
technology to the development of a surface contaminant detector,
of D. J. Sharp, who suggested the use of tin oxide for organic
vapor detection and performed many of the early experiments, of D.
W. Tipping, and of V. D. Nogle, who assisted in the construction
of the detectors. All are with Sandia Laboratories, Albuquerque,
New Mexico. This work was supported by the U.S. Department of
Energy (DOE), under contract AT(29-1)789, and was performed at
Sandia Laboratories, a U.S. DOE facility.

REFERENCES

1. R. E. Cuthrell and D. W. Tipping, Rev. Sci. Instr. 47, 595
    (1976).
2. Figaro Report, April 1975, Figaro Engineering Inc., 3303
    Harbor Blvd., Suite D-8, Costa Mesa, California 92626, USA.
3. Supported sensors are also available from Japan Electronics
    Manufacturing Agency, Inc., 7550 N. Kolmar, Skokie, Illinois
    60076, USA.
4. R. E. Cuthrell, "The Quantitative Detection of Molecular
    Layers with the Indium Adhesion Tester," SC-DR-66-300, Sandia
    Laboratories, Albuquerque, New Mexico 87185, USA, July 1966.

5. R. F. Tylecote, Br. Weld. J. 1, 117 (1954).

6. R. R. Sowell, R. E. Cuthrell, D. M. Mattox, and R. D. Bland,
    J. Vac. Sci. Technol. 11, 474 (1974).

7. L. C. Jackson, (1978), personal communication, Bendix Corp.,
    Kansas City, Missouri, USA.

8. Al Pshaenich, Application Note AN-735 (1974), Motorola Semicon-
    ductor Products Inc., P.O. Box 20912, Phoenix, AZ 85036, USA.

9. R. E. Cuthrell and D. W. Tipping, J. Appl. Phys. 44, 4360
    (1973).

10. R. E. Cuthrell and D. W. Tipping, IEEE Trans. Parts Hybrids
    Packag. PHP-10, 4 (1974).

EXTRACTION METHODS FOR MEASUREMENTS OF IONIC SURFACE CONTAMINATION

J. Brous

Alpha Metals, Inc.
57 Freeman Street
Newark, New Jersey 07105

One of the most sensitive techniques for measurement of surface contamination is electrical conductivity measurement of solutions of aqueous extracts of the surface. For complex mixtures of ionic and non-ionic contamination, water, alone is not sufficient for complete ionic extraction due to encapsulation of ions in the non-ionic material. Mixtures of alcohol and water have been found to be more efficient in solubilizing both types of surface residues to enable the conductivity measurement of the ionic contaminants. A number of specific techniques are available for ionic extraction. Most of these are static extraction and measurement. These methods, while simple, may not extract all of the surface ionic materials thereby giving erroneous measurements. A dynamic extraction method utilizing recirculating alcohol - water solutions can provide many advantages over static extraction.

1. Method is self-indicating of completeness of ionic removal.
2. Atmospheric carbon dioxide interference is eliminated.
3. Self-cleaning system eliminates equipment cleanliness variables.
4. Allows highest sensitivity of measurement.
5. Method indicates rate of ionic extraction.

A discussion of the theory and practice of dynamic and static measurement of ionic residues is presented.

INTRODUCTION

The electrolytic conductivity of ions in solution is a property which can, in special situations, be of great practical value. This property has, for example, been widely used to specify and control the quality of distilled or deionized water. It is also commonly used for measurement of higher ionic levels such as the salinity of sea water and mineral content of boiler water. The value of this measurement may be expressed in units of conductivity, resistivity or as an effective concentration of a specific salt such as sodium chloride in solution.

An important characteristic of electrolytic conductivity is its extreme sensitivity capability. Pure water, which is itself a weakly ionized medium, has approximately 10^{-7} moles/ liter of H^+ and OH^- in solution. The conductivity of the purest water is of the order of .05 micro Siemens/cm (20MΩ-cm resistivity). Additions of traces of other ions as low as 10^{-8} moles/liter can measurably affect the conductivity. The readings obtained for high purity water are therefore sensitively affected by ions which are brought into solution deliberately or inadvertently.

If a solid object is introduced into a pure water medium, the conductivity of the solution will be changed to a degree determined by the amounts of ions dissolved from the object. Generally, objects with higher levels of ionic contaminants on the surface will effect greater changes in the conductivity of the extracting solution. It is therefore possible to use the conductivity of an extracting medium as an indication of the degree of ionic contamination.

APPLICATION TO ELECTRONIC MANUFACTURING

This type of measurement has found important applications in the manufacture of electronic circuits and components. In many electronic applications both the initial behavior as well as the service reliability of circuits and components can be affected by the presence of traces of ionic contamination on insulating surfaces. This can be particularly troublesome in the presence of atmospheric moisture which can be absorbed by hygroscopic contaminants providing mobility to these surface ions. These mobile ions can be responsible for leakage currents across insulating surfaces which could seriously degrade the electrical characteristics of the circuit. In addition, these ions can have corrosive effects on metallic portions of the circuit.

It is therefore necessary to limit the ionic concentrations on electronic insulating surfaces to levels which are safely below those which may cause deterioration of circuit performance over its service life. No specific limit of ionic contamination can be designated which will be equally applicable to all circuits since they differ in geometry as well as sensitivity to leakage. General limits for ionic contamination have, however, been established under military specifications[1,2] which experience has shown represent safe operating levels in the majority of cases.[3] The ability to measure ionic levels of surfaces of electronic circuits is therefore a most important requirement.

STATIC EXTRACTION METHODS

One of the simplest methods for measurement of surface ionic density is by immersion of the object in a fixed (or static) quantity of deionized water as shown in Figure 1. The conductivity of the solution will be changed to a degree determined by the amounts of ions dissolved from the object. At low levels of ionic concentration, the conductivity increase of the resulting solution is linearly proportional to the concentration of ions as indicated in the equation:

$$\Delta L = k\Delta C = K \frac{\Delta g}{V}$$

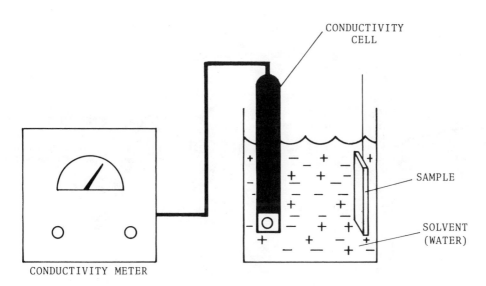

Figure 1.   Egan's method for static ionic extraction.

Where L is the conductivity, C, the concentration of ions, in
solution, g, the total weight of ions and V, the volume of solution. If all the ions which were originally on the surface of
the sample have been extracted into the solvent, then the value
Δg, represents the total initial ionic contamination of the
sample. This is also the product of the average surface ionic
density and the area. Therefore:

$$\Delta L = \frac{KA \; \Delta \; D}{V}$$

Where A is the area of the solid surface being measured and D
represents the average surface ionic density.

Such a method was described by T.F. Egan[4] of Bell
Laboratories in 1973. After a fixed period of bathing of the
sample, the conductivity is compared with the water conductivity
prior to immersion of the sample. The volume of the solvent
which is used is in a fixed ratio to the surface area of the
sample, therefore the conductivity change indicated is directly
proportional to the change of surface ionic density in the
extraction.

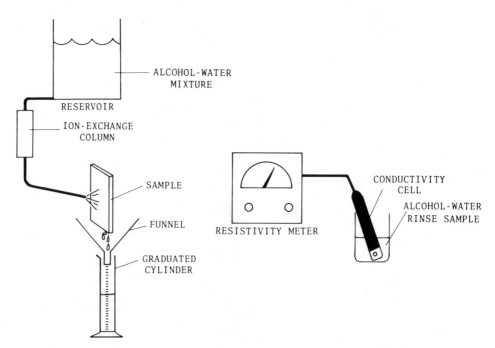

Figure 2.   Extraction method of Hobson and De Noon.

   Adaptations of this **method** were made by others[5-12] who
applied it specifically to measure ionic residues from electronic
circuits. One important modification of this method was made by
W.T. Hobson and R.J. De Noon[5] of the Naval Avionics Center,
Indianapolis, Indiana, who used mixtures of isopropanol and water
to extract the ions from surfaces of the assembly. (Figure 2.)
In their method, a stream of deionized alcohol - water mixture is
directed against the surfaces to dissolve a portion of the ions
present. The solution is collected in a graduated cyclinder to
measure the volume of the rinse solution used. Here too, the
ratio of the solvent volume to surface area is held constant so
that the ionic concentration can bear a relationship to the ionic
density of the surface. The rinse liquid is then measured
separately for resistivity. The use of alcohol - water mixture
greatly enhances the ability of the solvent to **thoroughly extract**
all of the ions for measurement. This is especially true if
nonpolar, as well as ionic materials are present. The presence
of nonpolar residues such as rosin fluxes can effectively mask
the removal of water-soluble ions trapped in and under this
material unless the solvent is capable of simultaneously
dissolving, to some degree, both types of contaminants. This
procedure is used in Military Specifications[1] for measurement
of surface ionic contamination.

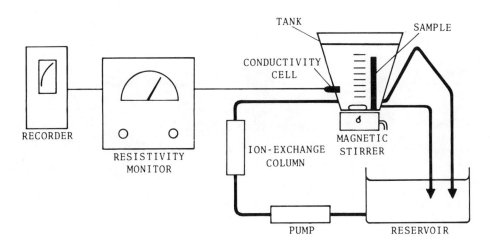

Figure 3.   Schematic representation of the OMEGA METER[(R)] system.

   Another variation of the static extraction method was
described by E.W. Wolfgram[6] of Kenco Chemical Company in a patent
describing a semi-automated system for ionic extraction from
electronic circuits. This equipment, manufactured under the name
OMEGA METER,(R) (Figure 3.), is designed to perform a static
measurement in a fixed period of time using alcohol-water
mixtures in a fixed ratio of volume to sample surface area.
Values of the solution resistivity are measured and indicated
on a recorder as well as a meter. After each measurement,
the mixture is recycled through ion-exchange columns to provide
a fresh supply of deionized mixture for the next measurement.
The pump is operated only to purge and introduce fresh solution
into the tank.

   Several modifications of static extraction methods were
made by others[7-11] all of which utilize similar procedures
with limited volumes of solvent, usually for fixed periods of
time.

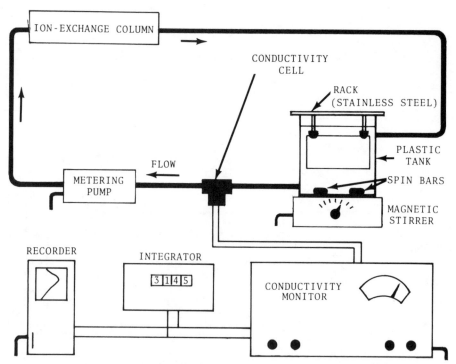

Figure 4. Schematic representation of the IONOGRAPH(R) system.

DYNAMIC IONIC EXTRACTION

A method and a system for quantitatively measuring all of
the extractable ionic contaminants from a circuit assembly was
developed in the Research Laboratory of Alpha Metals.[12,13]
This method employs the dynamic extraction of ions into a theoretically infinite volume of solvent. (Figure 4.) An alcohol-water
mixture is pumped continuously by a metering pump through a mixed
bed ion-exchange column, then into a tank in which the sample is
immersed for ionic extraction. The liquid is then flowed
through a conductivity cell and back to the pump to form a
closed loop. The conductivity of the mixture within the measurement cell is monitored continuously by a conductivity monitor.
This provides a DC voltage which is linearly proportional to the
solution conductivity. This voltage is fed to a meter and a
chart recorder which shows the time function of conductivity of
the effluent solution from the tank. The signal is also fed to
an electronic integrator which generates a reading proportional
to the time integral of the conductivity. With the system filled
with alcohol-water mixture and the pump operating, the constant
recirculation of the solvent through the ion exchange column will
purge it of all ionic salt content. With no sample in the tank,
the conductivity will fall until it reaches a small but finite
level value. This established the background ionic level
corresponding to the "base line" on the recorder curve, which
is respresented in Figure 5a.

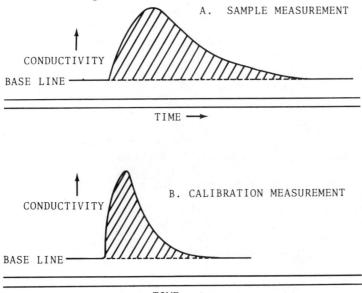

Figure 5. Typical appearance of IONOGRAPH[(R)] recorder curves
obtained by a. sample measurement and b. calibration measurement.

This value represents the sum of two conductivity factors:

   a. The natural ionization of water to the extent determined
      by its ionization constant in the alcohol-water mixture.
   b. The contribution of carbonic acid formed by a dynamic
      equilibrium between the CO_2 absorbed from air at the
      tank surface and the purified liquid flowing through
      the tank.

   When a contaminated sample is introduced into the tank, the
conductivity of the solution will rise rapidly and will be
indicated by the conductivity monitor and recorder. As the ions
are stripped from the sample and purged through the system, the
conductivity will fall, gradually returning to the initial base
line. When the conductivity has returned to the base line it can
be concluded that all of the extractable ionic material has been
removed from the sample and tank. The conpletion of ionic
removal is very sensitively indicated since the measurement near
the end point (return to the base line) is made at the lowest
concentration of ions so that the slightest addition of ionic
content is most easily discernable relative to the low background
level. Completion of ionic removal can therefore be indicated
at the highest level of sensitivity of the conductivity equipment.

   The entire amount of ionic material removed from the sample
can be related to an integration of the conductivity function over
the time of the extraction i.e. the area of the curve above the
base line: At any time, t, the number of moles of ionic material
within the conductivity cell, n_t, is $n_t = V_c C_t$ where C_t is the
concentration of ions and V_c is the volume of the cell, a constant. Over an infinite amount of time, the total number of
moles of ions, N, passing through the cell will be:

$$N = \int_0^\infty n_t\,dt = V_c \int_0^\infty C_t\,dt$$

Since we are dealing with very low ionic concentrations ($<10^{-4}M$),
we can assume complete ionization, therefore:

$$conductivity = L = kC$$

(assuming one salt to be present for purpose of clarity).
Substituting C in the equation above, it can be seen that the
total amount of extracted salts is proportional to the integral
of the conductivity readings above the base line:

$$N = kV_c \int_0^\infty L_t\,dt$$

If the conductivity monitor and recorder responses are linear with
respect to L and the pumping rate is constant, the area under the
conductivity-time curve is a function of the total amount of ions
removed from the sample. The area above the base line can be
integrated using the electronic integrator. The instrument can
be calibrated by measuring the response obtained by injecting
into the tank, a known amount of a standard sodium chloride
solution. The ionic extract level from a sample can then be
represented, as is often done for water purity measurements, as
a conductivity factor expressed as sodium choride. While not
accurate in terms of actual concentrations of the unknown ions,
this method enables us to represent ionic levels in a situation
where many unknown ion types may be present. Concentrations of
all ions of various mobilities are therefore normalized to a
function which is more directly related to surface electrical
leakages on a circuit board.

Figure 5b. represents the recorder curve obtained by
injecting a quantity of sodium chloride solution directly into
the tank as in calibration. It shows the rapid build-up of
conductivity followed by an exponential decline as the salt is
purged through the system. Comparison of this curve with Figure
5a. shows a more rapid build-up of the conductivity to a peak
value and a more rapid decline to the base line. In the case of
the second curve, the entire ionic content had been injected into
the solution at the start of the measurement. In the first case
the ionic material was extracted at a slower rate from the sample
surface. The dynamic extraction measurement can therefore provide an indication of the rate and completeness of ionic removal.

COMPARISON OF THE EXTRACTION METHODS

A. Extraction Completeness

In the static extraction methods, the final reading of ionic
content is made with the sample bathed in a solution at its
highest ionic level. Indications of resistivity or conductivity
changes due to final increments of extracted ions are least
sensitive, particularly if higher instrument multiplier scales
must be used to obtain a reading. In various applications of
the static method, the reading is usually terminated after a
predetermined period of time. Thus a static extraction can be
incomplete for the following reasons:

   a. Equilibrium between sorbed materials on the surfaces
      and the relatively high concentrations of ions in the
      solution can prevent further removal of ions.

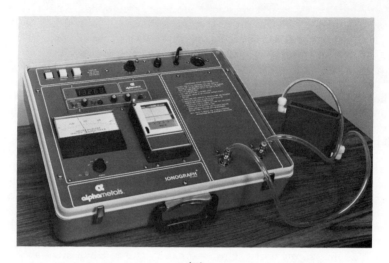

Figure 6.   Ionograph$^{(R)}$ with integrator.

b. Limited solubility of specific salts (such as lead
   sulfate or lead chloride) in the alcohol-water mixture
   could limit the amount of material which is removed
   in a static solution which has become saturated.
c. Diffusion rates of ions from surface porosity and out of,
   and from under components mounted on the circuit will
   limit amounts of ions removed in a finite period of time.

For the dynamic extraction method, the sample is bathed in
a theoretically infinite volume of deionized solvent for as long
a period as is necessary to completely remove all of the available ions. Since the completeness of ionic removal is self-
indicating, the measurement is not terminated until the completeness is indicated. The entire ionic content is measured despite
weaker solubilities of surface deposits, absorption or slow
diffusion rates of ions out of the surface, if the measurement
is run until the curve returns to its base line.

B. Atmospheric Carbon Dioxide

Another problem encountered in running a static extraction
is the contribution of absorbed carbon dioxide to conductivity.
This will be read as a finite conductivity added to the conduc-
tivities of the extracted ions being measured. Even if attempts
are made to subtract the carbon dioxide contributions, any
differences between initial and final CO_2 content can give rise
to errors which can be significant. In the case of the dynamic

extraction method, a dynamic equilibrium is established between the atmospheric CO_2 and the solvent flowing through the tank so that a stable base-line value is obtained. The subsequent integrated reading represents the conductivity function above the base-line contributed by the sample only. Background levels due to air and system contributions are therefore automatically subtracted.

C. Equipment Cleanliness

In applying the static method, it is necessary to employ great care to assure ionic cleanliness of all equipment used which may contact the solution. Furthermore, if any materials are present in the system which evolve ions at some slow, constant rate, this will also affect readings.

The closed-loop dynamic system is, however, self purging. Any foreign contaminants in the system at the time the pump is turned on will be removed by the ion exchange column in the initial pump-down to the base-line. Evolution of ions from materials within the system, if evolved at a constant rate, will contribute to the base line level. This will automatically be subtracted from the sample reading.

D. Experimental Comparison of the Two Methods

In order to compare the relative effectiveness of a time-limited static process to a dynamic extraction, the following procedure was followed:

1. An IONOGRAPH$^{(R)}$ filled with 1:1(by volume) isopropanol: water was run without sample to establish a stable base line. The integrator was then zeroed against this base line.
2. The pump was then stopped so that a static volume of the mixture was left in the tank. The magnetic stirrer was left on.
3. A sample printed circuit board which had been fluxed, soldered and cleaned was then immersed in the tank.
4. After a static soaking for 10 minutes, the sample was removed and the system pump was once again started.
5. After several minutes, all of the ionic material was flushed out of the tank and was indicated as a count on the integrator. The conductivity once again returned to its original base line. This value was recorded and is indicative of the amount of ionic material removed in the 10 minute static immersion.

6. With the pump still operating, the same circuit board was re-immersed into the tank to allow the further extraction of any additional ionic material still remaining on the board after the initial static extraction.

7. When the conductivity once again reached the base line value, the integrator was read again. This final reading is proportional to the total extractable ionic material which was initially on the board.

The ratio of the first (static extraction) reading to the final value is then the fraction of the ions which were removed in the ten minute static extraction. In running this test on a group of ten boards it was found that the percentages of the total ionic contents which were removed in the static extraction ranged from 67% to 88%. This indicated the failure of the static extraction to remove all, or even a constant fraction of the available ions in a fixed, ten minute period.

APPLICATIONS OF IONIC CONTAMINATION MEASUREMENTS

The ability to measure surface ionic contamination implies the ability to determine various factors contributing to surface contamination and its removal. In the case of electronic circuits, studies can be made of contaminating factors such as atmospheric dust, fingerprints, fluxes, soldering process, etchants, plating salts and other chemicals used in circuit manufacture. In addition, the effectiveness of different surface cleaning process variables can be measured such as cleaning method, equipment, chemicals, time, temperature and agitation.

Studies such as these can contribute much to our understanding of the causes and removal of ionic contamination to enable us to produce electronic circuits to the highest levels of reliability. Much work in these areas has already been done. With the wider use of methods for surface ionic measurement, we can expect many improvements in materials and processes for cleaner, more reliable and economical electronic products.

REFERENCES

1. MIL-P-28809 - Military Specification - Printed Wiring Assemblies March 21, 1975.
2. R.E. Martz, Materials Research Report No. 13-77, U.S. Naval Avionics Center, Indianapolis, Indiana.
3. W.B. Wargotz, IPC Technical Review, Issue No. 204, 9(Jan.1978)
4. T.F. Egan, Plating, 60(4), 350(1973)

5. W.T. Hobson and R.J. De Noon, Materials Research Report
    No. 3-72, U.S. Naval Avionics Center, Indianapolis, Indiana.
6. E.W. Wolfgram, U.S. Patent - 4,023,931, May 17, 1977.
7. H.E. Phillips, Electronic Packag.Prod, 13(9), 177 (1973).
8. J.W. Dennison, Jr., Corrosion/74 Paper No. 27, International
    Corrosion Forum, National Assoc. of Corrosion Engineers,
    Chicago, Illinois, March 1974.
9. W.G. Kenyon, in "Proceedings NEPCON West/East - 1977"
    p.313, Ind. and Scientific Conference Management Inc,
    Chicago, IL.
10. T.O. Duyck, Insulation/Circuits, 38, (Oct. 1978).
11. Data Sheet No. 10.031 "On Line Test Procedure for Determining
    Level of Cleanliness," London Chemical Co., Bensenville, IL.
12. J. Brous, Weld. J. Res. Supp., 444S (Dec. 1975)
13. J. Brous, U.S. Patent No. 3,973,572, Aug. 10, 1976.

CHARACTERIZATION OF BONDING SURFACES USING SURFACE ANALYTICAL

EQUIPMENT

C. E. Garrett and E. F. Good
Lord Kinematics Incorporated
Erie, Pennsylvania 16512 and
Purdue University
West Lafayette, Indiana 47907

Over the years, Lord Kinematics has evaluated several types of surface analytical equipment which aid in defining substrate conditions for producing quality elastomer-substrate bonds. In this study, evaluation was made of the removal of three cutting oils from steel and aluminum surfaces as an initial step in providing a clean, active surface for bonding. Evaporative rate analysis, a non-destructive analyzer of the cleanliness and activeness of a surface; reflection infrared spectroscopy, a destructive gross surface composition analyzer; and scanning electron microscopy, an analyzer of a surface topography, were used to characterize the effectiveness of degreasing, alkaline cleaning, and their combination in removing the cutting oils.

The results of this study may be summarized as follows:
1) Evaporative rate analysis demonstrated the cleanliness and activeness of the surfaces under study.
2) Scanning electron microscopy provided vivid photographs of the conversion coating growth and illustrated the results of the various cleaning, activation and growth parameters studied.
3) Reflection infrared spectroscopy was totally ineffective in determining the surface contamination because they were surface monolayers.

INTRODUCTION

The fabrication of a high quality, high performance reliability product requires that the surface conditions of the components be well defined. In the bonded rubber-to-metal industry, a contaminated substrate surface can be extremely detrimental to proper performance of the adhesive system. It is therefore imperative that adequate surface cleaning techniques be invoked to remove all undesirable contaminants prior to adhesives application. Of equal importance is the ability to measure how well these undesirable surface contaminants have been removed.

Research emphasis has been toward the development of sophisticated laboratory oriented surface analytical techniques, for observing, evaluating and understanding various surface phenomena. To date, however, few of these sophisticated techniques have found their way to the production floor to assist in defining in-process surface conditions. Having the capability of measuring as-processed surface conditions of production parts enables the fabricator to provide the quality necessary for proper part performance. The degree of measurement extends to the surface atomic monolayers where it is understood that bond performance emanates.[1]

To demonstrate the effectiveness of detailed surface analysis and the validity of our surface cleaning steps, a study of the cleanability of three cutting oils from the surfaces of steel and aluminum was undertaken. Sample coupons of 1020 cold-rolled steel and 2024 aluminum were machined in the environment of a water soluble, partially water soluble and a water immiscible cutting oil. After machining, the coupons were either degreased, alkaline cleaned or degreased and alkaline cleaned, followed by the application of a conversion coating. Various surface analytical techniques were used to determine the surface cleanliness and activity and the quality of the conversion coating.

EXPERIMENTAL

Steel and aluminum samples were prepared for this study by an initial degreasing and alkaline cleaning. The cleaned samples were then machined under the influence of each of three types of cutting oil - water soluble, partially water soluble and water immiscible. Those samples to be subjected to evaporative rate analysis and reflection infrared spectroscopy were cleaned by either:
1) Degreasing in Perchlorethylene
2) Alkaline cleaning in IRCO 51108 (steel) or Ridoline 53 (aluminum)
3) Degreasing and alkaline cleaning

Evaporative rate analysis is envoked to detect the amount of chemisorbed material remaining on the atomic surface after the cleaning operation. The active metal sites, not previously in-activated by the chemisorbed material, adsorb the radiochemical used in this analytical technique. A three-minute monitoring of the decay rate provides an indication of the surface cleanliness and activity.[2]   Typical results are illustrated in Figure 1. A high count indicates a clean, active surface (Curve A) while a rapid decay, caused by an absence of active sites, denotes a clean but inactive condition (Curve B). The intermediate decay curves indicate a surface coating (Curve C) or a non-volitile residue (Curve D) is present. The latter two curves represent the least desirable surface conditions for high bond strength.

Infrared spectroscopy, utilizing the reflection of the inci-dent light beam source, provides a spectrum for analysis of non-transmissible surface composition. This analytical technique was used in an attempt to identify either a surface coating or a non-volatile residue, as indicated from the evaporative rate analysis of the machined samples.

Additional samples from each cleaning procedure were either phosphatized (steel) or alodized (aluminum) and the resulting surface viewed under high magnification. Scanning electron

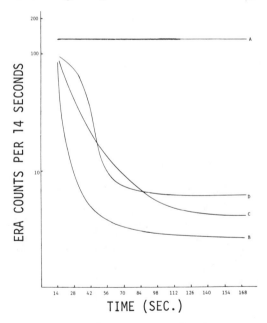

Figure 1. Typical evaporative rate analysis curves. (A) clean, active surface; (B) clean, inactive surface; (C) surface coating; (D) non-volatile residue.

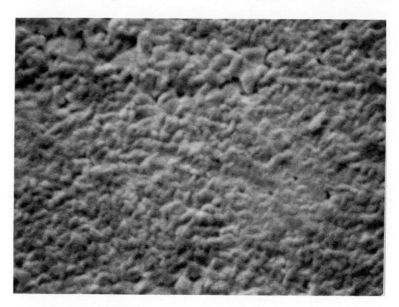

Figure 2. Scanning electron photomicrograph of a standard phosphate conversion coating of 1020 cold-rolled steel. 1200 X magnification.

microscopy provides a topographical overview of the macromolecular structure of the viewed surface.[3] Figure 2 is a scanning electron micrograph of a standard steel phosphatized surface at 1200 X magnification. The coating is complete, uniform and contains medium-sized nodule growth. A standard alodized aluminum sample, as seen in Figure 3, is a smoother coating than phosphatized steel, but once again contains a complete, uniform, medium-sized nodule growth coating.

RESULTS AND DISCUSSION

The steel samples machined in the water soluble cutting oil produced evaporative rate analysis results as shown in Figure 4. The degreasing operation left the surfaces clean but inactive while alkaline cleaning or degreasing and alkaline cleaning produced a clean, active surface ready for further processing. Figure 5 and 6 show similar results for the steel samples machined in the partially water soluble and water immiscible cutting oils respectively.

Scanning electron photomicrographs of the phosphatized steel surfaces after each cleaning operation show the following results:

Figure 3. Scanning electron photomicrograph of a standard alodized coating of 2024 aluminum. 1200 X magnification.

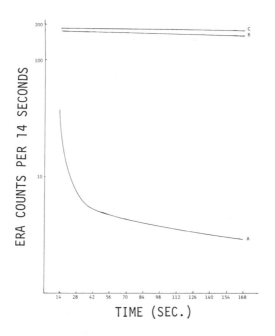

Figure 4. Evaporative rate analysis curves of 1020 cold-rolled steel machined in water soluble cutting oil. Curve A cleaned by degreasing; curve B cleaned by alkaline cleaning; curve C cleaned by degreasing and alkaline cleaning.

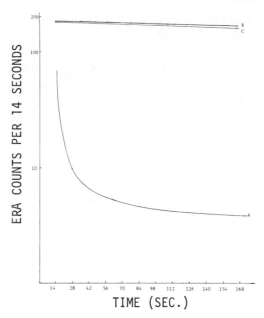

Figure 5. Evaporative rate analysis curves of 1020 cold-rolled
steel machined in partially water soluble cutting oil. Curve A
cleaned by degreasing; curve B cleaned by alkaline cleaning;
curve C cleaned by degreasing and alkaline cleaning.

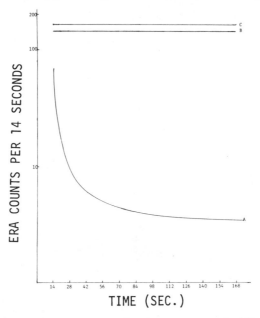

Figure 6. Evaporative rate analysis curves of 1020 cold-rolled
steel machined in water immiscible cutting oil. Curve A cleaned
by degreasing; curve B cleaned by alkaline cleaning; curve C
cleaned by degreasing and alkaline cleaning.

1) Degreasing and alkaline cleaning as shown by the evaporative rate analysis results leaves the steel surface clean and active for each of the three types of cutting oils studied. In Figures 7, 8 and 9, the phosphate growth uniformly covers the surface, has a medium nodule growth and most closely approaches the standard phosphatized surface seen in Figure 2.

2) Alkaline cleaning only, activates the surface, yielding complete phosphate coating coverage. However, the nodule growth, as seen in Figures 10, 11, and 12, is smaller than the standard. The confidence level for producing high-quality rubber-to-metal bonds is not as high as with those surfaces using the degreasing and alkaline cleaning steps to remove the cutting oils.

3) When degreasing only, the inactive surface, as indicated from evaporative analysis does not provide for complete phosphate coating coverage. Figures 13, 14 and 15 indicate that in each case, the nodule size is also larger due to the fewer number of nucleating sites on the inactive surface. The rubber-to-metal bonds from these surfaces are the least reliable of the three cleaning operations.

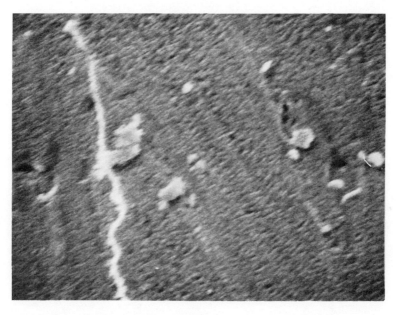

Figure 7. Photomicrograph of phosphate conversion coating on 1020 cold-rolled steel machined in water soluble cutting oil and cleaned by degreasing and alkaline cleaning. 1200 X magnification.

Figure 8. Photomicrograph of phosphate conversion coating on
1020 cold-rolled steel machined in partially water soluble cutting
oil and cleaned by degreasing and alkaline cleaning. 1200 X
magnification.

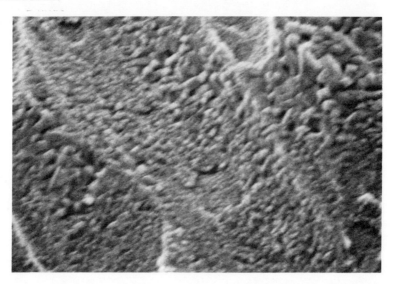

Figure 9. Photomicrograph of phosphate conversion coating on 1020
cold-rolled steel machined in water immiscible cutting oil and
cleaned by degreasing and alkaline cleaning. 1200 X magnification.

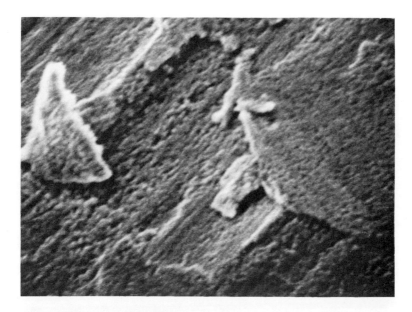

Figure 10. Photomicrograph of phosphate conversion coating on
1020 cold-rolled steel machined in water soluble cutting oil and
cleaned by alkaline cleaning.   1200 X magnification.

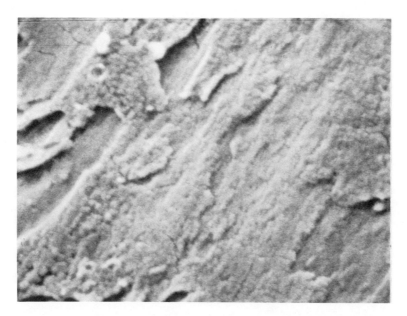

Figure 11. Photomicrograph of phosphate conversion coating on
1020 cold-rolled steel machined in partially water soluble cutting
oil and cleaned by alkaline cleaning. 1200 X magnification.

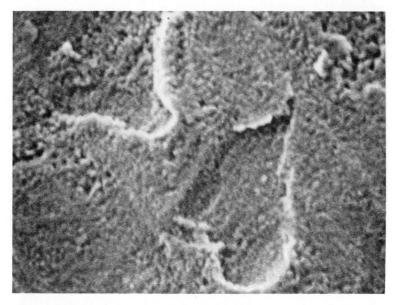

Figure 12. Photomicrograph of phosphate conversion coating on
1020 cold-rolled steel machined in water immiscible cutting oil
and cleaned by alkaline cleaning. 1200 X magnification.

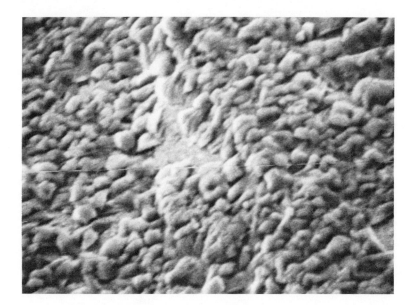

Figure 13. Photomicrograph of phosphate conversion coating on
1020 cold-rolled steel machined in water soluble cutting oil and
cleaned by degreasing. 1200 X magnification.

Figure 14. Photomicrograph of phosphate conversion coating on
1020 cold-rolled steel machined in partially water soluble cutting
oil and cleaned by degreasing. 1200 X magnification.

Figure 15. Photomicrograph of phosphate conversion coating on
1020 cold-rolled steel machined in water immiscible cutting oil
and cleaned by degreasing. 1200 X magnification.

The results of the aluminum sample evaluations are less defini-
tive than the steel samples, as indicated by the evaporative rate
analysis. Aluminum samples, machined in the water soluble cutting
oil, shown in Figure 16, indicate that a surface coating remains
after each cleaning operation. Scanning electron photomicrographs
illustrate the differences in alodize coating coverage as a result
of the various cleaning operations. Figure 17 shows that the
degrease and alkaline cleaning operation produces an alodized surface that best duplicates the standard alodized surface of Figure 3.
Figures 18 and 19 illustrate alodized surfaces after degreasing and
alkaline cleaning respectively.

Evaporative rate analysis results of the partially water
soluble cutting oil machined aluminum samples is shown in Figure 20.
After degreasing, a clean inactive surface results. However, in
Figure 21, the surface after alodizing is markedly different from
the standard surface of Figure 3. When aluminum samples from this
cutting oil were alkaline cleaned, a non-volatile residue was left
on the surface; refer to Figure 20. The alodized surface, Figure 22,
is more complete and uniform, but somewhat different in appearance
from the standard. When both degreasing and alkaline cleaning were

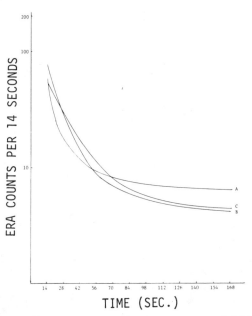

Figure 16. Evaporative rate analysis curves of 2024 aluminum
machined in water soluble cutting oil. Curve A cleaned by
degreasing; curve B cleaned by alkaline cleaning; curve C cleaned
by degreasing and alkaline cleaning.

Figure 17. Photomicrograph of alodized coating on 2024 aluminum machined in water soluble cutting oil and cleaned by degreasing and alkaline cleaning. 1200 X magnification.

Figure 18. Photomicrograph of alodized coating on 2024 aluminum machined in water soluble cutting oil and cleaned by degreasing. 1200 X magnification.

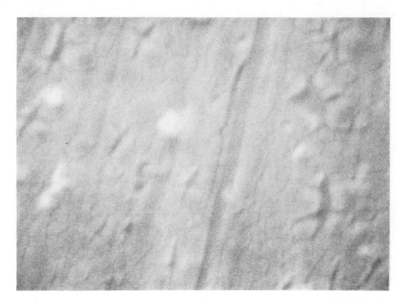

Figure 19. Photomicrograph of alodized coating on 2024 aluminum
machined in water soluble cutting oil and cleaned by alkaline
cleaning. 1200 X magnification.

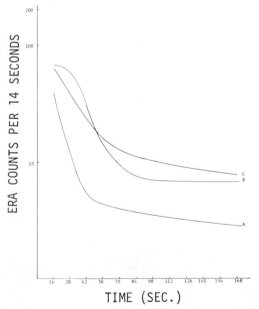

Figure 20. Evaporative rate analysis curves of 2024 aluminum
machined in partially water soluble cutting oil. Curve A cleaned
by degreasing; curve B cleaned by alkaline cleaning; curve C
cleaned by degreasing and alkaline cleaning.

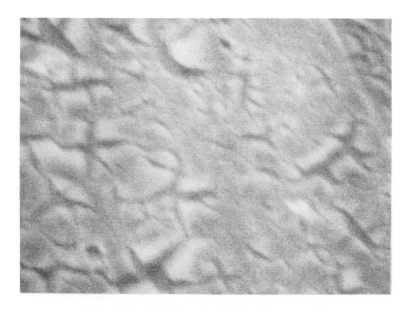

Figure 21. Photomicrograph of alodized coating on 2024 aluminum machined in partially water soluble cutting oil and cleaned by degreasing. 1200 X magnification.

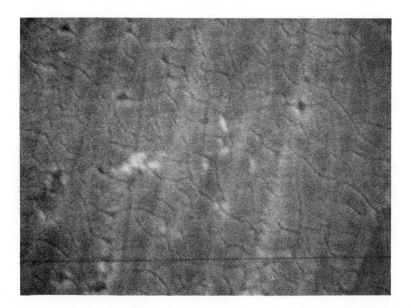

Figure 22. Photomicrograph of alodized coating on 2024 aluminum machined in partially water soluble cutting oil and cleaned by alkaline cleaning. 1200 X magnification.

Figure 23.   Photomicrograph of alodized coating on 2024 aluminum
machined in partially water soluble cutting oil and cleaned by
degreasing and alkaline cleaning.   1200 X magnification.

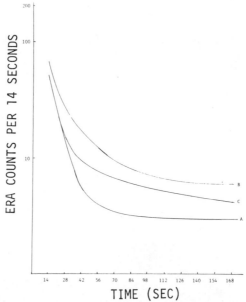

Figure 24.   Evaporative rate analysis curves of 2024 aluminum
machined in water immiscible cutting oil.   Curve A cleaned by
degreasing; curve B cleaned by alkaline cleaning; curve C cleaned
by degreasing and alkaline cleaning.

used to clean the partially water soluble cutting oil, from the machined aluminum sample, a surface coating remained. As Figure 23 indicates though, the resulting alodized surface most closely resembles that of the standard.

The aluminum samples machined in the water immiscible cutting oil have similar results to those of the other two types of cutting oils. After degreasing and degreasing and alkaline cleaning, the evaporative rate analysis in Figure 24 indicates that the surface is clean but inactive. Scanning electron photomicrographs of the alodized surfaces, Figures 25 and 26 indicate that both cleaning procedures provide similar alodized surfaces. After alkaline cleaning only, Figure 24 indicates a surface coating remained which, as seen in Figure 27, resulted in an alodized surface different from the standard of Figure 3.

An excellent bond quality record has resulted from steel and aluminum parts cleaned using degreasing followed by alkaline cleaning after machining. It is evident from the scanning electron microscopy results that not using both cleaning operations does leave a different appearing surface than a standard unmachined surface.

Results of reflective infrared spectroscopy are not presented because they were completely negative in this study. It is theorized that the surface coating or residue detected in the evaporative rate analysis is contained in the top few atomic

Figure 25. Photomicrograph of alodized coating on 2024 aluminum machined in water immiscible cutting oil and cleaned by degreasing. 1200 X magnification.

Figure 26. Photomicrograph of alodized coating on 2024 aluminum machined in water immiscible cutting oil and cleaned by degreasing and alkaline cleaning. 1200 X magnification.

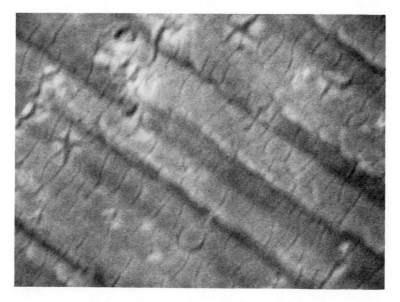

Figure 27. Photomicrograph of alodized coating of 2024 aluminum machined in water immiscible cutting oil and cleaned by alkaline cleaning. 1200 X magnification.

monolayers of the surface. Reflective infrared spectroscopy is simply not capable of detecting these thin surface compositions. Recently, there have been better monolayer characterization techniques developed which we intend to utilize in our further investigations of these surface conditions.

CONCLUSIONS

The three cutting oils tested - a water soluble oil, a partially water soluble oil and a water immiscible oil - are removed by proper cleaning techniques. The steel sample results were conclusive as evidenced by the evaporative rate analysis and scanning electron microscopy results. Although the aluminum sample results were somewhat less conclusive, adequate cleaning of the surface can be accomplished.

Based on the results of this study, it was recommended that degreasing followed by alkaline cleaning should be used to clean any of the cutting oil types tested from the machined surfaces of parts. This current practice has proven to be quite successful.

Evaporative Rate Analysis is a valuable analytical technique for determining the cleanliness and activeness of any surface. It is a non-destructive test which can be utilized on production parts. It is a valuable tool for routine production determination of how well a surface is being prepared for future operations.

Scanning Electron Microscopy is a valuable analytical technique for determining the actual macromolecular structure of a sample surface. It is useful as an in-process audit of how well a processing technique is performing.

Reflective Infrared Spectroscopy is not an adequate analytical technique for determining thin (less than 8 to 10 angstroms) surface compositions.

REFERENCES

1. N. M. Bikales, Editor, "Adhesion and Bonding," p.p. 35-59, John Wiley and Sons, New York, 1971.
2. C. S. Pietras, "Detection of Adsorbed Monolayers by Evaporative Rate Analysis," ACS. Div. Organic Coatings and Plastics Chemistry, Preprints, Vol. 33 (No. 1), p.p. 599-606, American Chemical Society, Washington, D. C., 1973.
3. G. Goldfinger, Editor, "Clean Surfaces: Their Preparation and Characterization for Interfacial Studies," pp. 97-114, Marcel Dekker, Inc., New York, 1970.

ION CHROMATOGRAPHY-QUANTIFICATION OF CONTAMINANT IONS IN WATER EXTRACTS OF PRINTED WIRING

W. B. Wargotz

Bell Telephone Laboratories, Incorporated
Whippany, New Jersey   07981

A new analytical technique, ion chromatography (IC) has been applied to the identification and quantification of various ionic species in samples of water digests of printed wiring. This analytical tool offers the advantage of providing a rapid analysis (30 minutes) of conductive and corrosive contaminants encountered on printed wiring board. Selectivity, sensitivity, calibration and repeatablility have been demonstrated.

INTRODUCTION

Ionic contaminants present on printed wiring boards (PWBs) surfaces can cause electical deterioration.[1,2] A surface contamination level (SCL) of 1 $\mu g/cm^2$ equivalent sodium chloride has been specified as the requirement for cleanliness of printed wiring board surfaces prior to application of a protective cover coat.[2] This requirement is measured by determining the conductivity of the deionized water extract of the printed wiring board. Non-compliance with this requirement informs the investigator that potentially corrosive water soluable ionic contaminants are present on the circuit surface. The measurement does not, however, provide information on the nature of the specific compounds (ions) which are causing the increase in conductivity. Knowledge of the identy and quantity of the corrosive ions present on the surfaces of printed wiring boards helps in determining where the contaminant originated, and remedial action can then be taken. In the past, identification of these potentially corrosive ions (anions or cations) in water extracts of printed wiring required time consuming elaborate analysis techniques. There was a real need for a reliable rapid method of analysis.

Ion Chromatography (IC) is a recently developed analytical technique for the analysis of anions and cations in chemical process streams, municipal water systems and biological fluids.[3-5] IC provides identification and quantification of various anions e.g., chloride, bromide, nitrate, sulphate, phosphate and simple organic acids in water. By employing a novel combination of ion exchange resins it is possible to separate ions in a background electrolyte used for elution. Ions are first resolved on one ion exchange resin column followed by neutralization of the background eluant and resolved ions in a second ion exchange resin column. The neutralized species in the effluent from the second column are then analyzed by a sensitive conductivity monitor (cell).

This paper reports in more detail on various facets on the use and limitations of IC applied to analysis of extracts from printed wiring boards. A brief review of the principles of IC will be followed by a discussion of results obtained on calibration, sensitivity, selectivity and repeatability of the method of analysis. The results from the IC analysis of extracts of printed wiring boards from various sources will demonstrate the utility of this analytical method of analysis.

Ion Chromatography

Ion chromatography (IC) is based upon three principles:[3]

Separation of Ions. The water solution to be analyzed is injected and deposited on a separation (analytical) ion exchange column. For anion separation the ion exchange resion would be represented as Resin $-OH^-$. Separation of ions is achieved by displacing the ions absorbed on the column with a dilute base or acid depending upon whether anions (-) (anion ion exchange Resin $-OH^-$ or cations (+) (cation ion exchange Resin $-H^+$) are to be separated. Table Ia and Table Ib list anions and cations which have been reported to be separable by IC.

Eluant Suppression. In the determination of anions by ion chromatography the effluent from the separation ion exchange column contains the resolved anions in a background of a dilute base eluant. The anions cannot be readily detected by a conductivity measurement in the presence of the base eluant due to the high conductivity of the eluant in relation to the anions of interest. The base eluant and resolved anions are then modified by passage of the effluent through a suppressor ion exchange column (Resin- H^+). The effluent from this column contains the resolved anions in a background of a

Table I. Ions Determined By IC*

<u>a</u> Anions   <u>b</u> Cations

Floride (F$^-$)	Formate	Sodium (Na$^+$)
Chloride (Cl$^-$)	Acetate	Ammonium (NH$_4^+$)
Bromide (Br$^-$)	Propionate	Potassium (K$^+$)
Iodide (I$^-$)	Chloracetate	Magnesium (M$_g^{2+}$)
Nitrate (NO$_3^-$)	Dichloracetate	Calcium (Ca^{2+})
Nitrite (NO$_2^-$)	Trichloracetate	Simple organic
Iodate (IO$_3^-$)	Glycolate	Amines (RNH$_2^+$)
Bromate (BrO$_3^-$)	Oxalate	
Sulphate (SO$_4^{2-}$)	Maleate	
Sulphite (SO$_3^{2-}$)	Fumarate	
Carbonate (CO$_3^{2-}$)	Succinate	
Chromate (CrO$_4^{2-}$)	Malonate	
Phosphate (PO$_4^{3-}$)	Itaconate	
	Benzoate	
	Ascorbate	
	Citrate	

---

*   Reference 4

weak acid of lower conductivity than the anions to be detected.
The reaction in the suppressor column can be depicted as follows:

$$\text{Resin-H}^+ \; + \; \text{Na}^+\text{HCO}_3^- \; + \; \text{M}_n^+\text{X}_y^- \; \rightarrow \; \text{H}^+\text{X}^- + \text{H}_2\text{CO}_3$$

Suppressor   Eluant   M=Na$^+$,Mg^{2+},Ca^{2+} carbonic
Column Resin   acid

X=NO$_3^-$,Cl$^-$,Br$^-$

n,y=0,1,2

Carbonic acid is the product from the ion exchange reaction of bicarbonate eluant with the suppressor column resin. The conductivity of this acid is much lower than that of hydrogen chloride or nitric acid the product from ion exchange of the salts of these acids with Resin $-H^+$.

Conductivity Detection. Hydrogen halide (H^+X^-) found in the effluent from the suppressor column in anion analysis or metal hydroxide (M^+OH^-) from cation analysis (see below) are then detected

e.g. Cation Analysis

$$\underbrace{Resin-OH^-}_{} + \underbrace{H^+NO_3^- + M_n^+X_y^-}_{} \rightarrow M^+OH^- + H_2O$$

Suppressor   Eluant
Column Resin

by a conductivity cell against the background of the weak acid (carbonic acid) or base (water). The response of the conductivity monitor is presented as a conventional chromatogram trace with measurable retention times (Figure 1). The retention time identifies the ion when compared to the retention time of standards. The retention time will vary based upon the ion exchange resin employed (separator column), the eluant employed and its flow rate through the columns, the temperature of column operation and packing of columns. The ion concentration can be quantitatively determined by converting the analog output of the conductivity monitor to a digital response (time integration) which is then processed in an integrator. The integrator then prints out the area or peak height of each chromatogram peak. By introducing standards of individual ions or mixtures,

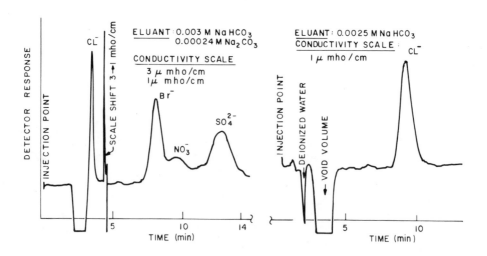

Figure 1.   Chromatograms of circuit (wave soldered/repair) extract.

a calibration curve can be obtained relating area or peak height of the chromatogram to concentration of ion. If a linear relationship exists between the response of the conductivity monitor and concentration of ion a proportional relationship is developed to obtain the concentration of the ion of interest.

EXPERIMENTAL

Ion Chromatograph: (Model 14-Dionex Corp., Sunnyvale, California)

A schematic representation of the IC is shown in Figure 2. General operating conditions for IC were as follows:

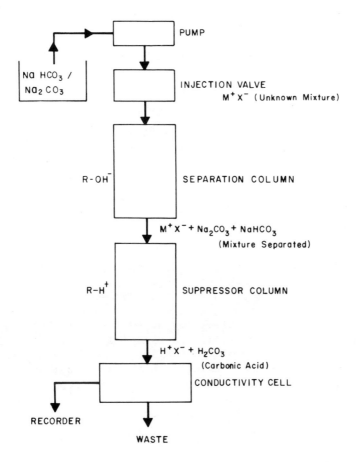

Figure 2.   Analytical system used for analysis of anions.

Analysis

| Anion | Cation |

Cl^-, PO_4^{3-}, SO_4^{2-}, Br^-, NO_3^- K^+, NH_4^+, Na^+

Eluant: 1. 0.004M $NaHCO_3$ 1. 0.006N HNO_3

         2. 0.003M $NaHCO_3$/0.0024M Na_2CO_3

Sample Loop Volume: 100 μl   100 μl
Flow Rate (Eluant):   92 ml/hr   230 ml/hr

Analytical Column (Separator)	2.8x500 mm	6x250 mm
Suppressor Column:	6x250 mm	6x250 mm
Conductivity Scale:	0.3-3 μmho/cm	0.3-1 μmho/cm
Regeneration Solution:	1N HNO_3	1N NaOH

Recorder: Hewlett-Packard Model 7132A Thermal Writing

Recorder Speed: 0.1 in/in
         Span: 5 mV

Integrator: (Minigrator: Spectra Physics, Santa Clara, California)

Parameters

Peak Width (PW) = 70. Peak Width (PW) - is the width, in seconds, of the narrowest peak of interest in the chromatogram trace at its half-height. Increasing the PW value provides more peak detection sensitivity.

Slope Sensitivity (SS) = 70000. Slope Sensitivity (SS) - is the peak detection threshold parameter and is used primarily as a noise rejection feature.

Baseline Test (BL) = 1,2 or 5. Baseline Test (BL) - number entered is number of times the integrator operation is performed and is proportional to the size of the PW value.

Tailing Peak Parameter (TP) = 40,60. Tailing Peak Parameter (TP) - used to identify tailing peaks and also for tangential level-ing of rider peaks. The minigrator identified a tailing peak as one exceeding the TP parameter value - which is the number of consecutive negative slope calculations before a positive slope.

Eluants and Standards

All eluants and standard solution of ions were prepared from

ACS reagent grade chemicals dissolved in deionized water. The
standards were prepared in glass volumetric flasks. These flasks
as well as other glassware had been cleaned by standing 24 hours
in 10% hydrochloric acid followed by rinsing with deionized water
until the conductivity of rinse after standing in glassware was
less than 1 µmho/cm.

Injection of standards and printed wiring board extracts onto
the IC separator column were made with disposable plastic syringes
(1 ml capacity). Adequate rinsing with deionized water of the instrument injection port was required to prevent cross-contamination
of samples.

Preparation of Extracts of Printed Wiring Boards

The extraction of printed wiring board specimens was conducted
as described by Egan.[6] The surface contamination level (SCL) on the
printed wiring specimens was determined[2] from the extract.

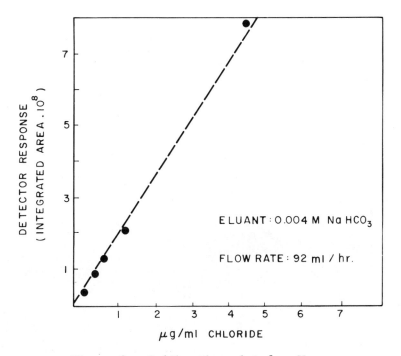

Figure 3. Calibration plot for CL-.

RESULTS AND DISCUSSION

Calibration, Sensitivity, Selectivity and Repeatability of IC

Calibration of IC Conductivity Monitor. Standard solutions of ions of 0.1-10 µg/ml concentration were analyzed by IC. Multiple analysis of each concentration were made, the integrated areas averaged and the area versus the concentration of ion (IC) plotted. In Figures 3 and 4 are the calibration curves for chloride (0.004M NaHCO$_3$ eluant) and phosphate (0.003M NaHCO$_3$/0.0024M Na$_2$CO$_3$ eluant) anions. The response of the conductivity monitor was linear with concentration. For sodium and potassium cations with 0.006N HNO$_3$ eluant the detector response was also linear. However, for ammonium cation it was not linear above 1 µg/ml concentration.

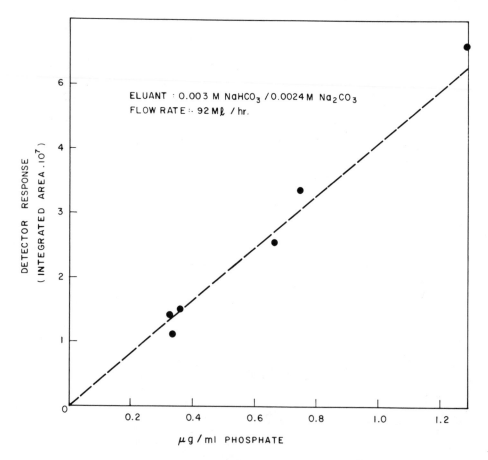

Figure 4. Calibration plot for PO$_4^{3-}$.

Ammonium ion deviation from linearity at high concentration has been observed by others and a corrective technique applied.[7]

Accuracy. In Table II are found the results from the determination of single ions. The data presented relates the percentage difference between the known concentration of ion (standard) added to the IC and that determined by IC analysis. For anion concentrations above 0.5 µg/ml with no apparent interference of ions with similar retentions times, the measurement is accurate to within 20%.

Selectivity. The selectivity of anion separation is shown in Figure 1. In Figure 5 are shown the separation of a mixture of cations. Good separation of Na^+, K^+ and NH_4^+ can be achieved.

Repeatability of Analysis. In Table III are found the area measurements and retention times for multiple analysis of chloride ion of 0.4 and 1.2 µg/ml concentrations. Representative chromatograms for the chloride concentrations are shown in Figure 6. The repeatability of peak area and retention times at high conductivity monitor sensitivity (0.3 µmho/cm for 0.4 µg/ml chloride) and lower sensitivity (1.0 µmho/cm for 1.2 µg/ml chloride) are considered quite good for these low concentrations of ions.

A standard mixture of anions was prepared in bicarbonate/carbonate eluant, and the selectivity, sensitivity and repeatability of the IC analysis determined. The data from multiple measurements of area and retention time for each ion in the mixture is found in the Table IV. A typical chromatogram for the mixture is shown in Figure 7 and

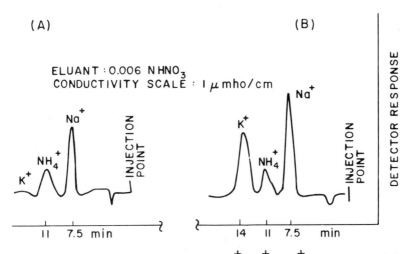

Figure 5.   Chromatograms of caption – K^+, Na^+, NH_4^+ from clean(a) and contaminated (B) PWB ·

Table II.   Accuracy of IC Results.

Chloride			Phosphate	
Std. μg/ml	% difference from IC		Std. μg/ml	% difference from IC
0.1	+34		0.4	−20
0.4	−25		0.74	− 8
0.7	− 5		1.2	−13
1.2	+ 3			
4.8	− 2			

Bromide			Nitrate	
Std. μg/ml	% difference		Std. μg/ml	% difference
0.4	−30		0.4	−12
0.7	− 6		0.75	+ 7
1.2	0		1.25	+13
10.3	+ 8		10.8	+16

Sulphate			Sodium	
Std. μg/ml	% difference		Std. μg/ml	% difference
0.44	+ 6		0.448	− 2
0.8	−15		0.8	+12
1.3	+10		1.1	− 8

Ammonium	
Std. μg/ml	% difference
0.44	+44
1.3	−28
11.4	−29

Table III. Repeatability of Area and Retention Times of Chloride[*]
Analysis by IC (Intergrator Printout).

0.4 μg/ml Chloride (0.3 μmho/cm IC Scale)		1.2 μg/ml Chloride (1.0 μmho/cm IC Scale)	
Area Count $\cdot 10^4$	Retention Time (Sec.)	Area Count $\cdot 10^4$	Retention Time (Sec.)
1464	280	712	272
1313	295	705	275
1247	294	826	279
997	275	727	276
1208	274	846	280
1432	279	902	287
1341	278	822	280
1039	294	794	286
975	281	659	281
1234	292	686	282

The 95% confidence limits for area:

$1101 \longleftrightarrow 1337 \cdot 10^4$

The 95% confidence limits for area:

$710 \longleftrightarrow 826 \cdot 10^4$

[*] Solution of ion prepared in 0.03M $NaHCO_3$/0.0023M Na_2CO_3 rather than deionized water.

should be compared to that of a mixture of anions in water without the bicarbonate/carbonate buffer, Figure 1. The improvement in resolution of the chloride peak (void volume and deionized water negative peak reduced in intensity) is readily apparent and indicates that selectivity has been improved through employment of the bicarbonate/carbonate solution. Use of the buffered solution simplifies IC analysis.

Analysis of Extracts of Printed Wiring Boards by IC

Fingerprint Contamination. Samples of control and handled printed wiring boards were available for analysis. The SCL was determined and found to be in range of 0.42-2.8 $\mu g/cm^2$ as equivalent sodium chloride. In Table V are shown the chloride ion surface

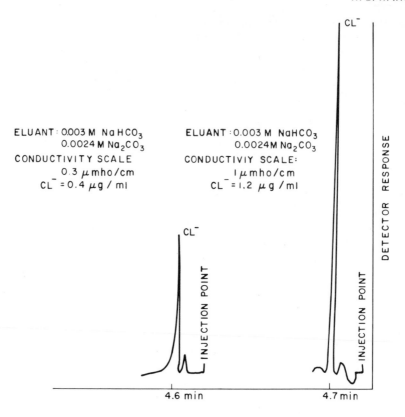

Figure 6. Chromatograms of chloride in bicarbonate/carbonate
          solution.

concentration in µg/cm^2 determined from the IC chloride ion analyses
and surface area of PWB extracted.

The results reveal that sample A was handled more extensively
than the others, as evidence by the high chloride level detected,
which correlates with SCL result. The SCL of samples B and C are
significantly lower than A. The cloride content appear equivalent
for B and C. The higher SCL exhibited by C relative to B might be
related to the higher relative sulphate concentration observed for C
on comparing the IC peak areas (not shown in Table). No sulphate IC
analysis was made to determine the concentrations from the B and C
extract.

Aged Printed Wiring Board. Printed wiring boards covercoated
with an epoxy solder mask were recovered after approximately three
years exposure in an office environment (35°–50°C/20–55% relative

Table IV. Repeatability, Selectivity and Sensitivity

95% Confidence Limits

Anion (conc. μg/ml)

	Area Measurement $\cdot 10^4$	Retention Time (seconds)
Chloride*, Cl^- (1.2 μg/ml)	870–968	274–285
Phosphate, PO_4^{3-} (0.4 μg/ml)	216–410	550–578
Bromide, Br^- (0.4 μg/ml)	351–411	677–693
Sulphate, SO_4^{2-} (0.9 μg/ml)	1462–1960	1196–1239

* The conductivity meter sensitivity scale was set at 0.3 μmho/cm except for chloride which was set at 1.0 μmho/cm.

humidity). These boards were inspected and a salt like contaminant was found surrounding the soldered component pin contact area of some boards.[8] Sections of boards which exhibited this contamination were submitted for analysis. In Table VI are found SCL results and IC analysis of deionized water extracts (SCL measurement) as well as IC results of bicarbonate/carbonate extracts of similar sections of two boards, KD and HG. The SCL results reveal a contamination level greater than 1μg/cm^2, a requirement that has been established

Table V. IC Analysis of Handled Printed Wiring Board ⋅

Sample	SCL μg/cm^2 as NaCl	Chloride Anion Concentration (μg/cm^2)
A	2.8	1.6
B	0.4	0.3
C	1.03	0.2

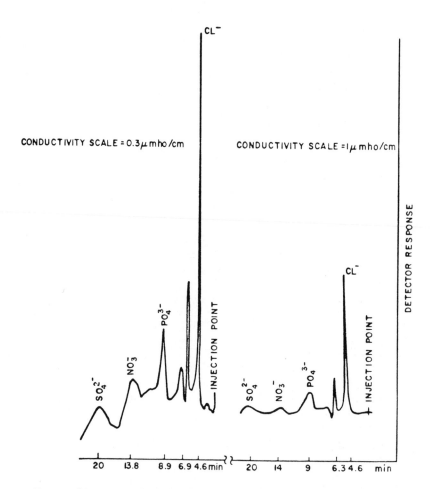

Figure 7.   Influence of conductivity scale setting on selectivity
of anion analysis in HCO_3^-/CO_3^{2-} solution.

Table VI. Aged Printed Wiring Board Extract Analysis

Equivalent Surface Concentration

$(\mu g/cm^2)$

Ion	KD (SCL = 3.7 $\mu g/cm^2$)		HG (SCL = 1.8 $\mu g/cm^2$)	
	DI Extract	Buffer Extract	DI	Buffer
Cl^-	1.10*	1.90	0.51*	1.30
Br^-	0.17	0.14	0.22	0.14
NO_3^-	0.50	0.30	0.25	0.11
SO_4^{2-}	0.85	1.24	0.96	0.70
	―――	―――	―――	―――
Total	2.62	3.58	1.94	2.25

* 0.004M $NaHCO_3$ eluant for chloride detection with DI
extracts of boards. All other anions and buffer extract
anions determined with 0.003M $NaHCO_3$/0.0024M Na_2CO_4 eluant.

for the cleanliness of processed printed wiring boards.[2] The IC
analyses reveal that the contaminates are predominantly chloride
and sulphate. The chloride can be associated with handling, while
sulphate may possibly arise from airborne pollutants. The small
amount of bromide detected is believed to arise from the decomposition of the bromide containing organic activator in the solder flux.
Bromide ion was also found on solder flux treated PWBs after wave
soldering.[9] (See Figure 1.)

Hard Gold Finger Plating Residual Salt Contaminant-Anion
Analysis. Contamination of the dielectric surface between gold plated
(conductor) fingers on printed wiring results in lower IR and may
lead to electrical shorts.[1] In the Table VII are found IC analysis
of the gold finger plated area of a PWB. The board was extracted in
a bicarbonate/carbonate solution. These boards had been stored for
two years in a laboratory environment and are replicates of samples
investigated earlier.[1]

The earlier work[1] has established qualitatively that chloride,
phosphate and potassium were the predominate contaminants, probably

Table VII. Hard Gold Plating Contamination.

		IC Analysis (Bicarbonate/carbonate Extract		
Sample	SCL	Chloride	Phosphate	Sulphate
		Cl^-	PO_4^{3-}	SO_4^{2-}
	$\mu g/cm^2$ as NaCl	$\mu g/cm^2$	$\mu g/cm^2$	$\mu g/cm^2$
Contaminated	9.0	0.3	1.1	0.1
Control	0.5	0.05	0.07	N.D*

* Not detected

present because of inadequate rinsing of the boards after gold finger plating. The IC analyses found in Table VII reaffirmed the earlier elemental analysis and in addition presents quantitative results. With adequate rinsing, the presence of corrosive contaminants is markedly reduced as evidenced from the results in the Table VII.

Soft Gold Plating – Measurement of Residual Salt Contamination.
In soft gold plating for printed wiring boards, ammonium citrate and potassium cyanide baths are employed. Inadequate rinsing of gold plated areas will result in the contamination of the PWB with these compounds. The potassium and ammonium cations can be detected in water extracts of soft gold printed wiring boards. In Table VIII are found the insulation resistance (IR) of contaminated Sample A (SCL = 7 $\mu g/cm^2$) and clean Sample B (SCL 1.1 $\mu g/cm^2$ boards) sections of the printed wiring board. The IR between gold finger pads (0.050 in. separation) readily differentiates between the contaminated (low IR, 10^8 ohms) and clean board areas. It is interesting to note that there does not appear to be any reduction in ammonium level (IC) on the clean boards over the contaminated. The presence of high levels of potassium suggests that the contaminant remaining on the board was potassium cyanide or potassium gold cyanide. An anion analysis was conducted to analyze for cyanide or citrate ion. These ions were not detected with a mixed carbonate eluant. Citrate has been detected, with another eluant, for which the retention time is greater than 30 minutes.[4]

Analysis of Process Rinse Water and PWB Extracts. The cleanliness of electronic devices or PWB is greatly dependent upon the efficiency of the final rinsing of the PWB. Analysis of the final

Table VIII.   Soft Gold Plating Contamination .

Sample	A	B
IR*	10^8 ohms	10^{13} ohms
SCL (as NaCl)	7.2 $\mu g/cm^2$	1.1 $\mu g/cm^2$
K^+ (IC)	9.0 $\mu g/cm^2$	0.8 $\mu g/cm^2$
NH_4^+ (IC)	0.8 $\mu g/cm^2$	0.9 $\mu g/cm^2$

* Measured at 35°C/90% RH/50 Vdc after conditioning in test
environment for 24 hours without bias applied.

rinse water of a process and water extracts of PWB processed with
that rinse water provides information on the efficiency of the
cleaning stage.

A sample of rinse water was removed from overflow of the final
rinse of a PWB process. The final rinse of the PWB follows a circuit copper etch/solder brightener process unit. Test PWBs were processed at the same time as product PWB and recovered for test. The
test PWB were extracted in deionized water for SCL measurement, and
the extract analyzed for cations and anions by IC. The final process
rinse water was also analyzed (Table IX).

Table IX.   Analysis of Process Rinse Water and Printed Wiring
Extracts .

Sample	IC Analysis $\mu g/ml$		
	Cl^-,	PO_4^{3-}	NH_4^+
Final Rinse Water	0.1	0.02	0.6
Test Circuit 1	0.4	0.09	1.2
2	0.2	0.05	0.9
3	0.04	0.7	1.0

The rinse water exhibited a low level of chloride (IC). The chloride and ammonium ions found in the rinse water are residuals carried over by the PWB from the etchant (ammonium chloride-ammonium hydroxide-sodium hyprchlorite). Phosphate ion arises from a dilute phosphoric acid rinse which follows the solder brightener treatment of the PWB. These treatments precede the final rinsing with the deionized water.

The results from the IC analysis of the PWB (0.015 in conductor spacing) are quite interesting. All ions measured in these extracts were found to be significantly higher in concentration ($\mu g/ml$) than in the final rinse water in which these PWBs had been processed. (Table IX) Insignificant amounts of sulphate were found in the rinse water and on the PWBs as compared to higher sulphate concentration on aged or handled PWBs. (Table IV) The concentration (surface) of chloride and phosphate ion in $\mu g/cm^2$ found for these PWBs, determined from IC results and surface area of PWBs extracted were 1-3 $\mu g/cm^2$. At this surface concentration one would expect a low insulation resistance. IR of test PWBs 1-3 were in range 10^6-10^7 ohms measured at 500 Vdc in a 35°C/90% RH environment. Test PWBs which exhibited low surface concentration of chloride < 0.5 $\mu g/cm^2$ had IR of 10^9-10^{10}ohms.

What effect PWB residence time in the rinse tank and water flow has on rinse water ion content and PWB contamination level, is unknown at present.

SUMMARY

Ion Chromatography (IC) has been shown to offer the advantage of providing a rapid analysis (30 minutes) of water soluble ionic surface contaminants on printed wiring boards and process rinse water. For chloride, phosphate, bromide (anions), potassium and ammonium ions (cations), contaminants which arise in chemical processing, soldering and handling, analysis can be completed in 15 minutes in separate anion and cation analyses. This rapid analysis technique can be automated as in conventional chromatographic methods and would provide a continuous analysis of rinse water from chemical processing of PWBs as well as analysis of contaminants from a large number of extracts from SCL measurements.

When compared to known ion standards, IC analysis is accurate to within 30% at concentrations of ions below 1 $\mu g/ml$ and within 10% in the range 1 μg-10 $\mu g/ml$. Reproducibility of retention times, a measure of separation efficiency of individual ions, is dependent upon eluant concentration, environment, and integrator parameters, and was found to be within 5% in this study. The error in area measurement (concentration) for mixture of ions approximates 15% for each ion at 1 µg/ml or less, except for phosphate ion which is 40% below 0.4 µg/ml concentration. Larger error for phosphate is due to similar separation characteristics as chloride. A high concentration of chloride (1 µg/ml) will cause overlap in area measurement at low concentrations of phosphate. A similar error in area measurement is found with low bromide ion concentrations in presence of higher concentrations of nitrate ion (Figure 1). It is believed that the separation characteristics of the chloride/phosphate and bromide/nitrate ions can be improved upon by slight changes in eluant concentrations with some sacrifice in total time for analysis.

REFERENCES

1. W. B. Wargotz, Preprints "Symposium on Deterioration of Electronic Devices," The Electrochemical Society Meeting, May 8-13, 1977.
2. W. B. Wargotz, Circuit Manufacturing, p. 40, (February 1978).
3. H. Small, T. S. Stevens, and W. C. Bauman, Anal. Chem., 47(11), 1801 (1975).
4. C. Anderson, Clinical Chem. 22, 1424 (1976).
5. Preprints "Symposium on Ion Chromatographic Analysis of Environmental Pollutants," EPA. Research Triangle Park, N. C., April 28, 1977.
6. T. Egan, Plating, p. 35, (April 1973).
7. S. A. Bouyoucos, Anal. Chem. 49(3), 401 (1977).
8. J. P. Mitchell and D. G. Denure, Unpublished Results, Bell Telephone Laboratories.
9. W. B. Wargotz, Unpublished Results, Bell Telephone Laboratories.

Part IV
Implications of Surface Contamination

EFFECT OF SURFACE CONTAMINATION ON SOLID PHASE WELDING - AN

OVERVIEW[*]

J. L. Jellison

Process Metallurgy, Division 5833
Sandia Laboratories
Albuquerque, New Mexico 87185

Surface contaminants hinder solid phase welding
by not allowing metallic surfaces to be brought sufficiently close together that short range interatomic
attractive forces operate. The effect of surface contamination on solid phase welding is worse than implied
by viewing contaminants simply as physical barriers
to adhesion because surface metallic atoms become
satisfied by forming bonds with contaminant atoms,
thus reducing attractive forces. How detrimental contaminants are depends on the process and metals involved because these factors govern which mechanisms
are available for elimination of surface films. Deformation welding is generally impaired by organic
films, whereas inorganic films constitute the major
barrier to diffusion welding. The detrimental effect
of organic contaminants on deformation welding
generally increases with increasing chemical activity,
molecular size, and film thickness.

---

[*] This work was sponsored by the U.S. Dept. of Energy
(DOE), under Contract AT(29-1)-789.
[**] A U.S. DOE Facility.

INTRODUCTION

Although no unified theory of solid phase welding has been developed, the generally agreed upon requisite for producing a weld is that the metallic surfaces must be brought sufficiently close together that short range interatomic attractive forces operate.[1] Intimate mating of surfaces is generally easily achieved by both deformation welding and diffusion welding. Therefore, it can be concluded that barriers such as oxide films and/or other surface contaminants are responsible in most instances when extensive metallic bonding does not result. Milner and Rowe have reviewed the literature regarding deformation welding and have found that complete mating of faying surfaces almost always occurs at levels of deformation significantly below that required for deformation welding.[2] Some threshold deformations for cold welding of common metals are summarized in Table I. These data illustrate that very high deformations are required to initiate welding, particularly in the case of indentation welding. Yet approximately 10% deformation is all that is required to bring faying surfaces into intimate contact.[1] Tylecote hypothesized that fragmentation of surface oxides controls welding and determines the threshold deformation,[3] but later found that experimental data are inconsistent with this proposal.[4] It is interesting to note that gold exhibits a threshold deformation for indentation welding of 24 to 30%, even though it is free of surface oxide.

Table I. Threshold Deformations for Cold Welding

Metal	Threshold Deformation, %	
	Indentation Welding	Roll Bonding
Aluminum	40	25
Copper	73	45
Gold	24–30	--
Iron	81	--
Lead	10	10
Silver	80	--

ROLE OF CONTAMINANTS IN PREVENTING SOLID PHASE WELDS

The thermodynamic driving force for solid phase welding is the lowering of surface energy; two free surfaces are replaced by one solid-solid interface. In the case of metal welding, the weld interface is energetically similar to a large angle grain boundary (for high temperature processes this weld interface may be destroyed by processes such as recrystallization and grain growth). The surface free energy of such an interface is about one-third that of a free surface. Consequently, substitution of a weld interface for two free surfaces, results in a reduction of surface energy to about one-sixth the original level. Although the welding of two metallic surfaces to one another is thermodynamically favorable, opposing surfaces must be brought within the range of their mutual short-range interatomic attractive forces for the normal interatomic cohesive forces of metals to result in strong adhesion at this interface. This distance is no more than about 1 nm (10 A) for most metals.[1] We can now begin to understand the role of surface contaminants. If the thickness of a surface contaminant exceeds 1 nm, the attractive forces between surfaces are inadequate to affect adhesion. Actually, the effect of surface contaminaton on welding is worse than implied by viewing contaminants simply as physical barriers to adhesion. The interatomic attractive forces are due to the fact that surface atoms on a clean surface desire to share electrons with additional atoms (possess unsaturated bonds). When a surface becomes contaminated, surface metallic atoms become satisfied by forming bonds with contaminant atoms. Thus, attractive forces at the surface (and consequently surface energy) are reduced. The marked effect that a thin contaminant layer can have on metal adhesion is illustrated by the work of Buckley on the role of oxygen in the adhesion of iron[6] (see Figure 1). He found that adhesion of clean iron was drastically reduced by as little as

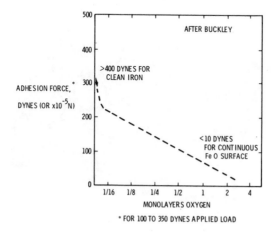

Figure 1. Effect of oxygen on the adhesion of iron.

one-half monolayer of oxygen. The arrangement of oxygen atoms
on iron for this condition is schematically shown in Figure 2.
Note that for as little as one-half monolayer of oxygen, each iron
atom has a near neighbor oxygen atom. Consequently, all the iron
atoms are partially satisfied with regard to electron sharing.

Clearly, unless removed or disrupted, most surface contaminants
will be sufficient to prevent metal-metal welding. The opposite
effect could also be envisioned to occur. Metal-contaminant bonds
might form so that contaminant inclusions could contribute to the
strength of a solid phase weld. Tylecote[1] and H. A. Mohamed and
J. Washburn[7] discount the contribution of metal-oxide bonding to
overall weld strength. However, Ludemann[8] extensively studied
metal-oxide welds and concluded that such welds can form under
diffusion welding conditions and contribute significantly to joint
strength. Ludemann summarized the requirements for a large metal-
oxide contribution to joint strength as: 1) both metals must bond
strongly to the oxide layer, 2) the oxide must be strong, and 3)
the oxide layer must be thin enough so that its brittle properties
do not influence joint strength. Good diffusion welds were obtained between aluminum and anodized beryllium, aluminum and
oxidized iron, copper and oxidized iron, and oxidized copper and
iron.[8] However, the strength of joints involving iron oxide were
limited to the strength of the oxide. Ludemann concludes that residua
oxide films in diffusion welds are not necessarily detrimental and,
in fact, in the case of some dissimilar metal welds, may be
beneficial by reducing intermetallic compound and/or void formation.
Metal oxide bonds are less likely to form under the conditions of
deformation welding than during high temperature diffusion welding.

Whereas oxide films and possibly other inorganic films might
contribute some strength to a solid phase weld, most organic
contaminants are either too weak or too weakly bonded to metals to

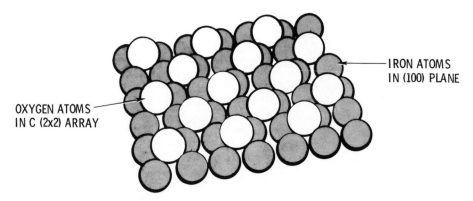

Figure 2. Schematic representation of one-half monolayer of
oxygen on iron.

make much contribution to joint strength. In summary, most
contaminants prevent metal-metal bonding and, because they are
intrinsically weak, comprise defects if not eliminated from the
weld interface.

CLASSIFICATION OF SURFACE FILMS

Inorganic Films

Contaminant surface films consist of two types: inorganic
films, principally oxides, and organic films. Fortunately, the
growth of oxide films on most metals occurs at a logarithmic
rate at room temperature; this results in limiting film thicknesses
of 2-10 nm. Small amounts of gases such as water, carbon dioxide,
and sulfur dioxide can significantly increase the limiting oxide
film thickness.[1] Similarly, oxide films formed by reaction with
chemical solutions or electrochemical reaction may be thicker or
of different structure. If sufficient oxygen is available, oxide
films formed at elevated temperatures are often much thicker than
those formed at room temperatures. Vacuum, inert, and reducing
atmospheres are often used during diffusion welding to reduce the
availability of oxygen.

Other inorganic films such as chlorides,[8] nitrides,[9]
carbides,[9] and phosphates can hinder welding. Nitrides form on
some metals (such as titanium and magnesium) by reaction with
air.[1] Chlorides and phosphates are known to form by reaction with
acids. Even degreasing solvents can react to form metal halides.[8]
Therefore, some treatments used to remove organic contaminants and
strip oxide films result in the formation of other inorganic films.

Although adsorbed films are generally thought of as organic
contaminants, inorganic gases can also be adsorbed. Water vapor
is a commonly adsorbed inorganic film. Other adsorbed inorganic
gases are carbon monoxide and carbon dioxide. Buckley observed
that carbon dioxide disassociated on iron surfaces to result in
iron oxide and carbon monoxide.[10]

Organic Contaminants

Organic contaminants can be further divided into two
categories: those that are adsorbed from the atmosphere and those
that are incurred by contact, e.g., oils, grease, and waxes.

Generally, organic atmospheric contaminants are relatively
small, volatile molecules. Also, the surface films resulting from
atmospheric contamination tend to be limited to monolayers because

once a monolayer of the organic contaminant is adsorbed, active
sites are no longer available. The resulting barrier films usually
do not exceed 4 nm in thickness and can be much thinner, as has
been determined by Auger electron spectroscopy.[11] Exceptions are
situations where a second species, perhaps even an inorganic like
water, attaches itself to the adsorbed monolayer of the first
species.

Organic contaminants resulting from direct contact often
result in surface barriers of considerable thickness. Obviously,
in the case of long chain polymers, the concept of a monolayer of
contaminant has no meaning. Thicknesses of nonatmospheric contaminants can easily exceed 5 μm. Aside from being thicker, nonatmospheric organic contaminants tend to differ from atmospheric
contaminants by consisting of larger, more strongly bonded molecules
and may exhibit cross-linked structures. It should be noted that
contaminants incurred by contact do not need to be thick to hinder
welding. An example of a thin contact contaminant is the common
fingerprint which typically includes sodium chloride, potassium
chloride, calcium chloride, lactic acid, stearic acid, sebacic
acid, and urea.

ROLE OF CONTAMINANT PROPERTIES

For both inorganic and organic contaminants, the effect of a
surface film as a barrier to welding depends on the properties of
the film as well as its thickness. Compared with metals themselves,
inorganic films tend to be more brittle, whereas organic contaminants are more mobile and tend to smear. However, variations
within each category exist and both types of contaminants are often
encountered together.

Inorganic Films

The influence of oxide films on deformation welding primarily
depends on how the oxide fragments and distributes itself along the
weld interface and the ability of the metal to extrude through breaks
in the oxide film. Although several workers,[3-5,12,13] have attempted
to elucidate the role of oxide films in deformation welding, a consistent correlation between weldability and the properties of
metal oxides has not been determined. Experiments in the field of
friction by Whitehead[14] and Wilson[15] indicated that the coefficient
of friction correlated with the relative hardness of the oxide to
the substrate metal. A hard oxide on a soft metal was found to
break more easily than soft oxides or hard oxides on hard metals.
Tylecote attempted to apply this concept to deformation welding,[3]
but experimental investigation by Tylecote, et. al., revealed
little correlation between the oxide hardness-to-metal hardness

ratio and weldability.[4] The correlation observed for friction
experiments probably breaks down for deformation welding because
most oxides fragment fairly easily at the high deformations
associated with welding. Surprisingly, even relatively thick oxide
films such as produced by anodization do not prevent welding
provided deformation is high enough to fragment and distribute the
oxide films. Vaidyanath and Milner[12] and Donelan[16] found that
anodized layers of 2 to 6 μm permitted aluminum welds to be made
which were nearly as good as those with thin oxides, if the parts
were first baked to remove adsorbed contaminants.

There is some evidence that thick, soft oxides such as can
form on copper are particularly detrimental to welding because they
tend to smear along the weld interface. Vaidyanath and Milner[12]
observed that, whereas two opposing thick oxide films on aluminum
generally fragment together during roll-bonding so as to provide a
high fraction of welded interface, copper oxide is sheared into
small fragments (less than the thickness of the oxide) so that the
total accumulative length of oxide fragments becomes more than
twice the original length of the specimen. Consequently, they
found that only 18% of the interface was made available for welding
at 60% deformation. Tylecote suggests that the poor weldability
of zinc is due to the softness of its oxide.[1]

Gilbreath[17] showed that the effect of gas adsorption on metal
adhesion correlated with the heat of adsorption for the particular
metal-gas reaction. Little loss of adhesion was experienced when
gases were physically adsorbed on metals (heats of adsorption less
than 5 kcal/mole) as for the adsorption of argon, hydrogen, and
nitrogen on aluminum or copper. Loss of adhesion in those metal-gas
systems having heats of adsorption above 20 kcal/mole (chemisorbed)
occurred at exposures of 10^{-4} to 10^{-2} Torr-sec. Gilbreath observed
that even in the case of gold, which does not form compounds with
most gases, chemisorbed species such as water, carbon dioxide,
carbon monoxide, and sulfur dioxide must be eliminated before
adhesion can be obtained at room temperature. Conversely, oxygen,
nitrogen, and noble gases which are only physisorbed on gold had
little effect on adhesion. Bowden and Throssell found that water
hindered gold welding far worse in the form of condensed water films
than when restricted to one or two chemisorbed monolayers.[18]

Organic Films

Much of the best work conducted to determine the effect of the
chemical and structural properties of organics on metal-metal welding
has involved adhesion and friction experiments, where the extent
of the plastic deformation is negligible compared to most industrial
deformation welding processes. While high plastic deformation is
known to reduce the effect of organic surface contaminants, the

qualitative observations made during adhesion experiments generally
would be expected to apply to deformation welding processes.

    Chemical activity is a dominant property of organic contaminants
with regard to their influence on adhesion. As indicated earlier,
Gilbreath observed that, for adsorbed contaminants, the effect of
environment on metal adhesion is related to the heat of adsorption.[17]
Organic gases are generally chemically adsorbed by metal surfaces
and consequently nearly always impair adhesion. The extent of this
impairment depends on the chemical and mechanical properties of
the adsorbed species. Buckley has extensively studied the effect
of adsorbed gases on metal adhesion, particularly with regard to
chemical activity.[6,10,19,20] The dependence of adhesion of iron on
the chemical structure of hydrocarbons as determined by Buckley is
summarized in Table II. The simple one and two carbon atom saturated
hydrocarbons, methane and ethane, had very modest effects on adhesion.
The more pronounced effect on iron adhesion due to the adsorption
of ethylene and acetylene is explained on the basis of the
progressive reduction in bond saturation for these hydrocarbons.[20]
Although the rigidity of the double and triple carbon atom bonds
(as in the case of ethylene and acetylene, respectively) as compared
with single bonds in saturated hydrocarbons probably contributed to
the difficulty in disrupting unsaturated hydrocarbons, the primary
effect is thought to be their greater chemical activity. An even
greater loss of adhesion was observed by the addition of chemically

Table II.   Effect of Various Hydrocarbons on Adhesion of
            Clean Iron, (011) Planes

Hydrocarbon	Projected Molecular Formula	Adhesive Force,* Newtons x 10^4
Methane	CH_4	40
Ethane	H_3C-CH_3	28
Ethylene	$H_2C \quad CH_2$	17
Acetylene	$HC \quad CH$	8
Vinyl Chloride	$H_2C \quad CHCl$	3
Ethylene Oxide	$H_2C \overset{O}{\quad} CH_2$	1

*Contact Load = 2×10^{-4} Newtons

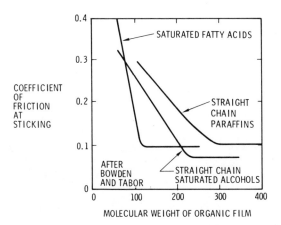

Figure 3.   Coefficient of friction (on steel) as a function of
molecular weight.

active atoms such as chlorine (halogenation) or oxygen to a
hydrocarbon, as in the cases of vinyl chloride and ethylene oxide,
respectively.[19] Buckley also conducted studies on the influence
of adsorbed saturated hydrocarbons on the coefficient of friction
of iron.[10] Ethane (2 carbon atoms) reduced the coefficient of
friction by more than a factor of two. As the number of carbon
atoms in the hydrocarbon chain was increased up to eight, the
coefficient of friction tended to decrease slightly.

Studies on boundary friction by Bowden and Tabor disclosed
that friction on steel is inversely dependent on the molecular
chain length of lubricants, which consisted of homologous series
of paraffins, alcohols, and fatty acids[21] (see Figures 3 and 4).
Also borrowing from the field of lubrication the effect of an

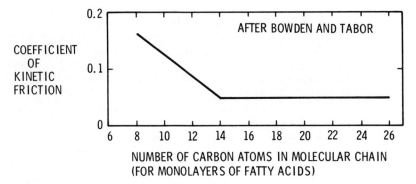

Figure 4.   Effect of molecular chain length on kinetic friction
(stainless steel sphere on glass surface).

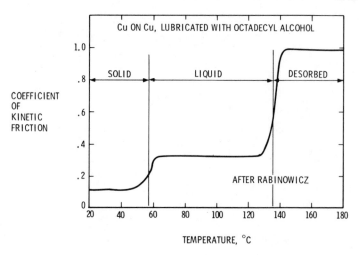

Figure 5. Effect of state of organic film on sliding friction.

organic contaminant on welding would be expected to be dependent
on the state in which contaminant exists on the surface. For
example, Rabinowicz found that the coefficient of friction for
copper lubricated with octadecyl alcohol underwent transitions
at the melting point and desorption temperature of the
lubricant[22] (see Figure 5). Similarly, a friction transition
temperature was observed by Bowden and Tabor for a steel surface
lubricated with fatty acids.[21] However, in this case the transition
temperature occurred at a higher temperature than the melting point
of the fatty acid (see Figure 6). Bowden and Tabor have shown that
the friction transition temperature for fatty acids occurs at
approximately the melting point of the metallic soap that would
be expected to form by reaction of the metal and the fatty acid.
The high chemical activity of the fatty acids accounts in part
for the greater dependence on molecular chain length than for the
other homologous series. Bowden and Tabor indicate that the strong
lateral attraction between molecular chains and high melting points
also contribute to the excellent lubricating properties of metallic
soaps.[21] Recall that fingerprints include fatty acids capable of
forming metallic soaps. These fatty acids account for the effectiveness of fingerprints in hindering deformation welding.

On the basis of adhesion tests between indium and platinum,
Cuthrell also observed that the coefficient of adhesion is inversely
proportional to the length of the hydrocarbons chain.[23] After
conducting adhesion tests on a large number of organic contaminants
as diverse as alcohols, halogenated hydrocarbons, and salts of
organic acids, Cuthrell further concluded that adhesion is very
sensitive to the contaminant layer thickness, as is graphically
illustrated in Figure 7.

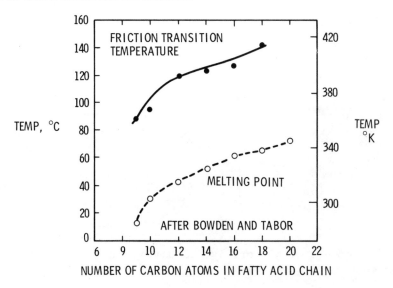

Figure 6.   Friction transition temperature of fatty acids on steel surfaces.

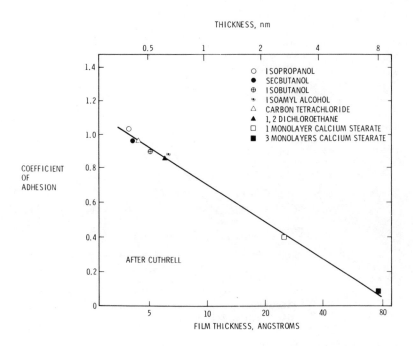

Figure 7.   Effect of organic film thickness on adhesion.

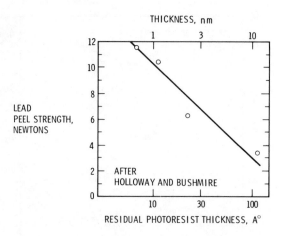

Figure 8. Effect of photoresist contamination on the thermocompression bonding of gold.

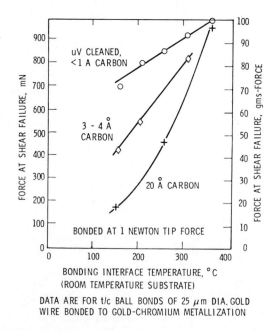

Figure 9. Effect of atmospheric (lab air) contamination on the
thermocompression bonding of gold.

A strong dependence on the deformation welding of gold on
contaminant layer thickness was shown in studies by Holloway and
Bushmire[24] and Jellison.[11] The effect of photoresist contamination
as observed by Holloway and Bushmire is shown in Figure 8. In
Figure 9, from Jellison's study, the dependence of weld strength on
welding temperature is seen to increase with increasing contaminant
layer thickness.

The effect of the chemical activity of contaminants on the
high deformation cold welding of aluminum, copper, nickel and
chromium was studied by Baranov[25] and Semenov.[26] Polar contaminants
(e.g., water, ethanol, oleic acid) were found to impede welding
much more than nonpolar materials (e.g., heptane and petroleum jelly).
Baranov also found that polar contaminants adhere more tightly to
chromium and copper than to nickel.

The effect of mechanical properties of organic films on solid
phase welding has not been systematically studied, and, in individual
experiments is difficult to separate from chemical effects.
Experiments involving thermocompression gold ball bonding to thin
film gold have shown that a strongly bonded photoresist impedes
solid phase welding more than comparable or even greater thicknesses
of contaminants adsorbed from the atmosphere.[27] Wetzel indicates
that strongly crosslinked organic films exhibit a threshold effect
on welding.[28] These hard layers prevent welding at lighter loads
but at higher loads fragment somewhat like inorganic film layers
so that welding becomes satisfactory. Wetzel's observation implies
that at high deformations highly crosslinked organics might actually
be less detrimental than weaker more mobile organics.

Whether synergistic effects involving oxide films in organic
contaminants exist has not been determined. Excluding charge
effects, organic materials are more weakly adsorbed on oxides than
on metals. This weak adhesion can facilitate removal of contaminants.
However, some investigators[8,12] suspect that an organic that is
more weakly adsorbed on the metal oxide than the metal itself,
might be particularly troublesome because it would be expected to
have high mobility on the oxide until it became readsorbed on the
higher energy nascent metal surface created by deformation.
Vaidyanath and Milner[12] suggest that high threshold deformations
are required for welding because adsorbed contaminants quickly
transfer to exposed nascent metal at lower deformation. Consequently, the small areas of freshly contaminated metal become
nearly as difficult to weld as those remaining covered with oxide.

Particulate Contaminants

Normally, contaminants are in the form of surface films.
However, particulate contaminants also can prevent welding. A

particle on a surface can prevent intimate contact over an area
much larger than its own cross section. The effect of particulate
contaminants depends on their hardness, the hardness of the metallic
surface, amount of deformation, and the number of particles.
Tylecote, et. al.,[4] studied the effect of alumina powder on the
deformation welding of tin. They found that alumina was far less
detrimental to welding than grease and that the effect of alumina
powder was small at high deformation where the particles became
imbedded in the metals (see Figure 10). Similar observations were
made by Williamson, et. al., in their study of the effect of
carborundum powders and polyethylene powders on the welding of gold.
They observed that the soft polyethylene particles were compressed
so as to permit greater metal contact than for carborundum powder.

MECHANISMS FOR ELIMINATION OF SURFACE BARRIERS DURING WELDING

As discussed earlier, welding is attained when surfaces are
brought into intimate contact with no barriers (non-metals) between
faying surfaces. Theoretically and in the laboratory these barriers
can be removed prior to welding. Although precleaning of surfaces
is common in industrial welding processes, surfaces are seldom
clean in the atomic sense and oxide films are generally unavoidable.
Therefore, contaminant barriers must be eliminated during the welding
process itself.

How detrimental contaminants are depends on the process and
metals involved, because these factors govern which mechanisms are

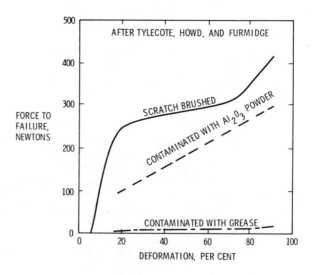

Figure 10. Effect of contaminants on the deformation welding of
tin.

available for elimination of surface films. Deformation welding
is generally impaired by organic films, whereas inorganic films
constitute the major barrier to diffusion bonding. Depending on
the amount of surface extension, the level of surface shear stresses,
the temperature, and the ability of the metal to assimilate surface
films by diffusion precleaning may only involve wiping the surface
with a dry cloth[25] or, at the other extreme, require procedures
capable of reducing films to less than a monolayer in thickness.[11]
At very low levels of plastic deformation, as in the case of
adhesion tests, a monolayer of oxygen may be sufficient to prevent
welding.[6] However, the limiting oxide film thicknesses on most
metals (typically 2 to 10 nm) generally do not prevent welding at
the high levels of deformation employed in most deformation
welding processes. Therefore, in considering the mechanisms which
overcome surface barriers to welding, we are mostly concerned with
the effect of organic contaminants on deformation welding processes
and oxide films on diffusion welding processes. Generally, surface
films are thought to be disrupted mechanically in the case of
deformation welding, whereas, for diffusion welding, thermal
mechanisms disperse the oxide film. These distinctions are not
very precise, since deformation welding often is performed at
elevated temperatures and, conversely, significant deformation
often accompanies diffusion welding.

Thermal Mechanisms

    Mechanisms which change the properties, reduce the amount,
or eliminate organic contaminants can be very effective in improving
both deformation welding and diffusion welding. As discussed
earlier, melting of organic compounds significantly increases
friction between metal surfaces and, therefore, would be expected
to aid deformation welding. Evaporation and dissociation of organic
contaminants can markedly improve weldability. Investigators of
adhesion[9,17,21,30] have often emphasized the inability to obtain
good adhesion on surfaces exposed to high temperature or high
vacuum. These observations hold for conditions of low deformation
because a monolayer of chemisorbed contaminant is usually sufficient
to prevent adhesion. However, the significant reduction in the
thickness of contaminant layers brought about by dissociation
and/or evaporation greatly aids welding processes employing high
deformation. This point is illustrated by the weldability of
gold. Under conditions of low deformation, gold welds can be very
difficult to obtain even for surfaces which have been heated to
greater than 1000°K in vacuum.[1,31-33] Tylecote showed that a
1173°K-one minute heat treatment was required in order to cold
weld gold at room temperature with 30% deformation.[33] Conversely,
Jellison found that atmospherically contaminated gold surfaces
could be restored to excellent weldability with regard to
thermocompression bonding by heating to 508°K in air.[11] Auger

electron spectroscopy revealed that a residual carbon layer of
approximately 2 Å remain (probably correspond to approximately
one monolayer of hydrocarbons,) which would be sufficient to
inhibit welding under low deformation conditions, but not for the
high deformations encountered in thermocompression ball bonding
(typically 70%). Where complete desorption on heating is possible,
as is probably the case for some organics adsorbed on metal oxide,
even greater improvements in weldability would be expected.
Vaidyanath and Milner found that the ability to roll bond aluminum
was much improved by a prebake in vacuum at 773°K followed by
cooling in a desiccator.[12] Since surface analyses were not
conducted, whether organic contaminants were desorbed or merely
reduced to tolerable levels is not known.

The principle mechanism for destruction of oxide films during
diffusion welding is assimilation of the films into the bulk metal
by diffusion. Some metals such as copper, iron, titanium, zirconium,
niobium, and tantalum dissolve substantial quantities of oxygen and,
consequently, are easy to diffusion weld. Bryant has reviewed
available data on hot isostatic welding (gas pressure bonding) and
has shown that the weldability can be predicted on the basis of
oxygen solubility, free energy of formation of the most stable
oxide, and the mechanical properties.[34] Bryant's graphical summary
of hot isostatic welding data is given in Figure 11. He has refined

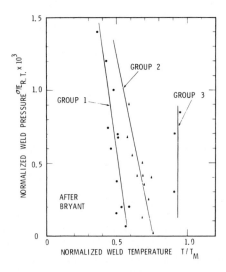

Figure 11. Minimum pressure-temperature conditions for hot
isostatic pressure welding.

the common rule-of-thumb that diffusion welding should be performed at a homologous temperature of one half or higher. For metals such as copper, silver and molybdenum, whose oxides are only modestly stable (free energy of formation - 100 kcal/mole or greater) or metals with high solubility for oxygen (greater than 1 atom %) an homologous temperature of approximately one half is adequate. At the other extreme, aluminum, which has a very stable oxide and negligible solubility for oxygen, requires a homologous temperature greater than .9 for diffusion welding.

As has been shown by Kinzel[35] and Fine, et. al.,[36] and Ham,[37] the time required for dissolution of the oxide can be readily calculated, if the diffusion constants are available. The perhaps surprising result of such a calculation is that, if the temperature is increased significantly above a homologous temperature of one-half for metals such as iron, copper, and titanium, the required diffusion time becomes only a few minutes. This results, of course, from the fact that diffusion rates are highly temperature dependent. The practical result is that much or all of the oxide is actually eliminated by diffusion during processes that might normally be thought of as hot deformation welding.

Another way in which oxide films are eliminated is by reaction with carbon from the substrate metal. Ham has calculated reaction times for reduction of molybdenum and iron oxides.[27] In one minute, 1 μm, 7 μm, and 100 μm of oxide can be removed from molybdenum at 673°K, alpha iron at 773°K, and gamma iron at 1183°K, respectively.

Finally, the possibility of the reduction of the oxide by a reducing atmosphere should not be forgotten. For metals with moderately stable oxides such as iron and nickel, addition of a reducing atmosphere to the process is entirely feasible, and often more effective than a vacuum or inert atmosphere. A pitfall of this approach is that an atmosphere which is reducing at welding temperature may actually be oxidizing to the same metal during heating. Therefore, the heating rate and time at welding temperature must accommodate the kinetics of the oxidation and reduction processes.

Jellison,[11,27] English and Hokanson,[38] and Hayasaka and Hattori[39] have observed a time-related phenomenon in the welding of gold which cannot be explained on the basis of assimilation of service films by diffusion because the solubilities of oxygen, carbon and hydrogen in gold are negligibly small. These workers each observed that gold-gold interfaces which were only partially welded by mechanical disruption of surface films would continue to grow in strength as a function of time at modest elevated temperatures (423-648°K), whether or not an external load was applied. In some cases weld strengths increased by nearly an order of magnitude following initial deformation (see Figures 12 and 13). As proposed earlier

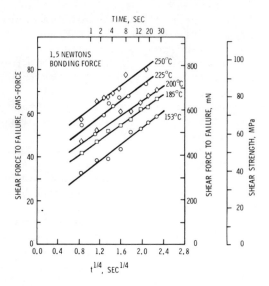

Figure 12.   Time-temperature dependence of the strength of gold ball bonds made to photoresist contaminated gold.

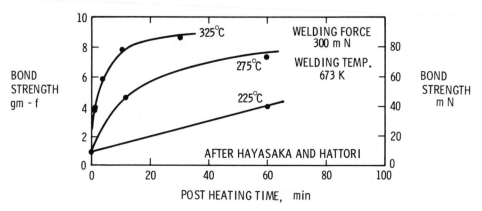

Figure 13.   Increase in gold beam lead bond strength on post heating.

by Jellison[11] the growth of the metal-metal bond interface during the second stage of welding or during post heating appears analogous to sintering phenomena. The metal-metal interfaces grow at the expense of the higher energy metal contaminant interfaces. In contrast to most sintering processes, however, the growth of gold-gold bonds appears to be far more stress dependent than temperature dependent. The rate of bond growth with an external load applied was found to be several times that without external loads. Also, within experimental accuracy, the rate of bond growth

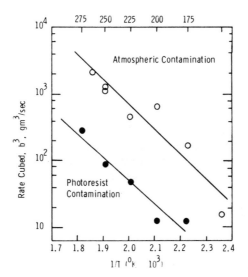

Figure 14. Arrhenius relations for the increase in strength of thermocompression gold ball bonds during post heating.

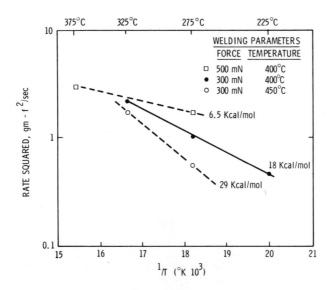

Figure 15. Arrhenius relations for the increase in strength of beam lead bonds during post heating.

under an external load does not appear to be temperature dependent. Conversely, the growth process was found to be thermally activated in the absence of an external load (see Figures 14 and 15). In plotting Figure 15, a parabolic rate law is assumed which seems more consistent with Figure 13 than the linear rate law assumed by Hayasaka and Hattori. The increased rate dependence on temperature in the absence of external load is consistent with a stress assisted thermally activated process such as low temperature creep of gold[40] and other fcc metals[41] where the activation energy decreases with increasing stress.

In summary, a sintering phenomenon appears responsible for the growth of welds following initial deformation. Whether this phenomenon is due to stress assisted thermally activated plastic flow as proposed by Jellison, bulk diffusion as assumed by Hayasaka and Hattori, or surface diffusion as suggested by Tylecote[32] requires further experimental and theoretical clarification. In any case, the process appears to be a potentially important way of improving weld strength.

Mechanical Mechanisms

On the basis of both theoretical considerations and experimental results, Ahmed and Svitak[42] recently summarized the fundamental parameters that influence bond formation in deformation welding as follows:

1) a centrally located nonplastic zone may exist where welding is difficult, 2) the magnitude of the normal stress, above that required for intimate contact, is unimportant, 3) mutual extension of two nominally clean surfaces in intimate contact beyond some threshold deformation is a sufficient condition for welding, 4) the required surface extension is greatly reduced by simultaneous application of an interfacial shear stress, 5) neither interfacial sliding nor high interfacial shear stresses result in welding unless surface extension occurs.

Ahmed and Svitak's experimental results were for gold thermo-compression bonding where a surface oxide is not encountered. Oxidized surfaces are generally expected to require greater surface extensions.

Vaidyanath, et. al.,[5] and Mohamed and Washburn[7] have shown that joint strength can be related to the amount of fresh metal surface exposed by the extension of surfaces. Both groups of workers predict the exposed metal (and consequently weld strength) on the basis of fragmentation of oxide films. Vaidyanath, et. al., treated the case of roll-bonding of anodized aluminum where opposing oxide layers tend to adhere and, therefore, break up as one, whereas Mohamed and Washburn calculated the strength of aluminum lapwelds on the basis of thin oxide layers fragmenting independently.

As indicated earlier, Ahmed and Vaidyanath[42] observed that simultaneous application of interfacial shear stresses with normal stresses reduces the required surface extension. Stapleton evaluated the role of interfacial shear in the deformation welding of gold and found that better welds could be obtained at 7% surface extension in the presence of high interfacial shear stresses than at 25% surface extension in the absence of interfacial shear stresses.[43] Both Anderson[44] and Ahmed and Svitak[42] believe that interfacial shear results in higher localized deformation leading to mixing of contaminants.

Ahmed and Svitak[42] emphasize the role of geometry in controlling the stress state at a weld interface. Geometrical and frictional restraint give rise to two conditions that influence solid phase welding: 1) a nonuniform distribution of normal stresses (pressure hill) and, 2) a nonplastic zone where, although isostatic compressive stresses may be high, the difference in principle stresses is insufficient to cause plastic flow. The pressure hill effect may result in insufficient mating of surfaces, particularly near specimen edges. Conversely, negligible surface extension occurs within the nonplastic zone with the consequence that contaminants are not mechanically disrupted during deformation welding. The nonplastic zone problem is particularly bad in the case of a thin film metallization attached to rigid substrates.[11,27,38]

Another factor which has been considered in deformation welding is interfacial sliding or relative movement between surfaces.[1,2,42,45-48] The general observation was that much stronger welds could be made for the same degree of bulk deformation. The effect of interfacial sliding and welding is thought to be analogous to the plowing effect observed in friction experiments, where contaminants are pushed aside.[21] Tylecote has commented that interfacial sliding would be expected to be more important for oxidized surfaces than for oxide free surfaces.[32]

SURFACE PREPARATION

Several good reviews have been written on surface preparation for solid phase welding.[1,2,8] The consensus of these reviews is that some form of surface preparation is generally required prior to welding. The amount and type of surface preparation required depends on the welding process to be employed. Diffusion welding often requires flatter surfaces than deformation welding but removal of contaminants may not be as important because contaminants can be assimilated by diffusion or reaction. The degree to which organic contaminants must be removed prior to deformation welding depends on the amount of surface extension, interfacial shear, and interfacial sliding.

The surface preparation methods that have been employed can
generally be classified as solvent cleaning, chemical pickling,
mechanical and atmospheric techniques. Solvent cleaning based on
degreasing solvents such as trichloroethylene or detergents may be
insufficient in some cases. Solvent cleaning normally leaves some
residual contaminants but for processes employing high deformation
or high temperatures, the residual level of contamination may be
tolerable. In addition to absorption of the solvent, the solvent
may react with the metal to form a compound, as in the case of
chlorinated solvents with copper.

Chemical pickling used by itself is not very effective as a
surface preparation with one possible exception; that of preparation
for diffusion welding. Tylecote[1] indicates that it is apt to
produce hydrated oxides which are even more detrimental to welding
than the thin oxides formed following machining. However, chemical
pickling may be necessary to remove heavy oxide scales from iron and
copper base alloys. Chemical pickling followed by heating[12,25]
(such as vacuum baking) affords better weldability.

Milner and Rowe's[1,2] review indicates that the most favored
surface preparation is degreasing followed by scratch-brushing
(rotary wire brushing). Vaidyanath and Milner[12] have shown that
this process is far more effective than degreasing following
scratch brushing (see Figure 16). Presumably, scratch brushing,
in addition to mechanically disrupting surface films, generates
sufficient heat to desorb organic contaminants. Following solvent
cleaning by dry machining to expose virgin metal also provides
weldable surfaces but not as effectively as scratch brushing.
(See Figure 16).

Aside from the ability of machining and scratch brushing to
remove contaminants, the effect of the surface roughness they
create must be considered. In the diffusion welding of high strength

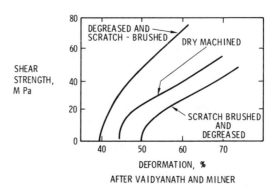

Figure 16. Effect of surface preparation on deformation welding.

alloys, excessive surface roughness can prove detrimental by preventing complete mating. Generally, however, some roughness is desirable. If significant interfacial slip occurs, rough surfaces aid in the plowing of contaminants. Also, since the magnitude of shear stresses are orientation dependent, a rough surface assures that part of the surface is oriented so that significant inter-facial shear stresses exist.

Finally, atmospheric cleaning approaches are available. Most common are vacuum baking or hydrogen firing. More recently, pro-cesses such as uV-ozone cleaning and rf plasma cleaning have been introduced. These processes have the potential of reducing oxygen contaminants to nondetectable levels[11,24,27] but must be used with care because they can also promote oxide film growth on base metals. Consequently, they have proved most suitable on gold and gold-platinum surfaces.

CONCLUDING REMARKS

The amount of contamination control required for reliable solid phase welding depends on the processes and metals involved. For most diffusion welding processes, solvent cleaning is adequate. Solvent cleaning followed by vacuum baking or rotary wire brushing to promote desorption of organic contaminants have proven highly effective for deformation welding processes. Processes such as friction welding, ultrasonic welding and multiple upset butt welding that result in very high deformation at the weld interface can tolerate more contaminants and, thus, require less stringent surface preparations. Conversely, many small parts used in the electronic industry, for which deformation is restricted by geometry, may require processes such as uV-ozone cleaning and rf plasma cleaning to obtain adequately clean parts.

The ability to eliminate contaminant layers either by pre-cleaning or during welding also is dependent on the metal involved. Some metals such as aluminum and gold possess little ability to assimilate contaminants by diffusion, whereas others such as iron or zirconium readily dissolve contaminants. In deformation welding, the ductility of the metal is extremely important, for ductility determines the ability to fragment oxide films, extrude the metal through gaps in the oxide, and deform the metal sufficiently to overcome other contaminants. In conclusion, successful solid phase welding must consider the properties (both chemical and mechanical) of the material(s) being joined in order to select the appropriate process. Then the contamination control must consider both the selected process and the metals involved.

REFERENCES

.1. R. F. Tylecote, "Solid Phase Welding of Metals," St. Martin's Press, New York, 1968.

2. D. R. Milner and G. W. Rowe, Metal. Rev., 7, 433 (1962).

3. R. F. Tylecote, Brit. Welding J., 1, 117 (1954).

4. R. F. Tylecote, D. Howd and J. E. Furmidge, Brit. Welding J., 5, 21 (1958).

5. L. R. Vaidyanath, M. G. Nicholas, and D. R. Milner, Brit. Welding J., 6, 13 (1959).

6. D. H. Buckley, NASA TND-5756, Lewis Research Cent., NASA Cleveland, OH, 1970.

7. H. A. Mohamed and J. Washburn, Welding J., 54, 302 (1975).

8. W. D. Ludemann, Lawrence Radiation Laboratory Report UCRL-50744, 1969, Livermore, CA.

9. W. P. Gilbreath, in "Adhesion or Cold Welding of Materials in Space Environments," STP No. 431, pp. 128-148, American Society for Testing and Materials, Philadelphia, 1967.

10. D. H. Buckley, NASA TND 4775, Lewis Research Cent., NASA Cleveland, OH, 1968.

11. J. L. Jellison, IEEE Trans. Parts, Hybrids & Packaging, 11, PHP-11, 206 (1975).

12. L. R. Vaidyanath and D. R. Milner, Brit. Welding J., 7, 1 (1960).

13. S. B. Ainbinder and F. F. Klovoka, Latvijas psr zinathu Akad Vestis, 87, 113 (1954).

14. J. R. Whitehead, Proc. Royal Soc., 20A, 109 (1950).

15. R. Wilson, Proc. Royal Soc., 213, 450 (1952).

16. J. A. Donelan, Brit. Welding J., 6, 5 (1959).

17. W. P. Gilbreath, NASA Report, NASA TND 4868, 1968.

18. F. P. Bowden and W. R. Throssell, Proc. Royal Soc., A209, 297 (1951).

19. D. H. Buckley, NASA TND 582, Lewis Research Cent., NASA, Cleveland, OH, 1970.

20. D. H. Buckley, NASA TND 5999, Lewis Research Cent., NASA, Cleveland, OH, 1970.

21. F. P. Bowden and D. Tabor, "The Friction and Lubrication of Solids," Clarendon Press, London, 1954.

22. E. Rabinowicz, "Friction and Wear of Materials," John Wiley and Sons, New York, 1965.

23. R. E. Cuthrell, "Quantitative Detection of Molecular Layers with the Indium Adhesion Tester," Sandia Labs Report SC-DR-66-300, 1966.

24. P. H. Holloway and D. W. Bushmire, in "Proceedings of the Twelfth Reliability Physics Symposium," pp. 180-186, Las Vegas, NV, 1974.

25. I. B. Baranov, "Cold Welding of Plastic Metals," OTS: 63-21785, JPRS: 19103, 1963, English Translation, Available from Office of Technical Services.

26. A. P. Semenov, Wear, 4, 1 (1961).

27. J. L. Jellison, IEEE Trans. Parts, Hybrids & Packaging, PHP 13, 132 (1977).
28. F. H. Wetzel, Discussion on Paper by P. F. Bowden, in "Adhesion and Cohesion," P. R. Weiss, Editor, p. 143, Elsevier Press, Amsterdam, 1962.
29. J. P. Williamson, J. A. Greenwood and J. Harris, Proc. Royal Soc., A237, 560 (1956).
30. K. I. Johnson and D. B. Keller, Jr., J. Appl. Phys., 38, 1896 (1967).
31. J. L. Jellison, Sandia Laboratories Report, SAND75-0053, 1975.
32. R. F. Tylecote, Gold. Bull., 11, 74 (1978).
33. R. F. Tylecote, Brit. Welding J., 1, 117 (1954).
34. W. A. Bryant, Welding J., 54, 433s (1975).
35. A. B. Kinzel, Welding J., 23, 1124 (1944).
36. L. Fine, C. H. Maak, A. R. Ozanich, Welding J., 25, 517 (1946).
37. J. L. Ham, Aerospace Engineering, 20-21, 49 (1961).
38. A. T. English and F. L. Hokanson, "Proceedings of the Ninth Annual Reliability Physics Symposium," pp. 178-186, Las Vegas, NV, 1971.
39. T. Hayasaka and S. Hattori, Review of Electronic Communication Laboratory, 23, 344 (1975).
40. R. R. Herrins and M. Meshii, Metal. Trans., 4, 2109 (1973).
41. K. A. Osipov, "Activation Processes in Solid Metals and Alloys," American Elsevier Publishing Company, Inc., New York, 1964.
42. N. Ahmed and J. J. Svitak, in "Proceedings of the 25th Electronic Components Conference, pp. 52-63, Washington, DC, 1975.
43. R. P. Stapleton, "Surface Elongation and Interfacial Sliding During Gold-Gold Thermocompression Bonding," Master's Thesis, Department of Metallurgy, Lehigh University, 1974.
44. O. L. Anderson, Wear, 3, 253 (1960).
45. F. P. Bowden and G. W. Rowe, Proc. Royal Soc., 233A, 429 (1956).
46. E. Holmes, Brit. Welding J., 6 29 (1959).
47. V. W. Cooke and A. Levy, J. Metals, 28 (Nov., 1949).
48. G. L. Knowles, Lawrence Radiation Laboratories Report, UCRL-50766, Livermore, CA, 1970.

CONTAMINATION AND RELIABILITY CONCERNS IN MICROELECTRONICS

A. V. Ferris-Prabhu

IBM General Technology Division

Essex Junction, VT 05452

This paper reviews the trend to greater density in
microelectronics and the correspondingly subtler
causes of reliability concerns, particularly those
due to contaminants. Methods for detecting contami-
ants are mentioned, reliability modelling is discussed, and several types of contaminants of actual
and potential reliability concern are described.

INTRODUCTION

The objective of this paper is to present a broad overview
of contamination and reliability concerns in microelectronics.
Reliability concerns arise when any initially functional unit
fails to function as designed. Reliability is to be differentiated from yield, which is a measure of the fraction of manufactured units which are functional from the time of manufacture.
Due to the economic cost, and more frequently the physical im-
possibility of testing every single unit, particularly in the
microelectronics industry, some non-functional units may escape
into use. The shipped product quality level is a measure of the
fraction of these escapees, which, when exercised at some later
time in the operational life of the unit, fail to function as
designed and thus mimic a reliability concern. It is therefore
customary to define reliability in terms of the time that an
initially functional unit can be expected to function within
design limits. Any time-dependent phenomenon which degrades the
effective functioning of a unit is thus a legitimate reliability
concern.

The reliability measure in terms of the expected time to
failure immediately introduces the concept of failure rate.
Noting that increased reliability means a decrease in the failure
rate, Figure 1 shows the increase in reliability of individual
storage cells or bits, over the last decade. There has also been
a rapid increase in the storage capacity of installed systems,
with the result that the increase in reliability at the total
system level, although considerable, has not been as great as
might be expected from the increase in reliability at the individual bit level. If the capacity of installed systems continues
to increase at the same rate, significant improvements are
essential in the reliability of individual units so as to ensure
acceptable reliability at the system level.

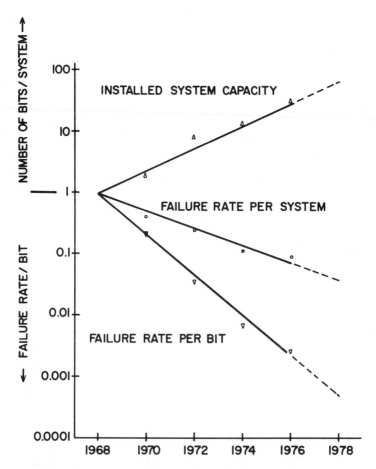

Figure 1. The failure rate per bit, the number of bits per
system and the failure rate per system over the decade 1968-
1978. All values have been normalized to unity in 1968.

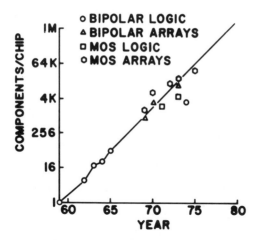

Figure 2. The number of components per chip from 1960 on.
If the linear increase continues, integrated circuits with
10^9 elements will be available in 20 years. (Courtesy of
R. N. Noyce, INTEL Corp.)

   The increased capacity of installed systems has been due to
the ability to make even smaller storage and logic devices.
Figure 2 shows the increase that has been achieved in the number
of components that can be packed onto a chip[1]. This trend becomes more meaningful when we consider some of the differences, as
shown in Table I, between the ENIAC, the first electronic digital
computer, and the Fairchild F8 minicomputer[2]. The reliability
has improved by a factor of 10^4 and the density has increased by
a factor of 3×10^5. The increased densities have been achieved

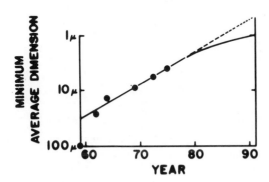

Figure 3. Projected minimum average dimension in advanced
integrated circuits. Minimum dimensions are projected to
decrease at a slower rate as optical lithographic limits
are approached. (Courtesy of R. N. Noyce, INTEL Corp.)

Table I. Comparison of Some Parameters of ENIAC with the Fairchild 8 (F8) microprocessor. (Adapted from, and courtesy of J. C. Linvill, Stanford University and C. L. Hogan, Fairchild Camera and Instrument Corp.)

ITEM	PARAMETER	ENIAC	F8*	COMMENTS
I	SIZE	3,000 CUBIC	0.011 CUBIC FEET	300,000 TIMES SMALLER
2	POWER CONSUMPTION	140 KILOWATTS	2.5 WATTS	56,000 TIMES LESS POWER
3	WEIGHT	30 TONS	<1 POUND	60,000 TIMES LIGHTER
4	CLOCK RATE	100 KILOHERTZ	2 MEGAHERTZ	20 TIMES FASTER CLOCK RATE
5	TRANSISTORS OR TUBE	18,000 TUBES	20,000 TRANSISTORS	ABOUT THE SAME
6	RESISTORS	70,000	NONE	FB USES ACTIVE DEVICES AS RESISTORS
7	CAPACITORS	10,000	2	5000 TIMES LESS
8	RELAYS AND SWITCHES	7,500	NONE	
99	ADD TIME	200 μSEC (12 DIGITS)	150 μSEC (8 DIGITS)	ABOUT THE SAME
IO	MEAN TIME TO FAILURE	HOURS	YEARS	MORE THAN 10,000 TIMES AS RELIABLE

by continued improvement in the techniques of fabricating minimum dimensions or linewidths. Figure 3 shows the trend of these minimum dimensions[1]. The limitations that appear to be over the horizon are due more to technological constraints, than to fundamental constraints imposed by the entropy or uncertainty principles. The limitations on device dimensions and the causes of these limitations, as shown[3] in Figure 4, support Richard Feynman's remark[4] that there is plenty of room at the bottom.

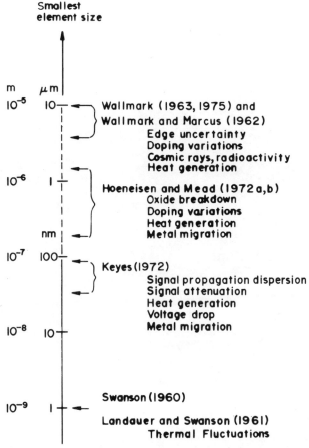

Figure 4. Smallest integrated device size derived by various
workers. The most stringent considerations are mentioned for
each author. (Courtesy of J. T. Wallmark.)

As devices become smaller, their effective functioning can
be affected by progressively subtler causes. An appreciation of
the size of microelectronic devices is obtained when it is
realized that a cerebral neuron is about 150 μm long[5] and
average device dimensions are less than 10 μm today[1]. In
Figure 5, the relative size of an amoeba relative to that of the
cells in a memory chip[6] illustrates the point well.

It is clear from these considerations, that the detection and
measurement of contaminant concentration are important concerns
with respect to evaluating the reliability of microelectronic
devices.

Figure 5. Photomicrograph of an 8K memory chip with an amoeba, the smallest single cell organism, superimposed. The amoeba covers about 40 memory cells, each about 4.65 μm^2. (Courtesy of R. Donlan, IBM General Technology Division.)

Methods for detecting and measuring the concentration of contaminants involve exposing the target to suitable radiation or particle bombardment, and examining the characteristics of the transmitted, emitted, or reflected radiation or particles, as illustrated in Figure 6. Details of these methods and of their applications are available elsewhere[8], and so will not be discussed here. A particularly informative illustration of some of these methods has been summarized by Wehner[9] as illustrated in Table II.

The material that follows discusses Reliability Modelling, followed by a section on Types of Contaminants, a categorization of types of contaminants and a description of several of them.

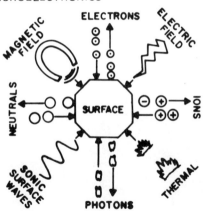

Figure 6. Pictorial representation of surface analysis techniques. Various combinations of probes in and particles out
determine the various surface analysis techniques. (Courtesy
of D. Lichtman, University of Wisconsin-Milwaukee.)

RELIABILITY MODELLING

Reliability concerns may be considered to be of basically
two kinds: intrinsic and extrinsic. By intrinsic, I mean defects
in the very structure itself due to irregularities in the fabrication process or to statistical fluctuations. By extrinsic, I
mean everything else, and this includes all forms of contamination.
The demarcation between the two is not sharp. For example, silicon
surfaces are oxidized in order to terminate the silicon lattice
in a more ordered way. However, the alterations of interfacial
structure at the silicon surface are still large enough to produce
interface state densities of the order of $10^{10}/cm^2.-eV$. These
interface states not only hinder performance of certain devices
by acting as a shielding layer between an applied signal and the
surface space charge region, but they also can act as traps for
surface contaminants, as a consequence of which an intrinsic
reliability concern can become an extrinsic concern. Despite the
inability always to make a sharp distinction, this paper will be
confined to contamination concerns and in particular to those
which affect the reliability by causing the performance to degrade
as time progresses.

It is therefore imperative to model each degradation
phenomenon so as to be able to predict, with at least reasonable
confidence, the expected reliable lifetime of a component. A
component functions reliably as long as each of its response
parameters, such as junction capacitance, current gain, resistance, etc., has values within design limits.   Each response

Table   II.   Survey of Compositional Analysis Methods based on
              physical phenomena. Methods for which commercial
              apparatus is on the market are underlined. (Courtesy
              of G. K. Wehner, University of Minnesota.)

	EXCITATION→			
	hυ	e	Ions	E
hυ	X-RAY FLOURESCENCE	ELECTRON MICROPROBE X-RAY APPEARANCE OR DISAPPEARANCE SPECTROSCOPY (APS, DAPS)	SURFACE COMPARISON ANALYSIS OF NEUTRAL AND ION IMPACT RADIATION (SCANIR) ION INDUCED X-RAYS (IEX)	
e	ELECTRON SPECTROSCOPY FOR CHEMICAL ANALYSIS (ESCA) OR SOFT X-RAY PHOTOELECTRON SPECTROSOPY (XPS) UV PHOTO-ELECTRON SPECTROSCOPY (UPS) AUGER ELECTRONS	AUGER ELECTRON SPECTROSCOPY (AES) IONIZATION SPECTROSCOPY	ION NEUTRALIZATION SPECTROSCOPY (INS)	
IONS			ION MICROPROBE OR SECONDARY ION MASS SPECTROSCOPY (SIMS) MASS SPECTROSCOPY OF SPUTTERED NEUTRALS ION SCATTERING SPECTROSCOPY (ISS) NECLEAR BACKSCATTERING	FIELD MASS SPEC-TROSCOPY

←—— EMISSION

parameter is itself a function of several other parameters, of
which some are the doping concentration, temperature, voltage,
humidity, and duty cycle. These parameters define a multi-
dimensional space within which each response parameter can assume
a number of possible states[10,11]. Some of these are "good"
or reliable states; all the others are "bad" or failed states.
For reliable operation, each of the response parameters must
simultaneously lie in one of the good states. As transitions

from reliable to failed states have been observed to occur over-
whelmingly more often than the reverse transition, it is reasonable
to assume that reliable states are metastable states, separated
from the stable failed states by potential barriers, to surmount
which, less energy must be provided than to make the reverse
transition. This is shown schematically in Figure 7. There are,
of course, failed states that are metastable, as anyone who has
attempted to determine the cause of an intermittent failure well
knows. However, it is not unreasonable to consider the pre-
ponderance of failed states to be stable states.

Any defect, whether intrinsic, such as a dislocation or
vacancy or interstitial, or extrinsic, such as the presence of
undesired impurities, that can be activated to surmount the
potential barrier between reliable and unreliable states, is a
potential reliability concern. Central, therefore, to all reliability modelling, is the need to determine the nature of the
contaminant that can be activated to undergo the undesired
transition, the amount of energy needed for the transition, i.e.,
the activation energy of the failure mode, and the time needed for
enough energy to be gained for the transition to occur, i.e., the
time-to-failure.

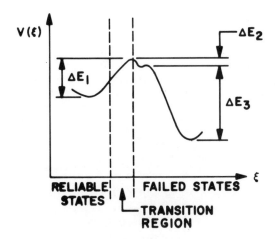

Figure 7. Potential energy diagram for reliability. The
transition from a reliable state to a failed state requires
less energy than the reverse transition. The narrower the
transition region between states, the more abrupt the failure.
Occurrence of intermittent failures can be explained in terms of
kinks in the potential energy curve to the right of the transition region.

Evaluating the activation energy and the time to failure involves the application of reaction rate theory. The Arrhenius, Delbruck or Eyring models are often used[10], [12], although other models may be needed to describe the particular failure mechanism involved[13]. Two examples of theoretical treatments pertain to the electromigration problem[14-16] and the metal depletion problem[17], [18]. In the former, high current densities in the metal stripes physically move the metal ions towards the cathode where they pile up, leaving voids and cracks, resulting in failures due to unduly high resistance. In the metal depletion problem for which an elegant theoretical formulation was first proposed by Rosenberg and Gniewek[19], [20] and extended by Ferris-Prabhu[21], contaminants already in the encapsulant, or entering it from the exterior environment, deplete the metallization of its high conductivity constituent by reacting with it. The increased resistance of the resultant reaction product and the remaining metallization results in failure.

However, fully theoretical treatments are rarely possible because of very incomplete knowledge of the dependence of the response parameter on the independent variables. As a consequence of the difficulty, semi-empirical methods are generally used[22], [23]. These usually involve measuring the change with time of a particular parameter, say the resistance, as the devices are maintained at several different but constant elevated values of the temperature and voltage. These are the operational variables that control the amount of energy available to activate the defect and thus cause failure. Acceleration of the aging process activates the failure mechanism within a reasonable time and hence such tests are called accelerated tests. Empirical fitting of the time to failure data to the voltage and temperature enables a so-called acceleration factor to be determined for use in calculating the expected time to failure under actual operating conditions. Applicability of the results of such accelerated tests depends strongly on the assumption that the aging process is only accelerated and not altered. This assumption is usually not wrong if the accelerated stresses are not much greater than the operational values.

TYPES OF CONTAMINANTS

There are so many different types of contaminants that can affect the reliability of microelectronic devices that there is no unique method of classifying them, let alone listing them all. One method of classifying the contaminants is by whether they are normally found in the bulk of the device, at, or near the surface, or enter from the environment. Another is in terms of whether the source of energy provided to activate the defect is mechanical, thermal, chemical, or electrical. Graft[24] has suggested the convenient classification shown in Table III.

Table III.   Contaminant Classification.

o   PARTICLE CONTAMINANTS
    −   DUST, HAIR, FIBERS, METALLIC MICROFRAGMENTS

o   ORGANIC CONTAMINANTS
    −   HUMAN OILS, SKIN FLAKES, GREASES, RESITUAL
        RESIST

o   IONIC CONTAMINANTS
    −   PLATING RESIDUES, WATER SOLUBLE ACIDS,
        METALLIC IONS

o   MICROBIAL CONTAMINANTS
    −   SPORES, BACILLI

     Particulate contaminants such as dust, hair, and fibers,
shown in Figure 8 - 10, can cause mechanical damage to the silicon
surface or to the metallization deposited on it. While dust, hair,
fibers, and other such particulates can be removed relatively
effectively, the problem of skin flakes is pernicious. Skin
flakes and oil from the human body are shown in Figure 11. Moore[25]
has shown that the shedding of skin flakes with associated body
oils, is a natural process over which there is no control. Because these biological products are insoluble in most, if not all,
common solvents, their removal presents a stimulating challenge.
If severe, these problems are yield detractors, but frequently
they become reliability concerns.

     Metallic impurities at the surface can diffuse into the
junction region, physically changing the location of the junction
and lowering the punch through voltage[27]. When in the bulk,
metallic impurities can diffuse preferentially along dislocations
and other structural irregularities, eventually precipitating out
as conduction "pipes" which lead to failure[28,29].

     Organically impure resins often release undesirable impurity
ions which can migrate to the surface of the device, leading
to degradation of its characteristics[26]. Hirsch and Koved[17]
found that the hydrogen generated by amine cured epoxies and some
silicones, reduces the palladium oxide in palladium silver re-

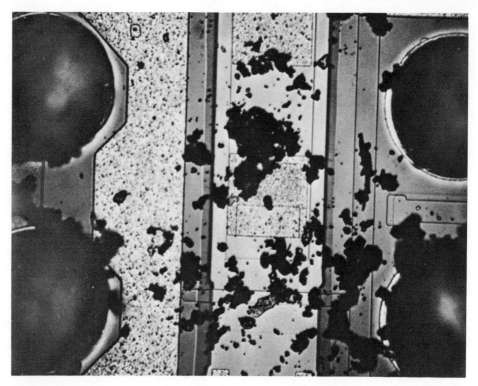

Figure 8. Conglomeration of different, unidentified dust particles.
(X200, reduced 50% for purposes of reproduction)

sistors to metallic palladium, degrading their resistance. Marks
and Stewart[30] found that in certain silicone foam encapsulated
encapsulants, polymerization of hydrosiloxane groups with polymeric hydroxy functional siloxanes releases hydrogen which diffuses through the silicone resin coating and is absorbed by the
metal. This results in loss of mechanical rigidity, change in
electrochemical potential, and increased resistance, all of which
age the device, decreasing its reliability.

     Sometimes the device surface is contaminated by cations.or
anions. These ions can drift along the surface due to operating
potentials, collect at the high bias electrodes, and when their
density is sufficient, effectively extend the area of the
electrode, altering the silicon surface potential in an undesired
way. The presence of moisture is well known to increase electrolytic corrosion, particularly when there is surface ionic contamination[31,32]. In some cases, the presence of alkali ions in the
oxide layer leads to their occupying weakly bound interstitial
positions which, under the excitation of the operating temperature

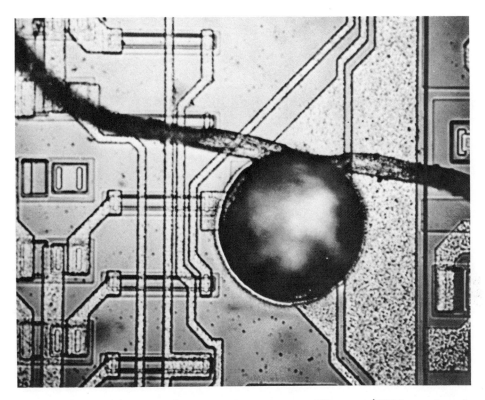

Figure 9. A flat, semitransparent cotton fiber. (X200, reduced 50% for purposes of reproduction)

and voltage, can move with the oxide, altering the surface potential of the adjacent silicon[33].

An interesting potential source of ionic contaminant has been suggested by W. Schuele[34], who has found bacteria on silica wafers, as shown in Figures 12 and 13. It is known that sodium and potassium ions may cause device failure due to excessive current leakage, and that phosphorus ions can dope silicon in undesired locations. Using the elemental composition of an average bacterium shown in Table IV[35], Schuele analyzed the salts on some typical wafers washed in deionized water. His results for the concentration of impurity atoms due to bacteria are shown in Table V. Concentrations of sodium on the order of 10^{10} atoms per cm^2 are known to affect leakage, and concentrations of phosphorus of 10^{16} atoms per cm^2 are used intentionally to dope silicon. The fact that skin flakes are densely colonized by bacteria, and Schuele's analyses, suggest the possibility of bacterial contamination as a potential reliability concern.

Table   IV.   Average Bacterial Composition.

CHEMICAL	% DAY WEIGHT
C	50.0
N	15.0
P	3.2
S	1.1
ASH	12.75
FIXED SALTS	7.25
FREE SALTS	5.5

Table   V.   Concentration of Impurity Ions due to Bacteria.
(Courtesy of W. J. Schuele, IBM General Technology
Division.)

ELEMENT	WT. OF ELEMENT PER BACTERIUM (g) $\times 10^{-15}$	NO. OF ATOMS PER BACTERIUM $\times 10^{6}$	CONCENTRATION OF ATOMS/cm^2 $\times 10^{15}$
Na	1.33	35	4.4
K	1.56	24	3.1
Na + K	2.89	59	7.4
P	5.8	112	1.4

Figure 10. A human hair, scaly on the surface.   The diameter
depends on which part of the body it comes from.

A totally different reliability concern has to do with alpha
particle irradiation.   The effect of high energy radiation on
semiconductors is well known, but the radiation has been caused
by environmental sources exterior to the semiconductor and its
packaging. May and Woods[36] have reported that various sealing
glasses such as oxides of lead, aluminum, etc., silicones,
epoxies, and packing materials used in semiconductor device manufacture contain amounts in the order of parts per million of
thorium and uranium.   The alpha particles emitted in the decay of
these radioactive contaminants, when they enter the silicon, can
ionize the silicon atoms sufficiently to produce enough electrons
to change the state of certain storage nodes in dynamic FET
memory devices, causing random, recurring, reversible failures.
They are random, because of the inherently probabilistic nature
of the radioactive decay; recurring, because once the storage node
has its state upset, a bad bit is delivered each time it is
addressed; and reversible, because no permanent damage is done,
the erroneous state lasting only until the correct information is
next written into the affected storage node.

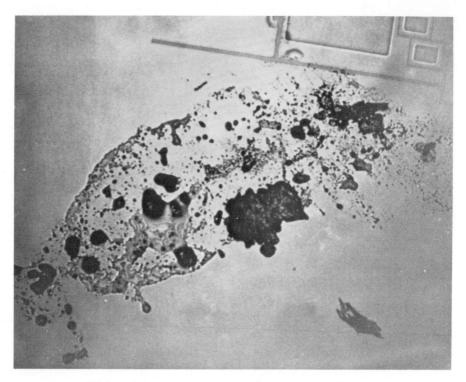

Figure 11. Skin flakes and oils from the human body. (X100, reduced 50% for purposes of reproduction)

        An alpha particle emitted by a uranium atom has an energy of
4.19 MeV. Despite losing energy while traversing the various
layers of metallization and oxides above the active silicon surface,
the alpha particle still has typically about 1.72 MeV left. This
is sufficient to produce 4.8×10^5 electrons, of which it is not
unreasonable to assume that at least half are collected by a
storage node.

        The smaller the storage node, the more susceptible it is to
alpha particle induced fails, as fewer carriers need to be generated near it to alter its state. For device densities compatible
with a sense amplifier sensitivity of about 30fC, less than
200,000 electrons are needed to cause an upset. An alpha particle
with 0.72 MeV energy is sufficient to produce this many electrons.
These observations together with the presence of radioactive elements
in the materials used in various stages of the manufacturing process,
as well as the trend in the industry towards greater density, i.e.,
smaller storage nodes, suggest that this reliability concern may
present new and interesting challenges in technology design as well
as in contaminant control.

Figure 12. Bacterium cell from deionized water, on a 0.2 μm Nucleopore filter (X20,000, reduced 50% for purposes of reproduction). (Courtesy of W. Schuele, IBM General Technology Division.)

SUMMARY AND CONCLUSION

This review has traced the trend in the microelectronic industry toward greater density and therefore, smaller devices. As a consequence, reliability - the concepts of which are briefly explained - can be affected by increasingly subtler causes. Many of these are due to the presence of contaminants, for the detection and measurement of which, various methods have been mentioned. The principles of reliability modelling, involving fully theoretical treatments as well as semi-empirical accelerated testing, have been discussed. The types of contaminants that lead to reliability problems have been listed and some recent potential contaminant concerns have been described.

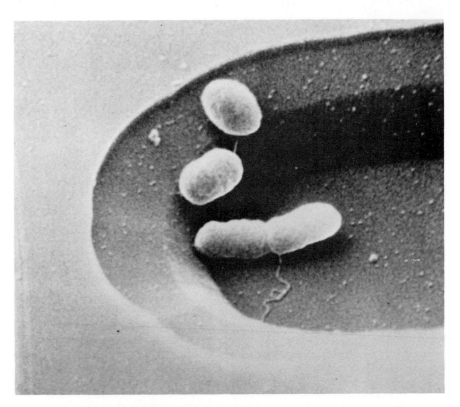

Figure 13. Bacteria cells on a silica wafer. Note the dividing
cell. (X20,000, reduced 40% for purposes of reproduction) (Courtesy of W. Schuele, IBM General Technology Division.)

ACKNOWLEDGMENTS

I wish to thank Dr. K. L. Mittal for inviting me to under-
take this review, M. Axelrod for encouragement, and P. Nerber for
supportive management which made it possible to accept. It is a
pleasure to acknowledge the assistance of Dr. B. Bertelsen,
D. Campbell, R. Donlan, E. Moore, J. Roderer, Dr. W. Schuele, and
W. Tice, in making some of their work available to me and for discussions from which I have learned much. For essential logistical
support, I thank L. Bickford, E. Delano, R. Gagner, and J. Knapp.
In particular, I wish to express my indebtedness to my wife Joy
and son Arjun for their generous sacrifice of the many weekends and
evenings spent in preparing this review.

REFERENCES

1. R. N. Noyce, Science 195, 1102 (18 March 1977).
2. J. G. Linvill and C. Lester Hogan, Science 195, 1107
    (18 March 1977).
3. J. Wallmark, in "Microelectronics," E. Keonjian, Editor,
    McGraw-Hill, New York, 1966.
4. R. Feynmann, in "Miniaturization," H. D. Gilbert, Editor,
    pp. 282-296, Reinhold, New York, 1961.
5. R. L. White and J. D. Meindl, Science 195, 1119
    (18 March 1975).
6. R. Donlan, Personal Communication, (1978).
7. D. Lichtman, in "Methods of Surface Analysis," A. W. Czanderna,
    Editor, pp. 39-74, Elsevier Scientific Publishing Co.
    Amsterdam, 1975.
8. See, e.g., "Methods of Surface Analysis," A. W. Czanderna,
    Editor, Elsevier Scientific Publishing Co., Amsterdam, 1975.
9. G. K. Wehner, in "Methods of Surface Analysis, A. W. Czanderna,
    Editor, pp. 5-38, Elsevier Scientific Publishing Co., Amster-
    dam, 1975.
10. R. E. Thomas and H. C. Gorton, Physics of Failure in Elec-
    tronics 2, 25 (1963), (Rome Air Development Center Series in
    Reliability.)
11. A. V. Ferris-Prabhu, in "Physics in Industry," E. O'Mongain
    and C. P. O'Toole, Editor, pp. 437-440, Pergamon Press,
    New York, 1976.
12. A. V. Pershing and G. E. Hollingsworth, Physics of Failure in
    Electronics 2, 61 (1963),(Rome Air Development Center
    Series in Reliability).
13. N. Johnson and K. Greenough, Physics of Failure in Electronics,
    2, 103 (1963), (Rome Air Development Center Series in
    Reliability).
14. R. S. Sorbello, J. Phys. Chem. Solids 34, 937 (1973).
15. J. R. Black, in "Proceedings of the 12th Annual Reliability
    Physics Symposium," p. 142 (1974).
16. R. Landauer and J. Woo, Phys. Rev. B 10, 1266 (1974).
17. H. Hirsch and F. Koved, Insulation,p. 51, (June 1966).
18. W. L. Clough and J. M. Collins, Microelectronics 3, 4 (1971).
19. W. Rosenberg and J. Gniewek, Personal Communication (1974).
20. W. Rosenberg, J. Gniewek and A. V. Ferris-Prabhu, Bull. Am.
    Phys. Soc. 20, 30 (1975).
21. A. V. Ferris-Prabhu, J. Appl. Phys. 47, 4078 and 4082 (1975).
22. See, e.g. "Semiconductor Reliability," Vol. 2, W. H. Von
    Alven, Editor, Engineering Publishers, Elizabeth, N.J. 1963.
23. John D. Venables, in "Proceedings of the 10th Annual Re-
    liability Physics Symposium," p. 159 (1972).
24. D. Graft, Microelectronics 3, 14 (1970).

25. E. J. W. Moore, "Contamination of Clean Room Operations by Human Dust (A Biological Problem)," IBM Technical Report 19.0052 (1969), IBM General Technology Division, Essex Junction, VT.

26. W. C. Miller and D. W. Crain, Microelectronics $\underline{5}$, 49 (1973).

27. M. C. Johnson, Physics of Failures in Electronics $\underline{2}$, 124 (1963) (Rome Air Development Center Series in Reliability).

28. G. Schwuttke, J. Electrochem. Soc. $\underline{108}$, 163 (1961).

29. W. K. Tice and R. Shasteen, Semiconductor Silicon, 639 (1973).

30. B. Marks and R. Stewart, in "Proceedings of the 8th Annual Reliability Physics Symposium," p. 101 (1970).

31. H. Koelmans, in "Proceedings of the 12th Annual Reliability Physics Symposium, p. 168 (1974).

32. D. B. Willmott, in "Proceedings of the 15th Annual Reliability Physics Symposium," p. 158 (1977).

33. R. P. Donovan, Physics of Failure in Electronics $\underline{5}$, 199 (1966) (Rome Air Development Center Series in Reliability).

34. W. Schuele, Personal Communication (1978).

35. I. C. Gunsalus and R. Y Stanier, in "The Bacteria - A Treatise on Structure and Function," Vol. 1, 15, Academic Press, New York (1960).

36. T. C. May and M. H. Woods, in "Proceedings of the 16th Annual Reliability Physics Symposium," p. 33 (1978).

EFFECT OF SURFACE CONTAMINATION ON HIGH VOLTAGE INSULATOR PERFORMANCE

G. Karady

Ebasco Services Incorporated

Forrestal Campus, Princeton, N.J.   08540

This paper presents a critical review and state of the art report on the surface contamination of high voltage insulators which reduce the flashover voltage of transmission lines and cause power outages.

The presently accepted theories of contamination build up, dry zone formation, and flashover mechanism are reviewed, together with a description of laboratory test methods used in the power industry for determination of insulator performance in contaminated areas. The basic criteria for insulator design and the type of insulators used in contaminated areas are described briefly. Finally, the paper concludes that in spite of the fact that over the last 50 years more than 800 technical papers have dealt with the problem, there is yet no comprehensive theory which is able to predict the performance of contaminated insulators. Even the results of laboratory tests of artifically contaminated insulators do not accurately represent the performance of the insulators in the natural condition. The major objectives of future research work are out-lined.

INTRODUCTION

Insulator contamination is a special type of surface contamination which is produced by airborne pollution. The major practical effect of this pollution is a reduction of insulator flashover voltage. For example, the flashover voltage of a standard suspension insulator under clean, wet conditions is 45 kV/unit. Light pollution, which corresponds to normal service conditions, reduces this value to 16 kV/unit; however, when the insulators are very heavily polluted the flashover voltage may be reduced to 6 kV/unit.

This problem has been studied since 1903 when the first high voltage transmission lines were constructed. Many papers have been written on the subject. A recently completed IEEE bibliography[1] contains more than 890 articles. In spite of the considerable research effort the number of flashovers and line outages caused by contamination have been on the increase. The problem is more severe for the Ultra High Voltage (500 kV and up) and DC transmission lines.

The purpose of this paper is to describe the state-of-the-art of insulator contamination research and to review the major problem areas associated with it.

BASIC FLASHOVER MECHANISM

It is observed that airborne pollution produces a deposit on insulator surfaces in dry weather and that later on, usually in the early morning hours when dew or fog wets the surfaces, flashover occurs. The flashover is preceded by surface arcing and heavy corona discharge.

The mechanism of this phenomenon may be described briefly as follows:

(a)   Air pollution produces a deposit layer on the insulator surface under dry conditions. This layer is not usually conductive and does not influence the flashover voltage.

(b)   Fog, dew, light snow and onset of rain wet the surface layer and produce an electrolyte which forms a conductive layer.

(c)   The voltage across the insulator drives a leakage current through the conductive layer which, in turn, heats the layer. The heating effect depends on the density of the current and the surface resistivity $(i^2\rho)$ on the insulator surface. The highest current density and heating effect are generally found near the electrode, e.g. near the pin.

(d)   The heating of the leakage current dries the surface layer and forms a dry zone near the pin. The resistance of this dry zone increases rapidly with time, with the result that the voltage distribution along the insulator surface changes and most of the voltage appears across the dry zone.

(e)   The high voltage across the dry zone produces a surface arc, which either extinguishes or evolves to the point that the insulator flashes over. The development of the arc is controlled mainly by the resistance of the rest of the surface. The leakage current in this case consists of intermittent bursts of impulses.

(f)   If the arc extinguishes and the dry zone is sufficiently wide to support the voltage across the insulators, the leakage current ceases altogether.

(g)   The fog and dew wet the insulator again, the described sequence of events starts again, and a partial arc appears after a few minutes near the pin. Generally, the formation of several partial arcs precedes the flashover of the insulator.

The periodic arcing near the pin of the insulator can be observed in the dark in any polluted area in the presence of fog and dew, or near the seashore during a sea-wind which drives humid air and sea-water droplets onto the insulator surface.

Flashover phenomena may be divided into the following four phases:

(A)   contamination process

(B)   wetting process

(C)   dry-zone formation and partial arcing

(D)   flashover or re-wetting

Each phase will now be discussed on the basis of the published literature.

A)   Contamination Process

The atmosphere contains particles produced by different sources.   The major sources of pollution are:

(1)   Industry.   In the vicinity of chemical plants and heavy industry, polluted air produces a deposit containing both conductive ($NaCl$, $CaCl_2$, $CaSO_4$, etc.) and non-conductive materials such as dust, tar, organic materials, SiO_2 and $CaCO_3$, etc.   Cement-type pollution producing a well-sticking, highly conductive deposit which can be found near coal-burning power plants and cement plants.

(2)   Highways.   In winter, the highways in Canada and the northern part of U.S.A. are salted and the traffic and wind produce a salt and dust deposit on nearby insulators.   This pollution can be observed at a distance of 0.5 to 1 mile from the highway.   The deposit layer takes the form of a white, easily removable powder that the rain washes down easily.

(3)   Desert.   Sand or clay contains 4-5% salt in desert areas where the wind produces a slow accumulation of deposit on the insulator surface during the dry period of the year.   The deposit is highly conductive when wetted.

(4)   Sea.   The sea wind carries small salt particles which are so fine as to be completely undetectable.   These particles deposit onto the insulator surface where they produce a thin transparent salt deposit which is very soluble.   Rain cleans the insulators easily.

The deposit layer on the insulator surface is influenced by several factors such as:

(1)   The gravitational force,

(2)   electrostatic forces and corona effect (electric wind),

(3)   wind (aerodynamic) force,

(4)   the chemical composition of the particles,

(5)   the surface of the insulators,

(6)   the insulator shape,

(7)   the surface condition of the insulators (dampness).

In calm weather, gravity deposits the larger particles on the top of the insulators. This effect may be significant near cement factories or coal-burning power plants when insufficient dust removal leads to the emission of large particles which fall on nearby insulators.

The electric field near the insulator is in the range of 1-2 kV/cm; e.g. in the case of a 20-kV distribution-type insulator, the measured maximum field strength is 0.8 kV/cm, which produces a force/weight ratio of 0.0005 for water particles at the insulator surface [2]. This example illustrates the fact that the effect of electric forces is negligible under ordinary working conditions at low voltages. However, the non-uniform voltage distribution may produce a higher electric field which increases the collection of particles. The electric force acting on the water particles depends on the particle size which in turn may vary from a few microns to millimeters. Close to the surface, this force should not be ignored, if the electric field is sufficiently high. Furthermore, the force which holds the water and contamination particles already on the surface is largely electrical in nature even under ordinary working conditions. This is particularly important with regard to high-voltage dc lines. In the case of distribution lines, the effect of electric fields is generally negligible even in the presence of mild corona.

The effect of wind is the most important of the forces that determine the amount and distribution of contamination on the insulator surface. The wind produces three types of air flow around the insulator. On the windward side, laminar flow is established while on the leeward side in the vortex sheet, high-speed eddies and, in the wake, slowly-moving eddies may be found. The air flow drives the particles to the insulator surface where they will either stick or be blown away.

The deposit formation by wind was studied in a wind tunnel by Thompson[2] who found the following:

- On conical vertical surfaces, the deposit is of variable thickness on the windward side, due to the direct impact of the particles. The deposit decreases with the angle of impact and practically no deposit was found in the tangential region,

- On the leeward side, a layer of more or less uniform thickness is formed,

- Horizontal surfaces under the sheds exhibit deposit patterns in conformity with the air flow in the wake of the neighboring vertical surface,

- Ribs arranged concentrically show deposits on the windward
  edges but remain comparatively clean where the curvature is
  in the direction of flow,

- Deep ribs show the same tendency around the edges but in the
  grooves a fairly uniform deposit is formed which decreases
  with the distance from the lip.

- The deposit is proportional to the amount of material in
  the air (number of particles per m^3),

- The wind velocity does not seem to affect significantly the
  amount of deposit. High-velocity wind cleans the insulator
  surface, particularly on the top. However, in the case of
  the pollution produced by the sea salt, the higher is the
  wind, the higher is the deposit on the insulators.

The shape of the insulator significantly influences the
deposit. A smooth disk shape, which reduces the formation of
vortices, greatly decreases the deposit density. This result
led to the development of aerodynamically-shaped insulators.
However, the overall performance of these insulators was
questioned and they have found only a few applications.

The deposit density is proportional to the surface roughness,
which is expressed in terms of the height of the surface
irregularities.

The relative humidity of the air does not influence the
deposit, as long as it does not produce water condensation on the
surface.

Dampness on the surface of the insulator increases the
deposit density nearly ten times.

The insulator material also influences the deposit density.
For example, the water-repellent, non-sticking surface of some
plastic materials reduces the thickness of the deposit layer.

The composition of the contaminant also has a significant
effect. Organic materials (tar) stick to the surface. Also,
cement-type materials form a well-sticking layer, difficult to
remove. Near the sea, an easily removable white salt layer is
formed on the insulator surface.

Rain cleans the insulators, the cleaning effect depending
on the shed shape. The slope of the upper surface increases the
washing effect because it allows the salt solution to run down
to the rim of the shed. There is a slight washing effect on the

bottom of the insulators mainly because of the water droplets splashing up from the shed below.

B) Wetting Process

Dry contamination is not normally conductive but becomes so when the insulator is wetted, which generally occurs in rain, fog and dew. However, there are other sources of wetting, all of which form a moisture film on the insulator surface. The following moisture films may be distinguished[4]:

(1) the surface film, present even in dry weather. This involves only a small amount of moisture and does not wet the contaminant significantly. The hygroscopic property of the contaminant does attract water from the air but this effect is negligible;

(2) the film produced by condensation as a result of temperature differences between the insulator and the ambient air. The insulator's thermal time constant is about one hour. A sudden increase in air temperature (warm damp wind) produces a temperature difference and, consequently, condensation; this wets the insulator surface more or less evenly, in protected as well as unprotected areas;

As the heat of condensation and the heat transfer from the fog to the insulator increase the temperature of the insulator, the rate of condensation decreases to zero. This phenomenon[5] can be described, in the first approximation, by heat balance and mass transfer equations.

The heat balance equation is:

$$\frac{dt_i}{d\tau} \ C \ M_i = A_i \ h \ (t_f - t_i) + A_i \ i \ M_w \qquad (1)$$

The mass transfer equation (rate of condensation) is:

$$M_w = h_D \ (p_f - \rho_i) \qquad (2)$$

The water accumulation per unit surface area during a period of time T is:

$$M = \int_o^T M_w \ d\tau \qquad (3)$$

where:

τ   time

t_i insulator surface temperature

t_f fog temperature

M_i insulator weight

A_i insulator surface area

C   specific heat of insulator material

h   heat transfer coefficient between insulator surface

        and vapor

i   heat of condensation

M_w rate of condensation per unit surface area

h_D diffusion coefficient

ρ_f saturation vapor density in fog at a distance from the

        insulator, at temperature t_f

ρ_i saturation vapor density near the insulator surface at

        temperature t_i

(3)   the film that forms on the insulator in mist and fog. When
       fog occurs, small (1-10mm diameter) liquid water particles
       are suspended in the air. These particles are driven by air
       streams into collision with the insulator where they produce
       wetting [5].   The air movement causes the fog to wet even
       protected surfaces between the ribs, although the wetting
       rate is smaller than on the upper surfaces; the calculation
       of the wetting produced by the collision of the droplets
       requirs a) the droplet size distribution (K); b) the liquid-
       water density (S_w) c) the speed of top droplets (v).

       The amount of water per unit volume carried by the droplets
with a radius between r and (r + dr) is:

$$d\rho = \frac{4\pi}{3} \rho_w r^3 \, K \, N \, dr \qquad (4)$$

where: r   radius of droplet

ρ_w water density
K   distribution curve
N   total number of droplets in unit volume of fog.

The total amount of liquid water in the fog per unit volume is:

$$m = \int_0^\infty d\rho = \frac{4\pi}{3} \rho_w N \int_0^\infty K r^3 \, dr \qquad (5)$$

Due to the combined effects of gravitation and air movement, the droplets move at a speed of v - c where v is the fall velocity of the droplets and c is the speed of the fog stream.

The ratio of flux of droplets with a radius between r and (r + dr) to the total amount of liquid water per unit volume is:

$$\frac{d\phi}{m} = \frac{K r^3 (v \pm c)}{\int_0^\infty K r^3 \, dr} \, dm \qquad (6)$$

This equation has to be evaluated using a step-by - step integration.

(4) surface wetting produced by rain and drizzle or light snow. Rain and drizzle contain large water droplets that fall in response to gravitational forces. The water droplets wet the upper surface directly but, at least in the case of light rain or drizzle, not the protected surfaces. In heavy rain and at higher wind velocities, the bottom surface is also wetted and flashover may occur in the first few minutes of rain. After a short period, the rain cleans the upper surface, thus preventing further flashover. The upper surface of the insulator is washed down within a short time during which the dry bottom surface provides sufficient insulation. The latter surface will be also wetted, by splashing water droplets; this happens fairly quickly but does provide enough time for the upper surface to become clean enough to withstand the voltage.

C) Dry-zone Formation and Partial Arcing

Dry zones are formed by the heating effect of the leakage current, which depends on the resistance, voltage stress and surface-loss distribution along the insulator leakage path.

1. Surface-Resistance Calculation: The interaction of the deposit and the water forms an electrolyte and a conductive

layer on the insulator surface which provides a path for the
leakage current. Supposing a uniformly distributed layer, the
surface resistance may be calculated using the following equation:

$$R(x) = \rho \int_{x_0}^{\ell} \frac{dx}{2\pi r(x)\delta} = \frac{\rho}{2\pi\delta} \int_{x_0}^{\ell} \frac{dx}{r(x)} \qquad (7)$$

where: ρ : resistivity of the deposit (Ω cm)
        δ : thickness of the deposit (cm)
        $r(x)$: radius of the insulator at leakage distance x (cm)
        x : leakage distance measured from the top electrode
                   (cm)
        L : total leakage distance (cm)

The potential distribution is determined by equation 8:

$$V(x) = I.R(x) \qquad (8)$$

Substituting equation 7 in equation 8 one can obtain
equation 9 which gives the potential distribution along the
surface:

$$\frac{V(x)}{V} = \frac{R(x)}{R} = \frac{\int_{0}^{x} \frac{dx}{r(x)}}{\int_{x_0}^{\ell} \frac{dx}{r(x)}} \qquad (9)$$

The voltage distributions of four different insulators were
calculated by integrating equation 9 numerically. The profile of
the insulators are shown in Figure 1. Figure 2 shows the
distribution of leakage distance; the normalized voltage
distribution is presented in Figure 3.

The following observations may be made following a study of
Figures 2 and 3.

(1) The largest voltage drop occurs near the pin. About 17-22%
     of the voltage is concentrated between the pin and the
     first corrugation (rib),

(2) The other critical area is that surrounding the cap. About
     6-18% of the voltage appears on a length equal to 4-8% of
     the leakage path near the cap.

(3) The voltage drop on the upper surface is about 20-30%, on
     the bottom surface about 70-80%. With regard to insulator
     design, it should be noted that the voltage drop across the
     ribs is important because the narrow gap between the ribs

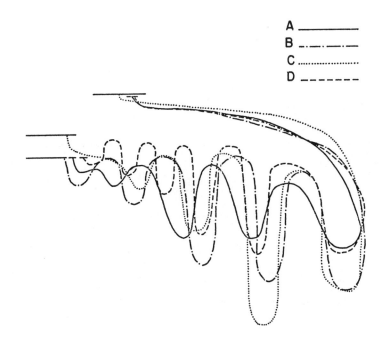

Figure 1. Profile of the tested insulators.

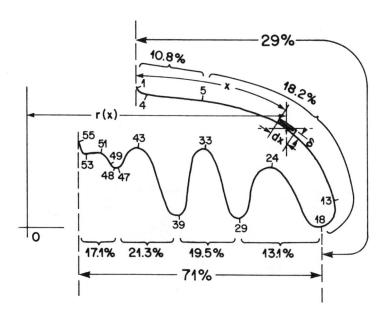

Figure 2. Voltage distribution along the surfaces of standard (No. 1) insulator ℓ = 30.8 cm, A = 1551 cm^2.

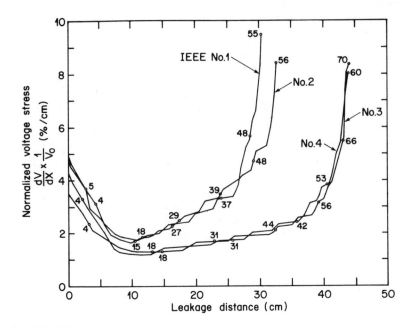

Figure 3.   Normalized voltage distribution.

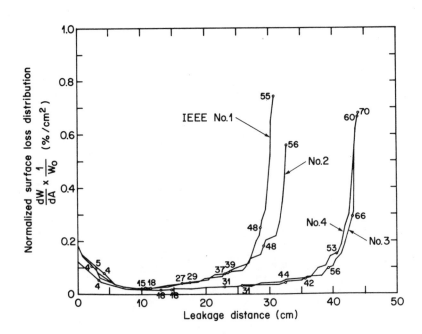

Figure 4.   Normalized surface loss distribution.

may flash over. In the case of uniform pollution, 17-23%
voltage appears across a gap of 1.2-1.5 cm, which is sufficient to withstand this voltage if the voltage across the
insulator is in the range of 10-15 kV.

From these results, one can predict that the dry zone will
first form near the pin, then near the cap. The dry-zone formation can be influenced and the non-uniform voltage distribution
improved by increasing the diameter of both the pin and the cap.

2) Surface-Loss Distribution: The leakage current heats
the insulator surface. The heat generated in the ring per unit
area (heat-generation density) is

$$\frac{dW}{dA} = I^2 \frac{\rho}{2\pi\delta} \frac{1}{2\pi[r(x)]^2} \qquad (10)$$

The current through the insulator is $I = V_o/R_o$; the total
loss is $W_o = V_o^2/R_o$. Substituting these values into equation 10 the
normalized loss density is calculated as shown in equation 11:

$$(\frac{1}{W_o}) \cdot \frac{dW}{dA} = \frac{1}{2\pi[r(x)]^2 \int_{x_o}^{\ell} \frac{dx}{r(x)}} \qquad (11)$$

Equation (11) was evaluated; the heat-generation density
rate is given in Figure 4.

It can be seen that:

(1) the rate of heat generation is greater in the vicinity of
     the pin. Between the pin and the first rib, the very severe
     heat generation may dry the contamination and the formation
     of a dry zone can be expected.

(2) the rate of heat generation is also greater near the cap
     than at any other part of the insulator, except near the
     pin. This indicates that a dry zone may also form around
     the cap.

The dry-zone formation changes the voltage distribution
along the surface. The surface-resistance variation on a standard insulator was measured by Tahasu[6];   the results are shown
in Figure 5 (where the pin is numbered 0, the cap 8). In this
figure, curve 1 is the voltage distribution for a wet, uniformly
contaminated insulator and curves 2,3,4,5 show the effect of
drying. It can be seen that the voltage drop between points 0
and 1 (first two ribs) increases very rapidly as a result of the

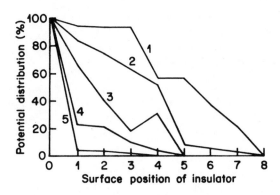

Figure 5.   Time variation of normalized voltage distribution.

drying effect; for example, about 92% of the voltage appears
across 0-1 in the case of curve 5.

D) Flashover and Rewetting 7, 8

    The leakage current forms a dry zone near the pin and/or cap;
the resistance of the dry zone increases rapidly as does the voltage across it. The increasing voltage may even arc over the dry
zone. This arc either extends, and the insulator flashes over, or,
if the dry zone withstands the arc is extinguished. In the latter
case, fog rewets the dried surface, and this leads to further
arcing. After several arcing periods the arc finally extends and
the insulator flashes over. The phenomenon may be followed on
the leakage current oscillogram, where the arc formation is indicated by a current burst (series of current impulses) showing
that the arc lasts several cycles. The current increases for a
few cycles during which time the arc and the dry band presumably
extend. The current then starts to decrease and the arc is extinguished. After several arcing periods, the current remains
constant on the increase and eventually, flashover occurs, which
is indicated by the collapse of the voltage.

    Owing to the phenomena outlined above, the flashover voltage
of contaminated insulators is much lower than that of clean ones.

    The criteria of flashover can be formulated to the first ap-
proximation by equation 12.

    The applied voltage V is divided between the arc voltage
(first term of equation) and the voltage drop across the wet contaminated layer.

$$V = (B.x) \, I^n + Ir \, (\ell-x) + V_e \qquad (12)$$

where: n   is an arc constant; typical value is 0.63
        B   is an arc constant; typical value is 63
        r   is the surface resistance per unit length
              of leakage distance
        x   is the arc length
        V_e is the sum of anode and cathode drop; typical
              value 80V

The insulator will flashover if the arc extends and bridges
the insulator surface which causing a continuous increase in
current accordingly the criteria of the flashover is

$$\frac{dI}{dx} < 0 \qquad\qquad (13)$$

Unfortunately the laboratory tests do not completely verify
this theory; even the other more sophisticated methods do not
correlate better with laboratory test results.

EXPERIMENTAL STUDY OF CONTAMINATION

Due to the limitation of the theory, field and laboratory
tests are used to determine the contamination level and to determine the flashover voltage as a function of contamination level.

A) A Survey of Insulator Contamination Levels

The contamination level forms the basis of the insulator
selection for a transmission line. There are two widely used
methods.

1. Leakage-Current Measurement: The leakage current is a
direct measure of the degree of contamination, and takes into account the accumulation of contaminant together with the wetting
effect of fog. The leakage currents of the energized insulators
in the test station are monitored continuously. In order to avoid
the dry-zone formation and flashover of the insulators, the voltage is applied repetitively, only for a few seconds at a time.
The leakage-current record is then evaluated by computer, which
correlates the data with the meteorological parameters.

The evaluation presents:

a) the leakage-current time-distribution curve for rainy,
   foggy, etc. days. From this, it is possible to distinguish
   the period when flashovers may be expected;

b) cumulative probability distribution curve of the leakage
current can also be determined. From this, the designer
may choose a characteristic value which is representative of the site.

Additional laboratory tests are required for determining the
salinity corresponding to the measured characteristic leakage
current. During this test the leakage current of artificially
polluted insulators is measured and plotted against resistivity or
equivalent salinity. When the leakage current of the artificially
contaminated insulators is the same as the measured leakage
current, simulation is achieved and the salinity which corresponds
to this leakage current value represents the actual pollution
level of the site. The described laboratory test is usually performed in a salt-fog chamber[3].

2. _Deposit-Density Measurement_: The deposit density may be
determined by periodically washing down the contamination from some
of the non-energized insulators of the test station with distilled
water. The conductivity of the obtained solution is measured and
the amount of soluble materials is expressed in terms of an equivalent quantity of NaCl which gives the same conductivity. Thus the
equivalent salt-deposit density is obtained as NaCl mg/cm^2.

Chemical analysis of the solution gives the type of pollution
and helps for its laboratory simulation.

The distribution of the contamination may be measured by determining the resistance between two electrodes on different parts
of the insulator surface when saturated by water. For the laboratory simulation of natural contamination, the insulators are contaminated according to the determined distribution. A practical
solution is to apply a different amount of contamination at the
top and the bottom of the insulators.

The frequency of collection of insulators for the washing
depends on the weather conditions and of the rate of contamination.
In Northern United States and Eastern Canada, a fortnightly collection of the contaminant proved to be sufficient. The deposit
density values obtained are statistically evaluated, and the
histogram and cumulative probability distribution curves are drawn.

B) Determination of the Reduction of Flashover Voltage

Contamination reduces the flashover voltage of insulators.
In order to determine the extent of this, the contamination is
artificially simulated. It is believed that contamination produced by the sea should be simulated by the salt-fog method, and
the industrial or urban contamination by the clean-fog method.

In the first case, the insulators are installed in the fog chamber and energized. Then salt-fog is applied for 60 minutes. If the insulators withstand the test, the voltage is increased and the test is repeated using new insulators. This process is continued until the insulators flashover; the complete test is repeated several times.

In the second case, the insulators are coated with a suspension which contains NaCl and different additional substances. The coated and subsequently dried insulators are wetted by clean fog and the flashover voltage is determined similarly as in the salt-fog test.

The results of this test give the flashover voltage of the insulators as a function of the salinity. The salinity is expressed as NaCl g/l in the case of the salt-fog test and as NaCl g/m^2 or surface conductivity in the case of clean-fog test. Typical flashover-voltage salinity curves of standard and fog-type suspension insulators are shown in Figure 6. The flashover voltages obtained in the laboratory have to be near to those occurring on naturally polluted insulators on the transmission line.

C)   Selection of Insulators

Flashover-voltage salinity curves are available for different types of insulator (standard, fog, etc.) and different insulator arrangements. The correlation of the contamination level with the flashover voltage produces several technically acceptable solutions.

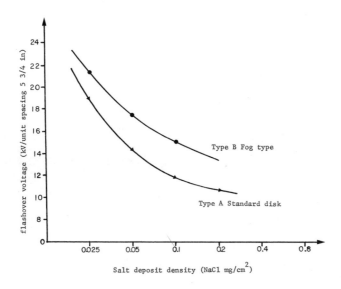

Figure 6: Flashover voltage of contaminated insulators.

For example, in the case of a 315-kV line and an equivalent of 0.2 NaCl mg/cm^2 salinity twenty standard insulators are needed to avoid the flashover according to Figure 6, whereas sixteen fog-type insulators will withstand the same voltage. The length of the chain in the first case is 9.6 ft., in the second case 7.3 ft. The difference in length influences the tower design as well as the economy of the line. It should be noted that a 15% safety factor was used in this calculation.

The method described gives a very conservative design; the flashover voltage measured in the laboratory does not correlate very well with the actual performance of the insulators.

In the case of major new transmission lines, it is advisable to also test the selected insulators under natural conditions. For this purpose an energized test station is used where all the insulators are energized at the full voltage.

INSULATOR DESIGN CONSIDERATIONS

On the basis of the data presented in the literature, the major factors influencing insulator performance in polluted areas may be summarized as follows:

(a) The pollution deposit can be reduced with a non-sticking water-repellent surface. This can be achieved either by coating the porcelain surface with a synthetic material (10) or by using entirely synthetic insulators. Both methods are as yet in the experimental stage but preliminary results are promising.

(b) The leakage distance is not the only determining factor and its increase does not necessarily improve the insulator performance although it does enhance the surface resistance. The latter effect may be lost, however, if the insulator diameter is large, no matter how much the leakage distance is increased[6].

(c) Insulator performance improves with increasing surface resistance which in turn is a function of the decreasing insulator diameter.

(d) Part of the insulator surface should be kept dry as protection against rain. The contamination rate is also lower on protected surfaces (e.g. ball-type insulators, or insulators with deep ribs on the bottom). On the other hand, with heavy dust contamination, narrow sheltered areas tend to fill up with dust, and, what is more, are not cleaned by rain. Sheltered areas usually prolong the time to flashover.

(e) The cleaning effect of the rain and wind will be more ef-
     fective if smooth, aerodynamic surfaces are used, for these
     reduce the rate of deposit.

     In the case of sea-salt contamination, John and Sayers [4]
     point out that a steeply sloped upper surface will allow
     the salt solution to run to the rim of the shed.

(f) Reduction of the leakage-current density near the pin and
     the upper electrode lowers the risk of dry-band formation
     and seems generally advantageous.

(g) The insulator performance may be improved by using a semi-
     conducting glaze which ensures a more uniform voltage dis-
     tribution and reduces or entirely eliminates the wetting by
     heating the insulator surface.

(h) The distance between the ribs should be sufficient to prevent
     arcing, presuming that arcing starts above a field strength
     of 3kV/cm on a polluted surface.

     Application of the above design criteria can be traced in
existing insulator design. It seems to be appropriate to describe
briefly the insulators most commonly used in contaminated regions.

(a) Fog Type Insulators. (Figure 3) are built with deep ribs on
     the bottom surface and with increased leakage distance. The
     ribs provide sheltered areas, which are not wetted. The
     longer leakage distance increases the flashover voltage.

(b) Semiconductive (RG) Type Insulators. Insulator performance
     under polluted conditions can be improved by covering the
     insulators with a semiconducting glaze. The resistance of
     a standard insulator covered with a semiconducting glaze is
     between 4 and 100 M Ω if measured with dc and 4-30 M Ω with
     ac. A constant leakage current flows through the glaze,
     heating the surface and keeping the insulators dry (except
     in heavy rain) and thereby minimizing condensation and
     wetting. Furthermore, in the case of heavy pollution and
     wetting, the developed dry band is shunted by the glaze
     resistance which reduces the probability of scintillation
     discharge and enables the insulators to withstand extremely
     heavy pollution.

     The disadvantage of the RG-type insulator is the possible
     deterioration of the glaze over fairly long periods of
     service due to the electrolytic effect. In addition, a
     semiconducting glaze has a negative temperature coefficient

which, in the case of overvoltage together with strong
radiation by the sun, may produce "thermal runaway." In
other words, the increase in leakage current increases the
temperature and this, in turn, reduces the resistance;
hence, further increases are caused in the leakage current.
This phenomenon finally destroys the insulator.

Another problem arises with the "cold switch-on" when the
insulator is suddenly energized after a fairly long period
out of service. During this time, contamination accumulates
on the surface and the insulator is wetted in the same way
as a standard insulator because the leakage current does not
heat the surface. When the insulator is suddenly energized,
therefore, it behaves as a normal insulator and, owing to
the contamination, may flash over.

(c) Synthetic Insulators. In recent years, a new type of insu-
     lator has appeared as an experimental device in the high-
     voltage transmission line. This insulator contains a fiber-
     glass core covered by sheds of a synthetic material such as
     silicon rubber, polytetrafluorethylene, cycloaliphatic resin,
     etc. Fiberglass ensures high mechanical strength while the
     synthetic material need for the shed, highly water- and dirt-
     repellent, ensures good electrical performance. The per-
     formance of new insulators of this type in contaminated
     areas is superior to that of porcelain insulators. However,
     the water- and dirt-repellent characteristic is affected by
     exposure to the weather which causes surface erosion. The
     extent of this effect is not known because only limited
     field experience is available as yet and the long-term per-
     formance of these insulators has not yet been assessed.

Finally, it is interesting to note that the rain cleans the
insulators in the horizontal position better than in the vertical
position. A further advantage with horizontal insulators is that
the probability of arc extinction is enhanced by the upward movement of the arc due to thermal effects. On the whole, the performance of horizontal insulators is superior to a vertical arrangement. This has led to the application of horizontal post-
insulators on distribution lines which, has significantly reduced
the need to wash the insulators in the Los Angeles area. The
insulators were not washed between 1959-1969 yet showed no contamination flashovers. Previously, a yearly washing had been
necessary to maintain trouble-free operation.

CONCLUSIONS

(1) Although theoretical studies have been made on each major
     area of the insulator contamination problem no comprehensive
     theory, which permits the calculation of flashover voltage
     exists.

The results of theoretical calculations correlate poorly with the test results. The development of a comprehensive flashover theory and the utilization of modern computational techniques is expected to lead to a method which permits the more accurate design of transmission line insulation.

(2) The flashover voltages predicted from the laboratory tests and the actual performance of insulators on a transmission line do not correlate well. This calls for the improvement of the testing methods.

(3) The results of insulator contamination surveys cannot be directly correlated with contamination levels used in laboratory tests. This indicates the need for the development of a more sophisticated survey method.

(4) The insulator contamination problem should be treated statistically. Methods which consider the statistical nature of the weather and of airborne pollution should be developed and related to the contamination problem.

REFERENCES

1. Institute of Electrical and Electronics Engineers, Inc., Bibliography on High Voltage Insulator Contamination IEEE, New York, No. 77BLO 100/8/PWR.
2. W. G. Thompson, IEEJ 91, Part II, 317 (1944).
3. P. J. Lambeth, J.S.T. Looms, M. Sforzini, R. Cortina, Y. Porcheron, and P. Claverie, IEEE Trans. on Power Apparatus and Systems, PAS-92, p. 1876, (November/December 1973).
4. W. J. John, F. M. Sayers, IEEJ, 629 (1935).
5. G. Karady, IEEE Trans. PAS-94, No. 2, p. 378, March/April 1975.
6. N. Takasu, Electrical Engineering in Japan, 83, No. 5, 63, (May 1963).
7. E. Nasser, IEEE Trans. on Power Apparatus and Systems, PAS-87, 957 (April 1968).
8. D. Jolly, J. Franklin Inst., 294, No. 6, 483 (December 1972).
9. G. Karady, IEEE Trans. on Power Apparatus and Systems, PAS-94, 378 (March/April 1975).
10. "Transmission Line Reference Book - 345 kV and Above." Electric Power Research Institute sponsored book, published by Fred Weidner and Son Printers, Inc., 1975.

ELECTRICAL CONDUCTION MECHANISMS OF ELECTRIC CONTACTS COVERED WITH CONTAMINANT FILMS

T. Tamai

Sophia University

Tokyo, 102, Japan

The contaminant films formed on the contact surface in the atmosphere harmfully affect the contact reliability. This is an important problem to be solved for electrical contacts. The electrical conduction mechanisms in the film covered contacts have not been clarified because of their complex phenomena. From these viewpoints, in the present study, Schottky conduction was found to be predominant conduction mechanism by examining the relation between the voltage and the current, and the temperature dependence of the current. It was also found that a sudden decrease in the contact resistance is caused by the voltage, namely, recover of the low contact resistance resulting from the electric breakdown of the film which is induced by decrease in the film resistance and thermal deformation of the contact area. Therefore, by appling the voltage which corresponds to the barrier height to the contact with the films, it can be possible to cause low contact resistance characteristics. From these standpoints, as a criterion of the contact resistance, the breakdown voltage of the films concerned with the barrier height and softening of metals may be recommended. However, for the cryogenic application of contacts, the contact resistance degrades at low temperature.

INTRODUCTION

Air pollution brings much damage to the industries as well as human health. It is well known that the reactive gases such as SO_2, H_2S, HCl, NO_2, and O_2 which are contained in the atmosphere corrode the metal surfaces. The gases from organic insulators or painted matters which were used in the electromechanical apparatus installation and from a volcanic or a spa region have been known formerly to affect the surface of contacts. Moreover, recently various kinds of gases exhausted industrially and artificially are contaminating the atmospheric environment in which the electrical contacts are used.

On the surface of the contacts which are exposed to the air, the reaction with gases in the air produces various kinds of contaminant films including oxides and sulfides. This is noticeable in base metals, and is also observed in noble metals. Even on Au, adsorbent films are produced, and their influence cannot be ignored[1,2]. Within these contaminant films produced at the contact surface and present at the interface of surfaces in contact with each other, the resistivity is much higher than that of bulk metal. Thus, the contact resistance increases, the contact reliability is greatly degraded, and the contact performance and the lifetime are impaired[3-6].

Further, in the small contacts, the electric energy level is low and the contact load is small. Accordingly, it is difficult to break the contaminant films by an electrical or mechanical means to eliminate the influence of the films on the contact resistance. Therefore, it is important to study the problems of these contaminant films and to clarify the conduction mechanisms of them.

It is generally known that for a sandwich constitution putting the thin insulating or semiconducting film between the metal electrodes, when the voltage is applied between them, the current flows through the film independently causing insulating or very high resistivity. Holm and Meissner[7] studied the tunneling effect in these films. However, it is known that the Schottky current flows through the film also, which consists of the electrons surmounting the potential barrier due to thermal energy. Therefore, the temperature dependence of the Schottky current dominates at higher temperature. It is known that the tunneling current dominates in the case of thin surface films of several tens of Å at room temperature, and the Schottky current dominates in the case of thickness of several hundred Å. Holm[8] and Price[9] examined the conduction mechanism of contact between a Cu ring with a Cu_2O film and a graphite brush, and showed that the Schottky current is greater than the tunneling current by a factor of 10^5 at room temperature. Holm[8] indicated that the two current components became comparable at a temperature of 170 K.

Therefore, in the case of relatively thick films leading to faulty contacts, taking into account the generation of Joule heat in the contact resistance due to the current flow, it is considered in this study that the Schottky current is dominant except for the case of the thin films of less than about 100 Å and very thick films.

In recent years, to improve the characteristics of electrical machines and electronic devices, various cryogenic methods to cool electric conducting materials to lower temperatures have been actively studied and developed[10]. These methods use essentially the positive temperature coefficient of resistivity of electric conductors, and the electric power loss is reduced by lowering the resistance by lowering the temperature and by superconducting the conductors.

For the low temperature applications, it is needed to clarify the basic phenomena and the characteristics of contacts at low temperatures. In contrast to the improvement of the characteristics of conductors realized by the cooling, the characteristics of contacts, especially the contact resistance at lower temperatures, are different from those at room temperature, and it has been reported that the contact resistance increases several times on Au, and over one hundred times on base metals[11]. Scince the Schottky current depends heavily on the temperature as mentioned above, it is expected that the contact resistance increases as the temperature is lowered. Therefore, in the contacts used for low temperature applications, the characteristics, especially the contact resistance characteristics, can be greatly degraded as compared to those at room temperature.

The present paper discusses the electrical conduction and breakdown mechanisms of the contacts covered with contaminant films such as oxides and sulfides which cannot be avoided in the atmosphere.

EXPERIMENTAL

Specimens

The simple metals shown in Table 1 were used as specimens, which are widely different from each other in physical and chemical properties such as softening temperature, hardness, conductivity and contamination resistance[3,6]. As the contaminant films, the oxide and the sulfide films of varying compositions, were formed on the metal surfaces. The contact material used is a wire of 1 mm (diameter) x 20 mm (long) with a purity of 99.9%. The wire was mirror-polished by a conventional method using alumina paste, and was degreased and cleaned ultrasonically using trichloroethylene, acetone and methyl alcohol. After the cleaning with distilled water, N_2 gas was blown against the wire for drying to obtain the clean surface. Then, in order to oxidize, the wire was heated in clean air by using an electric furnace. The film thickness was controlled by the heating time amd the temperature, was measured by the electrolytic reduction method, and was evaluated by the Arrehnius equation[12].

Table 1.   Contact specimens

contact specimen	Pb	Sn	Ag	Cu	Ti	Ni	Mo	W
softening temp.(°C)	190	100	180	190	490	520	900	1000
hardness (Kg/mm^2)	5.0	6.0	85	80	70	250	280	450
film	PbO	SnO	Ag_2S	CuO $+Cu_2O$ Cu_2S	TiO_2	NiO	MoO_3	WO_2

Moreover, since Ag and Cu can be easily sulfided, the sulfide films were formed on them in 1000 ppm H_2S gas atmosphere at 25°C and 50%RH.

The composition of the films were identified by electron beam diffraction as shown in Table 1.

Measurement procedures

To determine the conduction mechanism of the film covered contacts, it is essential to measure voltage–current characteristics and temperature dependence of the current characteristics. The experiment was conducted by using a cryostat with a double-walled Dewar to change the temperature[13].

The contact was so arranged that the crossed-rod contact is obtained under the load of 10 g. The temperature of the test chamber was controlled over a range from liquid nitrogen temperature (77 K) to room temperature (300 K), and arbitrary temperature within this range was obtained through the electric heater control. By using a Cu-Constantan thermocouple the temperature was measured. The inside of the cryostat was evacuated down to 10^{-4}-10^{-5} Torr by using a sorption pump and a rotary pump to prevent the freezing of water vapor on the contact specimen at low temperatures. Thus, the contact under test is placed in a clean vacuum. The contact can be made and broken from outside the cryostat.

First, the voltage-current characteristics of the contact with the surface films were measured at the temperature within the range from 300 K to 77 K. Then, the current characteristics of the contact with the contaminant films were measured at low temperature. The conduction mechanisms through the contaminant films were examined, and the properties of the contact resistance at low temperature were studied. Moreover, the breakdown voltages of the films were measured at room temperature for various thicknesses of the films.

VOLTAGE-CURRENT CHARACTERISTICS

A non-linear relation between the voltage drop and the current between contaminated surfaces was indicated by the measurement, and it was found that the contact resistance is not ohmic. Typical voltage-current relationships of Cu specimen with a film at 300 K are shown in Figure 1 for various film thicknesses. It should be noted here that the voltage-current characteristics were found to be independent of the electric polarity since the contact is made by bringing two specimens with surface films into contact with each other. For the other specimens, similar conditions obtained. Therefore, from the viewpoint of the Schottky effect as mentioned above, the voltage-current relationship as shown in Figure 1 shall be examined by applying the following equation for the Schottky conduction[14-17]:

$$I_S=(1-r)AT^2s\cdot\exp(-\phi/kT)\cdot\exp((q^3/\varepsilon d)^{1/2}E^{1/2}/kT) \tag{1}$$

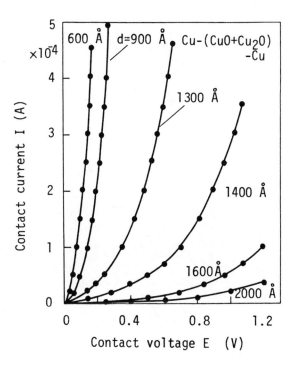

Figure 1.   Typical voltage-current characteristics for Cu with oxide films.

Where, I_s is the Schottky current, E is the voltage applied across
the contact, A=120 (A/cm^2/K), q is the electron charge, s is the
true contact area, ϕ is the barrier height, r is the ratio of electron reflection by the potential barrier, k is the Boltzmann constant, T is the temperature, ε is the dielectric constant of the
film and d is the film thickness.

Equation 1, by assuming m and n to be constants, can be reduced to the following

$$\log I_s = m \cdot E^{1/2} + n \ . \tag{2}$$

The non linear relation between the current and the voltage
shown in Figure 1 can be replotted as shown in Figure 2 by taking
$E^{1/2}$ as the axis of abscissa and log I as the axis of ordinate
according to the Equation 2. A linear relation is obtained in this
way between $E^{1/2}$ and log I. For the other specimens, the linear
relation is indicated as shown in Figure 3.

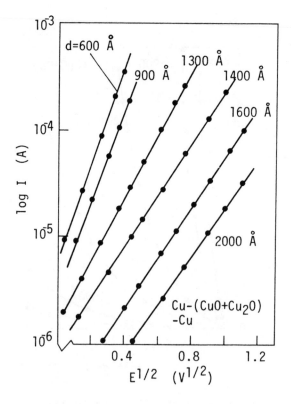

Figure 2. Log I - $E^{1/2}$ characteristics.

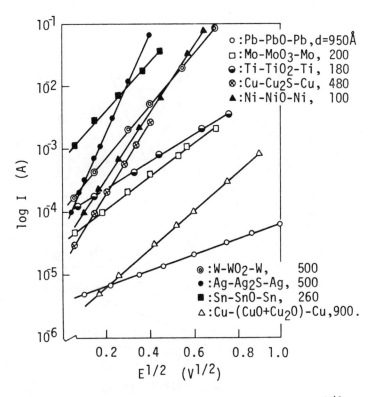

Figure 3.   Verification of the relationship log I - $E^{1/2}$ for various specimens.

From above results, it can be concluded that the conduction mechanisms of the contact film with a considerably large contact resistance is governed by the Schottky effect.

TEMPERATURE DEPENDENCE OF ELECTRICAL CONDUCTION

As a typical example, voltage-current relationships for Cu contacts with contaminant films are shown for various temperatures in Figure 4. Furthermore, the contact current as a function of the temperature is shown in Figure 5 for the contacts with contaminant films of different thicknesses. the figures show that the contact resistance increases greatly as the temperature is lowered, and finally approaches a saturation value. The following two conduction mechanisms may be possible. The first one explains the conduction mechanism in the temperature range where contact resistance changes

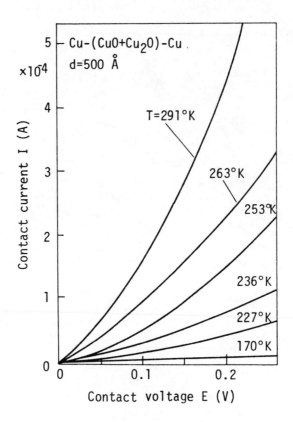

Figure 4.   Temperature dependence of voltage-current character-
            istics.

with temperature, and the second one corresponds to the temperature
range where no temperature dependence of contact resistance is seen.
     From Equation 1, a relationship between $d(\log I) / d(E^{1/2})$ and
$1/T$ should be a straight line with a constant slope. Namely, the
following equation holds assuming y a constant.

$$d(\log I) / d(E^{1/2}) = y \cdot (1/T) \qquad\qquad (3)$$

Therefore, the temperature dependence of the voltage-current relationship as shown in Figure 4 was examined by utilizing Equation 3,
and characteristic curve shown in Figure 6 was obtained. The straigt
line portion with a constant slope indicates the Schottky conduction region.
     Viewed from the conduction mechanism at 300 K as discussed

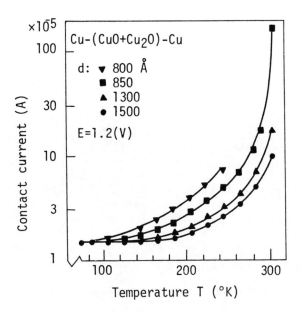

Figure 5.   Temperature dependence of the current through film covered contacts.

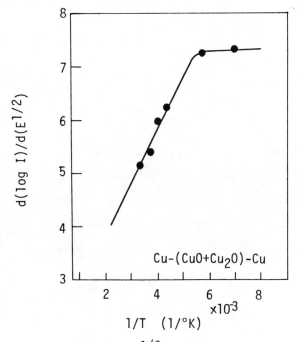

Figure 6.   d(log I) / d(E$^{1/2}$) - 1/T characteristic.

above, the Schottky conduction is most probable when one considers
the strong temperature dependence. Since the saturation region at
lower temperature is seemingly attributable to the other conduction
mechanism, by assuming this saturation current to be I_{const}, the
net Schottky current is given by $I_s = I - I_{const}$. From Equation 1, the
following relation holds:

$$\log(I - I_{const})/T^2 = b(1/T) + c \tag{4}$$

Where b and c are assumed to be constants.

   Thus, utilizing Figure 5 and Equation 4, and taking $\log(I - I_{const})/T^2$ as the axis of ordinate and $1/T$ as the axis of abscissa,
the straight line with the slope f should be obtained. Figure 7
shows in fact a straight line as expected, indicating the Schottky
conduction. Here, the slope f of the straight line is (from Equation 1) given by

$$f = (q^3/\varepsilon d)^{1/2} E^{1/2}/k - \phi/k \tag{5}$$

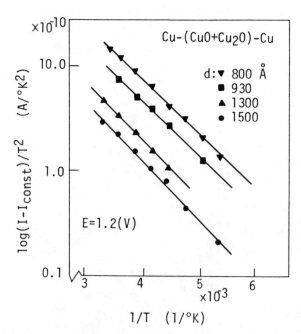

Figure 7. $\log(I - I_{const})/T^2$ - $1/T$ characteristics.

The barrier height ϕ is obtained by substituting the slope of the straight line shown in Figure 7 into Equation 5. The barrier heights for various specimens in Table 1 were obtained by this procedure as shown in Table 2. The barrier height of Cu-Cu_2O-Cu contact agreed very well with data by Holm[8] and Henisch[18]. Available data on barrier heights for the other contacts were not found in literatures. However, these measured barrier heights fairly agreed with differences of work function[19,20] between metals and films. This fact can be regard measured barrier heights as appropriate.

Table 2. Barrier Heights

contacts	Pb PbO Pb	Sn SnO Sn	Ag Ag_2S Ag	Cu O. orS. Cu	Ti TiO_2 Ti	Ni NiO Ni	Mo MoO_3 Mo	W WO_2 W
barrier height(eV)	0.38	0.25	0.36	0.45 0.34	0.18	0.19	0.19	0.22

Now, in the case of the second conduction mechanism, since there is no temperature dependence of current, the tunneling conduction is most probable. However, according to the Fowler-Nordheim equation[3], the tunneling current becomes negligibly small for the film thickness of 800 Å, E=1.2 V. The film thickness through which a sufficiently large tunneling current flows to allow measurement is about 200 Å. Therefore, the tunneling current cannot be a possible explanation for the films discussed here.

On the other hand, impurity conduction is possible. In this case, even when the thermal energy is not present, the wave functions of impurities overlap each other, the holes or the electrons are transmitted via these overlapping, and the so-called hopping process[15] results. The details are, however, left for future study.

MECHANISMS OF ELECTRIC BREAKDOWN OF THE FILMS AND EVALUATION OF BREAKDOWN VOLTAGE

The breakdown voltage of the film can be considered a criterion of the contact resistance because of recovery of low contact resistance characteristics caused by the electric breakdown of the films, measurements of the contact resistance and application of contacts[3,6,21-23]. Therefore, it is important to understand the breakdown voltage of the films. From these standpoints, the mechanisms of the electric breakdown of the films were investigated and the possibility of evaluation of the breakdown voltage for the contact with the films were studied.

In the first place, it is necessary to know the relation between true contact resistance and breakdown voltage. A relation between them was obtained by taking measured breakdown voltage as the axis of ordinate and true contact resistance which was measured by extremely low electric condition as axis of abscissa. Some typical examples are shown in Figure 8 and Figure 9. The true contact resistance R_k is given by Equation 6[3].

$$R_k = R_c + R_f = (\rho_c/2) \cdot (\pi H_c)^{1/2} W^{1/2} + (\rho_f \cdot d \cdot H_f) W^{-1} \tag{6}$$

Where, R_k is the contact resistance, R_c is the constriction resistance, R_f is the film resistance, W is the load, ρ_c is the resistivity of the metal, H_c is the hardness of the metal, ρ_f is the resistivity of the film, d is the thickness of the film and H_f is the hardness of the film. Therefore, if contact conditions such as metals and contact load are given, the constriction resistance R_c is determined, the thickness d of the film directly relates to the film resistance R_f. Thus, since the contact resistance R_k corresponds to the film thickness d, the thickness d is indicated with the true contact resistance as shown in Figures 8 and 9. In the case of thicknesses of a few hundred Å, since the Schottky conduction is dominant as mentioned above, the film resistance R_f mainly depends

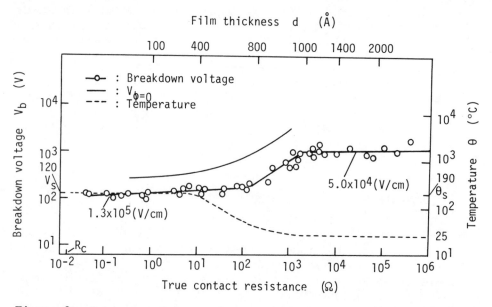

Figure 8.   Contact voltage and temperature for breakdown of film covered contact. Cu contact with $CuO+Cu_2O$ film.

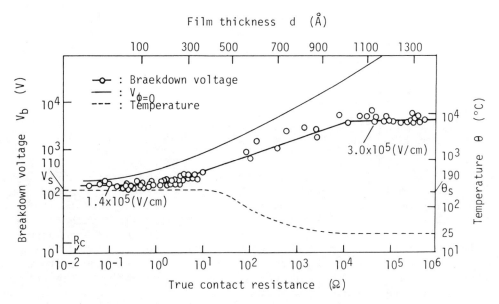

Figure 9. Contact voltage and temperature for breakdown of film
           covered contact. Pb contact with PbO film.

on the barrier height changes with applied voltage. Therefore, from
Equation 1, the decrease in the barrier height ϕ caused by the applied voltage is given by Equation 7.

$$\Delta\phi = (q^3/\varepsilon d)^{1/2} E^{1/2} \tag{7}$$

Therefore, when the height of the barrier is lowered to the Fermi
level of the substrate metal by the applied voltage, the film resistance R_f will disappear, and then the constriction resistance R_c
remains as the contact resistance R_k. Furthermore, with decreasing
film resistance R_f an increase in current occurs. This raises the
Joule heat in the contact area. Then, the film resistance R_f, which
is sensitive to temperature as shown in Equation 1, decreases like
a snowslip. And finally when the temperature reaches the softening
point θ_s of the metal, the true contact area should grow and the
contact interface containing the film must be broken, because the
softening point of the film is generally higher than that of the
metal.
     From these standpoins, since the breakdown voltage may be re-
lated to the contact resistance, evaluation of the breakdown voltage was investigated as follows. In the first place, voltage $V_{\phi=0}$
which lower the barrier height to the Fermi level of the metal was

derived from Equation 7. A relationship between the voltage $V_{\phi=0}$ and the true contact resistance was shown in Figures 8 and 9 compared with the measured breakdown voltage. The voltage $V_{\phi=0}$ indicates the upper limit of the breakdown voltage V_b.

On the other hand, the temperature at the true contact area is given by Equation 8[24].

$$\theta = \frac{1}{2}\left[\left\{\frac{V_k^2}{4\mathcal{S}_\infty}Y_a + \left(\frac{\mathcal{S}_0}{\mathcal{S}_\infty}T_0\right)\right\}^{1/2} + \left\{\frac{V_k^2}{4\mathcal{S}_\infty}Y_k + \left(\frac{\mathcal{S}_0}{\mathcal{S}_\infty}T_0\right)\right\}^{1/2}\right] - \frac{\mathcal{S}_0}{\mathcal{S}_\infty}T_0 \qquad (8)$$

Where V_k is the voltage of the contact, $\mathcal{S}_0$ is the Lorenz constant for room temperature, $\mathcal{S}_\infty$ is the Lorenz constant for high temperature, T_0 is room temperature, Y_a and Y_k are coefficient of deviation of the temperature due to Kohler effect for a anode and a cathode. The temperature in the true contact area at the time of the breakdown of the film was evaluated by Equation 8 and shown in Figures 8 and 9 as dotted lines. The temperature in which the breakdown voltage can be explained by the voltage $V_{\phi=0}$ is in agreement with the softening temperature θ_s.

The mechanisms of the electric breakdown of the film can be understood from above. However, the breakdown mechanism for higher contact resistance R_k than $10^1 \, \Omega$ is different from above and may be due to Zener type breakdown considering the strength of electric fields and the thickness of the film. For lower R_k values, as the mechanisms of breakdown of the film, the process of decrease in the film resistance due to the Schottky conduction and the softening of substrate metal must be considered. The voltage corresponding to the barrier height ϕ for the contacts with the films and the softening temperature can be considered as factors in predicting the breakdown voltage as a criterion in contact applications.

CONCLUSIONS

The electrical conduction mechanisms of the contacts with contaminant films were studied from the viewpoint of the recovery of low contact resistance characteristics by electric conditions and for the cryogenic applications.

It was found that Schottky conduction is dominant in the contaminant films when the film is thick enough to cause faulty contacts. Thus, it is reasonable to expect that low contact resistance can be recovered by applying voltage which corresponds to the barrier height. This voltage is the breakdown voltage of the films and can be recognized as a criterion for the contact resistance applications.

However, the contact resistance increases at low temperature due to the Schottky effect. Therefore, the performance of contacts at low temperature, especially when the contaminant film are present, is a important subject to be studied and the problem is more serious than at room temperature.

REFERENCES

1. C. A. Haque, IEEE Trans. Parts, Hybrids, and Packaging, P.H.P.-9, (1), 58, (1973).
2 S. Yamazaki, T. Nagata, Y. Kishimoto and N. Kanno, in "Proc. 8th Int. Conf. Electrical Contact Phenomena", Tokyo, 282, (1976).
3. R. Holm, "Electric Contacts" 4th ed., Springer, Berlin, 1976.
4. W. H. Abbott, IEEE Trans. Parts, Hybrids, and Packaging, P.H.P. -5, (4), 195, (1969).
5. M. Antler, IEEE Trans. Parts, Materials, and Packaging, P.M.P.-2, (3), 59, (1966).
6. T. Tamai and K. Tsuchiya, Trans. Inst. Elec. Engrs. Japan, 93-A, (6), 237, (1973).
7. R. Holm and W. Meissner, Z. Phys., 74, 787, (1932).
8. E. Holm, in "Proc. 4th Int. Conf. Electrical Contact Phenomena", England, 12, (1968).
9. M. J. Price, Proc. Inst. Mech. Engrs., 182, (t3A), 394, (1967-8).
10. Special issue on applications of superconductivity, IEEE Proc., 61, (1), 1, (1973).
11. A. Kawashima and S. Hoh, Cryogenics, 14, 381, July, (1974).
12. O. Kubaschewski and B. E. Hopkins, "Oxidation of Metals and Alloys", Butterworths, London, 1953.
13. T. Tamai and K. Tsuchiya, in "Proc. 1977 Holm Contact Conf". (Illinois Institute of Technology, Chicago), 151, Chicago.
14. J. G. Simmons, J. Appl. Phys., 35, (9), 2472, (1964).
15. J. G. Simmons, J. Appl. Phys., 35, (9), 2655, (1964).
16. T. J. Coutts, "Electrical Conduction in Thin Metal Film", Elsevier, Amsterdam, 1974.
17. A. A. Milgram and C. Lu, J. Appl. Phys., 37, (13), 4773, (1966).
18. H. K. Henisch, "Rectifying Semiconductor Contacts", Clarendon Press, Oxford, 1957.
19. E. H. Rhoderick, "Metal-Semiconductor Contacts", Clarendon Press, Oxford, 1978.
20. D. E. Gray, Editor, "American Institute of Physics Handbook", McGraw-Hill, New York, 1972.
21. J. Pullen, Proc. Instn. Elect. Engrs., 109, 220, (1962).
22. E. Holm, IEEE Trans. Power Apparatus and Systems, P.A.S.-84, 404, (1965).
23. T. Tamai and K. Tsuchiya, Trans. Inst. Elec. Engrs. Japan, 94-A, (9), 373, (1974).
24. H. Höft, Wiss. Z. Tech. Hochsch. Ilmenau, 12, (2), 1, (1966).

REACTION OF ANTHRANILIC ACID WITH CUPRIC ION-CONTAINING

HYDROXYAPATITE SURFACE

D. N. Misra and R. L. Bowen

American Dental Association Health Foundation
Research Unit at the National Bureau of Standards
Washington, D. C. 20234

Equal amounts of anthranilic acid (o-aminobenzoic acid) are irreversibly adsorbed on hydroxyapatite surface with or without pretreatment of the surface with cupric ions. Cupric ions enhance the strength of the bonding between anthranilic acid and the hydroxyapatite surface by effecting chemisorption of the acid molecules at their sites. This result, in itself, is of considerable interest to the chemistry of adhesion; but it is also shown that the chemisorption process is followed by a desorption of cupric anthranilate molecules and the formation of its crystals as a separate phase. A kinetic analysis of the latter process is of interest, because it is consistent with a model in which the rate is controlled by the concentration of the unreacted cupric ions remaining on the surface of the hydroxyapatite.

INTRODUCTION

This paper describes research devoted to development of an improved adhesion between tooth surface and restorative materials. As such, it does not deal directly with surface contamination. However, the phenomena described here are similar to those which may be encountered in instances of interest to this conference. The study concerns interactions between a model compound, anthranilic acid (o-aminobenzoic acid, $o-NH_2C_6H_4COOH$) containing a strong surface-active moiety and the surface of hydroxyapatite crystals and the effects of cupric ions on these processes.

983

The first step in the process is rapid chemisorption of the anthranilic acid at the cupric ion sites on the hydroxyapatite surface. Cupric ions exchanged for calcium ions on the surface do not affect the irreversibly adsorbed amount, but they do increase strength of the bonding. The second step is desorption of the cupric anthranilate, Cu $(\underline{o}\text{-}NH_2C_6H_4COO)_2$, molecules from the surface to form a crystalline precipitate as a separate phase. The kinetics of the latter process, a reaction between anthranilic acid and the cupric ions on the surface of hydroxyapatite, follows a rate law that seems to depend on the concentration of unreacted cupric ions on the surface.

EXPERIMENTAL

Materials*

Hydroxyapatite and its treatment with cupric ion. Certified "tribasic" calcium phosphate was used (Fisher C-127); its chemical formula is approximately $Ca_{10}(OH)_2(PO_4)_6$. It was repeatedly washed with boiling water before use. The physical and chemical details of its preparation and its analysis as hydroxyapatite have been reported.[1] The washed apatite (50 g) was treated with 50 mL of an aqueous solution (pH = 3.5) of 0.15 mol/L of cupric nitrate for 10 min, filtered, and washed with 200 mL of distilled water. This treated hydroxyapatite was slightly bluish. Under these conditions, cupric ions (3.00×10^{-4} mol/g) replace the surface calcium ions.[1,2] The copper content of the treated hydroxyapatite was determined to be $(3.00 \pm 0.05) \times 10^{-4}$ mol/g by atomic absorption spectrophotometry. No new phase was discernible microscopically. The amounts of adsorbed water on pure and cupric ion-treated hydroxyapatite were found to be $(3.25 \pm 0.05)\%$ and $(3.80 \pm 0.05)\%$ respectively in specimens equilibrated at 95% relative humidity and evacuated (100 N/m^2) at 105°C. The surface area, 41 m^2/g (BET, N$_2$), of the hydroxyapatite does not change after the treatment with cupric ions.

Preparation of cupric anthranilate. To 30 mL of one molar solution of anthranilic acid in ethanol (95%) was added slowly 15 mL of one molar solution of aqueous cupric nitrate.[3] The resulting precipitate was repeatedly washed with ethanol, then with water and finally rinsed with ethanol. It was dried in vacuum at 60°C. The product was composed of microscopically fine, slightly greenish

---

*Certain commercial materials are identified in this article to specify the experimental procedure. In no instance does such identification imply recommendation or endorsement by the National Bureau of Standards or by the American Dental Association Health Foundation, or that the material identified is necessarily the best available for the purpose.

crystals that decompose above 285°C. Elemental analysis was satisfactory; calcd for $Cu(C_7H_6O_2N)_2$: C, 50.1; H, 3.6; N, 8.4; Cu, 63.5;
and found: C, 50.0; H, 3.7; N, 8.4; Cu, 63.8.

Reagent grade anthranilic acid (Eastman Organic Chemicals,
recrystallized) was used without further purification. It had a
melting point of 146.5°C.

Reagent grade ethanol (95%, U.S.P. 190 Proof, Publicker
Industries, Inc.) was used as a solvent. The methylene chloride
was Fisher certified reagent grade.

Methods

Reaction kinetics. Cupric ion-containing hydroxyapatite
samples, 1.000 g each, were shaken with 10 mL aliquots of various
concentrations of anthranilic acid in 95% ethanol for various
periods (Figure 1). The supernatant solution was then filtered
through a medium-pore fritted disk by applying a light suction
initially. The concentration of anthranilic acid remaining in
the filtrate was determined from its absorbance at 335 nm on a
double-beam spectrophotometer (Coleman model 124D) and a standard
Beer's law plot. Cupric anthranilate is very insoluble, and its
relative contribution towards total absorbance of a solution at
any time is insignificant. The color of the apatite after this
reaction was slightly greenish.

The rates of reaction were studied from four solutions of
anthranilic acid having starting concentrations of 100, 80, 60
and 40 mmol/L. The individual rate points were reproducible to
within 1%.

Adsorption on pure hydroxyapatite. The adsorption of anthranilic acid on pure hydroxyapatite from solutions in ethanol (95%)
and methylene chloride were determined by the methods previously
reported. [4,5]

Desorption and reversibility. The irreversibility of the
interaction was studied for various runs by repeatedly washing
or desorbing the apatite with about 100 mL ethanol after a given
reaction period and analyzing the filtrates for anthranilic acid.

The reversibility of adsorption of anthranilic acid on pure
copper-free hydroxyapatite from solutions in ethanol and methylene
chloride was studied in a similar fashion. Null absorption of the
final aliquots of the filtrate showed the completeness of desorption process.

RESULTS

The concentration vs. time plot for the reaction of anthranilic acid with cupric ion on hydroxyapatite surface is presented
in Figure 1 for four starting concentrations. The initial amount
of anthranilic acid that is irreversibly removed from the solution may be calculated by extrapolating the four curves in Figure
1 to zero time. The initial amount that is removed from solution
and is calculated from the difference between the extrapolated
initial concentration and the starting concentration in each of
the four cases is $(17.2 \pm 0.1) \times 10^{-5}$ mol/g.

The irreversible nature of interaction, which is due to the
insolubility of cupric anthranilate, is shown by the amount of the
reacted anthranilic acid, which is very nearly the same (Table I)
before and after washing the apatite.

After the reaction of cupric ion-treated hydroxyapatite with
anthranilic acid, the crystals of cupric anthranilate are clearly

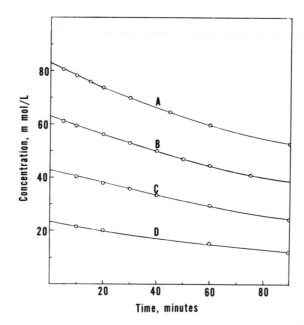

Figure 1.   Concentration vs. time at 23°C for the reaction of
anthranilic acid with cupric ions (3×10^{-4} mol/g) on hydroxyapatite; starting concentrations of (A) 100, (B) 80, (C) 60 and
(D) 40 mmol/L. It is to be noted that the extrapolated initial
concentrations of the solutions are different from the starting
concentrations because of the removal of 17.2×10^{-5} mol/g of
anthranilic acid by "instantaneous" chemisorption.

visible under the microscope by matching the apatite with an oil having a refractive index of 1.620. The crystals show characteristic birefringence of cupric anthranilate and are very small (typically 2 by $0.5 \mu m$).

It was found that the adsorption of anthranilic acid on pure hydroxyapatite from solutions in ethanol (95%) follows the Langmuir adsorption isotherm and is reversible in nature, whereas the adsorption from methylene chloride solutions is irreversible and the amounts of adsorbate, $(17.20 \pm 0.05) \times 10^{-5}$ mol/g, were the same for the four concentrations that were investigated. The details of these studies will be published elsewhere.[6]

Table I.   Irreversible Nature of Reacted Anthranilic Acid

---

1 g of Cu-OHAP[a] with 10 ml of anthranilic acid in ethanol (95%)

Time min	Amt. of acid in soln .	
	Before washing $\times 10^5$, mol	After washing[b] $\times 10^5$, mol

---

	Starting conc. 100 mmol/L	
10	78.1	78.2
30	69.9	69.9
60	59.8	59.8
90	52.7	52.3

	Starting conc. 60 mmol/L	
10	40.3	40.3
30	35.7	35.7
60	29.5	29.6
90	24.0	24.0

---

[a]The amount of cupric ion on hydroxyapatite is 3×10^{-4} mol/g.

[b]After repeatedly washing with ethanol. The total volume of filtrate was 100 mL.

Experiments showed that the final amounts of anthranilic acid
eventually removed from 10 mL solutions having concentrations of
200, 150 and 100 mmol/L by 1 g samples of cupric ion-containing
(30 x 10^{-5} mol/g) hydroxyapatite are the same, i.e. (60 $\pm$ 0.4) x
10^{-5} mol. This is in accord with the stoichiometry of cupric
anthranilate.

DISCUSSION

The interaction of anthranilic acid with cupric ion-treated
hydroxyapatite is a chemical reaction as distinguished from adsorption or chemisorption alone. This is shown by a microscopic examination of the apatite after the reaction. Distinctive crystals
of the product are observed. The optical properties and the stoichiometry of the product are identical to those of cupric anthranilate crystals.

The specific treatment of hydroxyapatite with cupric ions,
the absence of any new mineral phases and a comparative study of
the interaction of anthranilic acid on pure and cupric ion-
enriched hydroxyapatites from ethanol and methylene chloride show
that the available surface calcium ions were replaced by cupric
ions, as is discussed subsequently.

The amount of anthranilic acid which is "instantaneously"
chemisorbed from ethanol solution on the cupric ion-treated apatite is the same as the irreversibly adsorbed acid on pure apatite from methylene chloride solution. The equality between the
two amounts shows that the treatment with cupric ions does not
change the number of active sites on the apatite surface. Furthermore, if the cupric ions had not completely substituted for
the available calcium ions on the apatite surface, the treated
apatite would possess some calcium ion sites where reversible
adsorption of anthranilic acid from ethanol solution could be
possible; however, no such reversible adsorption was observed
on the cupric ion-containing apatite surface.

The effective area occupied by an anthranilic acid molecule
on the apatite surface may be calculated from the maximum adsorbed
amount (17.2 x 10^{-5} mol/g) and the surface area (41 m^2/g; BET, N_2)
of the apatite; it is 39.6 $\mathring{A}^2$. If the anthranilic acid molecules were
to adsorb in an upright position, with the carboxylate and amino
groups directed towards the surface, a rotating molecule oriented
in this manner would effectively occupy an area of $2\sqrt{3}(3.4 \mathring{A})^2 =$
40.0 $\mathring{A}^2$. The effective area of a cupric ion on the treated apatite
or an exchangeable calcium ion on hydroxyapatite surface is 23 $\mathring{A}^2$.[1,7]
Therefore, there are more cupric ions on the surface than could
interact on a one-for-one basis with the irreversibly adsorbed
molecules.

   A theory may now be developed for the kinetics of the initial
phase of the reaction of anthranilic acid with the surface cupric
ions on hydroxyapatite. It may be assumed that, so long as the
surface remains covered with anthranilic acid, the rate of change
of the acid concentration (C mol/L) with time is proportional to
the concentration of the unreacted cupric ions on the surface, and
to the initial concentration of solution (C_i mol/L); this may be
represented as:

$$- \frac{dC}{dt} = kC_i [Cu^{2+}]_{surface} \qquad\qquad (1)$$

where k[L/(mol min)] is the rate constant. The distinction between
the "starting concentration" and the "initial concentration" should
be emphasized. The starting concentration, C_o, refers to the concentration of anthranilic acid before it is added to the apatite,
whereas the initial concentration, C_i, refers to the concentration
of the acid after it has been added to the apatite and "instantaneous" chemisorption has removed an amount M from the solution. The
two concentrations are related by an equality: $VC_o = (VC_i + M)$,
where V is the volume of the acid used. If V liter of anthranilic
acid having a concentration of C mol/L at time t is in contact with
1 g of treated hydroxyapatite containing mmol/g of cupric ions,
the unreacted cupric ion concentration on the surface may be calculated as:

$$[Cu^{2+}]_{surface} = 0.5[(2m + M - VC_o) + VC], \qquad\qquad (2)$$

where the constants m, M and V have respective values equal to
3.00×10^{-4} mol/g, 17.2×10^{-5} mol/g and 1×10^{-2} L in the present
case. The integrated form of Equation (1), after substituting
$[Cu^{2+}]_{surface}$ from Equation (2) and separating the variables,
becomes:

$$- \frac{kC_i t}{2} = \frac{1}{V} \ell n[(2m + M - VC_o) + VC] + constant. \qquad\qquad (3)$$

   The value of the constant may be evaluated from the initial
condition. When t = o, the amount of acid in solution is
$VC (=VC_o-M)$, and the constant becomes equal to $V^{-1} \ell n(2m)$. Equation (3) now becomes:

$$- \frac{kC_i V}{4.606} t + \log (2m) = \log [(2m + M - VC_o) + VC]. \qquad\qquad (4)$$

   If the model is valid, then a plot of $\log [(2m + M - VC_o) + VC]$
vs. t should be linear, and the slope and intercept should be equal
to $-kC_i V/4.606$ and $\log (2m)$, respectively. The plots are shown in
Figure 2 and are reasonably linear. The intercept is the same for

all four curves, as it should be according to Equation (4). The
rate constants are given in Table II and agree well within the
limits of experimental error (within 10%).

This demonstrates the reasonableness of the assumed model for
the initial phase of the reaction. The reaction was not studied
for prolonged periods. The assumptions of the model would not
apply then, and other factors (e.g., diffusion) would probably
play a significant role.

In light of the foregoing, it may be expected that if an
anthranilic acid moiety were the surface-active part of a coupling
agent, its interaction with hydroxyapatite will probably be physical and reversible in nature. Therefore, any such coupling agent
may be less than ideal for a permanent chemical bonding with
unmodified tooth or bone surface. On the other hand, the interaction of such a coupling agent with cupric ion-modified hydroxyapatite surface will be irreversible in nature. If so, it may

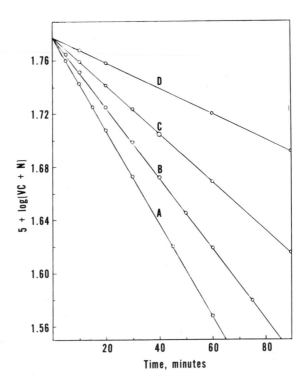

Figure 2. Rate law plots at 23°C for the reaction of anthranilic
acid with cupric ion-treated (3 x 10^{-4} mol/g) hydroxyapatite having starting concentrations of (A) 100, (B) 80, (C) 60 and (D) 40
mmol/L $\left[N = 2m + M - VC_o \right]$.

Table II. Rate Constant of the Reaction of Anthranilic Acid
with Surface Cupric Ion on Hydroxyapatite

Starting acid conc. mmol/L	Slope[a] x 10^4 L/min	Rate constant (k)[b] L/mol min
100	35.08	19.51
80	26.50	19.44
60	18.11	19.49
40	9.67	19.54

[a]Obtained from Figure 2.

[b]$k = 4.606$ slope/$C_i V$. The initial concentrations (C_i) are
82.8, 62.8, 42.8 and 22.8 mmol/L, respectively.

form a strong and durable bond with the surface. The replacement
of two hydrogens on the nitrogen atom (of anthranilic acid) by the
attached copolymerizable groups might prevent sequestration and
salt formation because of steric effects. These concepts are now
being tested.

ACKNOWLEDGEMENTS

This investigation was supported, in part, by Research Grant
7 R01 DE05129-01 and 7 S07 RR05689-10 to the American Dental
Association Health Foundation from the National Institutes of
Health – National Institute of Dental Research, and is part of the
dental research program conducted by the National Bureau of
Standards in cooperation with the American Dental Association
Health Foundation.

REFERENCES
1. D.N. Misra, R.L. Bowen and B.M. Wallace, J. Colloid Interface 51, 36 (1975).
2. D.N. Misra and R.L. Bowen, J. Biomed. Matls. Res., 12, 505 (1978).
3. H. Funk and M. Ditt, Z. Anal. Chem. 93, 241 (1933).
4. D.N. Misra and R.L. Bowen, J. Colloid Interface Sci. 61, 14 (1977).
5. D.N. Misra and R.L. Bowen, J. Phys. Chem. 81, 842 (1977).
6. D.N. Misra, unpublished data (1978).
7. M. Kukura, L.C. Bell, A.M. Posner and J.P. Quirk, J. Phys. Chem. 76, 900 (1972).

SURFACE-ACTIVE CONTAMINATION ON AIR BUBBLES IN WATER

Andrew Detwiler

Department of Atmospheric Science
University at Albany
Albany, New York   12222

     Studies of small air bubbles in water reveal
that adsorption of surface-active material from the
bulk water modifies both the rise speed of the
bubble and the ejection heights and sizes of the
small droplets produced when the bubble bursts. At
present observations of rise speeds and droplet-
ejection kinetic energies give a qualitative indication of the activity and concentration of surface-
active species in the water. Gaps in the theory
available to describe rising and bursting bubbles
at present prevent truly quantitative inferences
from being made.

INTRODUCTION

In natural bodies of water there are many surface-active
species which find an air-water interface an energetically more
attractive location than the bulk fluid. Although many inorganic
ions show weak surface effects in their distribution, the
strongest surface activity is generally shown by dissolved organic
ions, particulate organics, and by some tiny organisms like certain
species of viruses and bacteria. (For larger organisms the
definition of air-water interface becomes ambiguous.) While this
activity plays an important part in living processes and in the
ecology of natural waters, this report will be concerned with
describing some of the ways in which the adsorption of such surface-
active materials modifies the rising and bursting of small air
bubbles. These bubbles have been found to be very important in the

993

transfer of vast quantities of material from freshwaters and the ocean to the atmosphere.[1,2,3]

It will be shown how relatively simple observations of bubble behavior can allow one to deduce qualitatively the amount of sur-face-active material present in such water and some of its properties. The gaps in current knowledge that have to be filled before quantitative inferences can be made will be described.

RISING BUBBLES

One parameter of rising bubbles that is relatively easy to measure is rise speed. It turns out that rise speed is directly affected by the adsorption of surface active material onto the bubble surface.

As a bubble rises through water containing small amounts (on the order of a few parts per million or less) of dissolved surface-active material, this material will be adsorbed onto the bubble surface. The flow of fluid around the bubble will carry this material, which during the adsorption process has become relatively insoluble, around to the rear of the bubble. The result will be surface-tension gradients which will oppose the motion of the fluid around the bubble surface. A rigid cap will typically grow from the rear stagnation point of the bubble as long as the surface-tension gradient, with low surface tension on the rear part of the rising bubble due to the adsorbed material and high surface tension on the cleaner parts of the forward surface, is sufficient to oppose the flow. The cap may grow to cover the entire bubble or it may reach an equilibrium size before this occurs. For a given adsorbed species bubbles smaller than a certain size will rise slowly enough so that their entire surface will be covered. On larger bubbles only a partial cap will form.

For bubbles less than about 100 micrometers in diameter in water the Reynolds number is one or less. The Reynolds number reflects the ratio of the inertial to the viscous forces involved in the bubble motion. When the Reynolds number is much less than one a particularly simple form of the hydrodynamical equations will describe the bubble motion. This is known as Stokes flow. Only viscous forces are important and the non-linear inertial forces can be neglected. Analytical solutions are possible for many boundary conditions.[4]

Savic[5] and Davis and Acrivos[6] considered bubbles rising in such a flow regime with rigid adsorbed caps forming around their rear stagnation point. Figure 1, which is derived from the work of Davis and Acrivos, shows how the drag on an air bubble in water

depends on the difference in surface tension between the clean
fluid surface and the surface covered by the rigid adsorbed cap.
The rise speed, U, is the terminal rise speed reached when the cap
of material has spread to its equilibrium size.

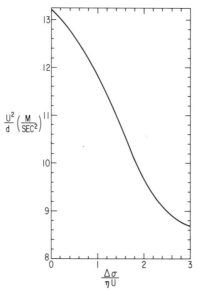

Figure 1.   (After Davis and Acrivos[6]) The drag on a rising bubble
increases as the difference in surface tension between the clean
fluid surface and the surface covered by the stagnant cap
increases.   "U" is bubble rise speed in m/sec; "d" is bubble
diameter in meters; η is the bulk-fluid dynamic viscosity; $\Delta\sigma$ is
the surface-tension difference.

        Saville[7] considered the case where the adsorbed material does
not have sufficient activity to form a rigid cap.   His work shows
results approaching those of Davis and Acrivos in the limit of an
almost stagnant bubble surface.   These results strictly apply only
to air bubbles in water considerably less than 100 micrometers in
diameter.   The terminal velocities of bubbles of this size always
seem to be those of rigid spheres.[8,9]   Apparently the decrease in
surface tension due to the adsorption of materials commonly found
in water is always large enough to produce a stagnant surface completely covering these slowly moving bubbles.

        Theoretical methods are available for treating the case of
bubbles larger than this, but the calculations are much more complex.   To my knowledge results are not available for the case of

a rigid cap on such bubbles, but only for the case of weak surfactants which form surface-tension gradients but not a rigid cap
along the bubble surface.[10] For these larger bubbles a dynamic
boundary layer forms which makes the drag more difficult to
evaluate. Another difficulty is that these larger air bubbles in
water start to be distorted from a true spherical shape.

    Despite the lack of a quantitative theory to describe the
behavior of these larger bubbles in water containing dissolved
surface-active material, observations of their rise speeds can
supply important qualitative information. Using the bubble aging
tube (Figure 2) developed by Blanchard,[11] it is possible to study
rising bubbles with diameters between about 400 micrometers and

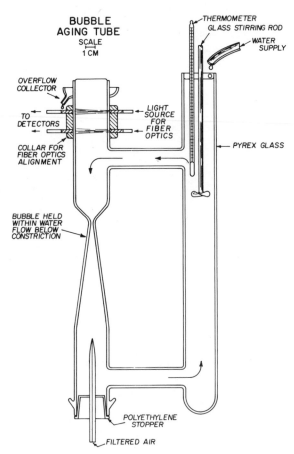

Figure 2. A bubble aging tube in which air bubbles can be suspended for any desired length of time in a downward-moving stream
of water. When the stirring rod is turned off, the bubble rises
through the fiber optic system and its rise speed is determined.

1500 micrometers. In the aging tube bubbles are suspended for
arbitrary periods of time in a downward moving stream of water.
When the stream is stopped the bubble is allowed to rise upward
into a quiet volume where its rise speed is determined.

It has recently been pointed out that for bubbles in this
larger size range terminal velocity is also almost always that of
a rigid sphere (or a sphere of equivalent volume if the bubble is
measurably non-spherical) in natural waters, tap water, and distilled tap water.[9,12] However these bubbles must rise a considerable length of time, or equivalently a considerable distance, in
order for this terminal velocity to be reached. This length of
time increases with bubble size in a given water sample and
increases with decreasing concentrations of surfactant for a given
bubble size.

Figure 3 gives an example of the decrease in rise speed as a
bubble spends greater amounts of time rising through the fluid.
The curve was derived by making one bubble at a time with the
capillary tip in the bottom section of the aging tube, suspending
it for a period, then releasing it and measuring its rise speed.[11]
The curve is then the result of single observations taken on many
identical bubbles aged for different periods of time.

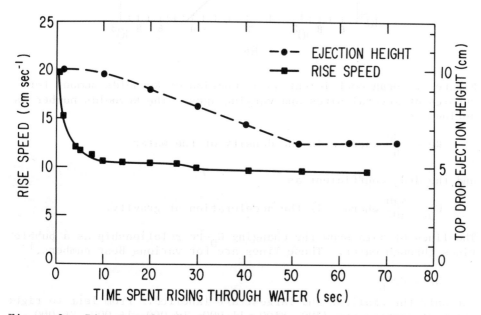

Figure 3.   Rise speed and top jet-droplet ejection height as a
function of age in a 3% NaCl-distilled water solution.   (See next
section for discussion of jet droplets.)

Figure 4 contains the results of seven series of observations on seven different bubble sizes. The Reynolds number has already been defined. The drag coefficient is the ratio of the total drag on the bubble to the kinetic energy of the water striking its horizontal cross section. One finds that a very young bubble behaves nearly as a fluid sphere while a relatively old bubble of the same size behaves nearly as a solid sphere.

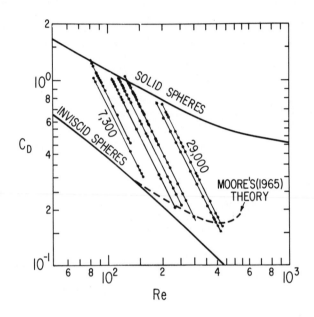

Figure 4. Drag coefficient as a function of Reynolds number for bubbles of several sizes and varying ages. The Reynolds number is defined as

$$Re = \frac{\rho U d}{\eta} \text{ where } \rho \text{ is the density of the water}$$

and the drag coefficient as

$$C_D = \frac{4gD}{3U^2} \text{ where } g \text{ is the acceleration of gravity.}$$

The lines of data show the changing C_D-Re relationship as a bubble rises through water. These lines are for various Best numbers,[13]

$$Be = ReC_D^2$$

but only the smallest and largest are labelled. From left to right the Best numbers are 7300, 8100, 11,000, 13,000, 16,000, 26,000, and 29,000. The solid sphere and fluid sphere lines are drawn from Harper.[8] The dashed line is from Moore.[14]

A spherical-cap model might be appropriate in describing the behavior of bubbles for which the Reynolds number is one or greater as well as for the case of small Reynolds number. A theory describing such a phenomenon is not yet available, however, and for these larger bubbles only qualitative inferences can be drawn.

We claim, then, that by making the approximation that a rising bubble passes through a succession of equilibrium states, the rising behavior of very small bubbles through water can give an indication of the instantaneous amount of surface-active material adsorbed onto them and the extent to which it reduces the surface tension. From such data bulk concentrations of dissolved surface-active material might be inferred and adsorption kinetics studied.

For bubbles in water much larger than a few tens of microns in diameter the Reynolds number is no longer much less than one and available quantitative theories cannot be applied. For these larger bubbles only qualitative judgements can be made. However, bulk concentrations of the order of a part per billion or less can be detected. For bubbles of a given size terminal velocity will be reached more quickly as the activity and/or concentration of dissolved surface-active material in the bulk fluid increases. For only weakly surface-active materials a terminal velocity may be reached which is greater than that of an equivalent solid sphere. We have never observed the latter in our work, though.

With air bubbles in water the problem is to prepare water which has only one surface-active species whose activity and concentration is to be measured. Bubble fractionation techniques[15,16] have been found most suitable in these investigations. We have yet to prepare truly pure, simple chemical systems in which results inferred by bubble rise speed measurements could be compared rigorously to surface-tension changes observed during the compression of an adsorbed surface layer in a more traditional apparatus, say a Langmuir trough. Further refinements in our technique are being made.

In order to make truly quantitative inferences of the activity and bulk concentration of surface-active species in water, it would seem most imperative to extend present theories to produce a theory relating the deceleration of rising bubbles to such parameters, rather than the terminal velocity. We have observed that for all air bubbles in water which do not depart radically from a spherical shape (of the order of a millimeter or less in diameter) the true equilibrium terminal velocity is always that of a rigid sphere of equivalent diameter. Such a result allows one to infer at best a lower limit for the equilibrium surface-tension gradient in the theory discussed above, and only for very small bubbles.

BURSTING BUBBLES

When single bubbles reach the surface they eventually rupture.
When this happens droplets are ejected by two mechanisms. The
first is the breakup of the bubble film which had extended above
the bulk water surface. We only mention these for completeness.
The second is the breakup of the bubble jet. The jet forms from a
layer skimmed off the inner surface bubble by a capillary wave
that propagates around the bubble from the initial point of
rupture.[17] This material is driven into the bottom of the collaps-
ing bubble cavity and erupts upward in a jet. See Figure 5.
Actually a pair of jets forms, one moving downward and one moving
upward. The upward moving jet becomes unstable and breaks up into
a set of five to eight droplets which is called the "jet set." They
are all about the same size. For bubbles in seawater the jet
droplets are roughly one tenth the diameter of the original bubble.
These droplets continue upward until drag by the atmosphere brings
them to a halt. For bubbles in seawater, again, the top droplet
is ejected to a height of roughly 100 bubble diameters.[11]

The rupture of a bubble and the breakup of its jet happen very
quickly. For bubbles around 100 micrometers in diameter the whole
process happens in a fraction of a millisecond. For bubbles
several millimeters in diameter the formation of the jet may be
slow enough to be observed by the eye.

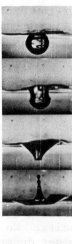

Figure 5.   Selected frames from a high-speed movie sequence showing an air bubble bursting at the bulk water surface.

Careful observation of adsorbed material contained in the jet droplet confirms the hypothesis that a bursting bubble acts as a microtome. The top droplet appears to contain material skimmed from the very inner surface of the bubble. None of the lower droplets contains as much adsorbed material as the top droplet. The available data indicates that the next highest concentrations are found in the second droplet, somewhat less in the third droplet, and so forth.[17,18,19] This would seem to indicate that the second droplet came from the layer a few tenths of a micron outward from the surface, the third droplet from a concentric layer just outside that, and so forth.

The concentrations in the top droplet of surface active material may be thousands of times that of the concentration of the same material in the bulk water. Dissolved organics, particulate organics, and some varieties of viruses and bacteria all tend to be concentrated in this way.[19]

The bubble microtome will also split electrical double layers which may exist near an air-water interface. This results in a net charge being transported away in the jet droplets.[11,20] In some cases the top droplet may carry a different sign charge from the lower droplets.[21]

As bubbles rise for longer periods of time through a bacterial suspension of about 10^6 Serratia marcescens per milliliter (in distilled water with trace amounts of nutrient solution) the number of bacteria transferred into the top jet droplet increases.[22] If the concentration of the bacteria in the bulk suspension is increased, the concentration of bacteria in the top jet droplet does not increase. Higher concentrations may show up in the lower jet droplets, however.[19] Experiments in suspensions of bacteria also indicate that concentration in the top jet droplet depends on droplet (and therefore bubble) size. For droplet diameters greater than 35 micrometers the ratio of bacterial concentration in the droplet to that in the bulk suspension decreases with increasing droplet size. The trend for diameters less than 35 micrometers is unclear.[23,24,25]

Unfortunately the way in which a bubble bursts and the jet forms and breaks up is only qualitatively understood. Quantitative calculations of fluid motion during the process are not available.[26] That adsorption of material which will lower the surface tension does modify the dynamics of bubble bursting is very clear. As bubbles rise for greater times through water containing such material ejection heights of the top droplets decrease and their sizes also decrease. Using the bubble aging tube described above one can make observations like those in Figure 6.

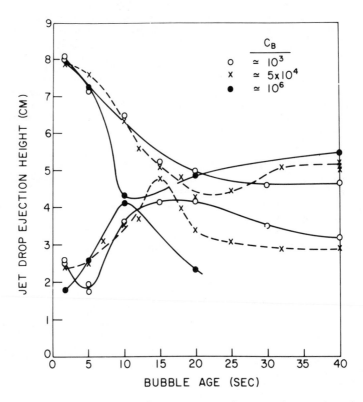

Figure 6.   Ejection height of the top and second jet droplets from a 700-micrometer-diameter air bubble rising through a suspension of S. marcescens as a function of bubble age and bulk concentration of bacteria.[22]

Knowing the ejection heights and droplet sizes one can cal-culate the initial kinetic energy the droplet must have had in order to have been ejected to the observed height.   The integration of the equation of motion for the jet droplet is made particularly simple if the heuristic expression for drag developed by Abraham[27] is used.   Figure 7 shows the results of such an integration.

For bubbles between 100 and 1000 micrometers in diameter bursting in distilled water at young ages one typically observes that the kinetic energy required to eject the jet set is of the order of 10% of the total surface free energy represented by the original bubble surface.   Within the jet set 70% to 90% of the kinetic energy goes into the top jet droplet and most of the remainder goes into the second droplet.

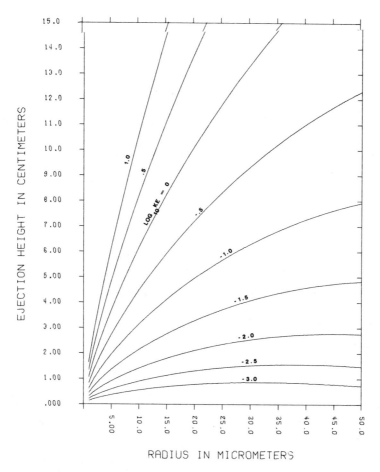

Figure 7. Final ejection height as a function of initial kinetic energy for rigid spheres of density 10^{-3} kg/m^3 moving initially upward (against gravity) in free fall in air at sea-level pressure. KE (kinetic energy) in units of 10^{-7} J.

Figure 8 shows the initial kinetic energy of the top and second droplets as a function of age from data contained in Figure 6.[22]   (The size of the second droplet was not measured, but an initial size was picked by comparison with other observations of a bubble of similar size and age, and the changes in size with age were assumed to be proportional to the changes observed in the top droplet.)   The combined kinetic energy decreases monotonically with increasing age and then starts to level off.   This decrease stops eventually, when the bubble surface reaches equilibrium with respect to the net adsorption of material.

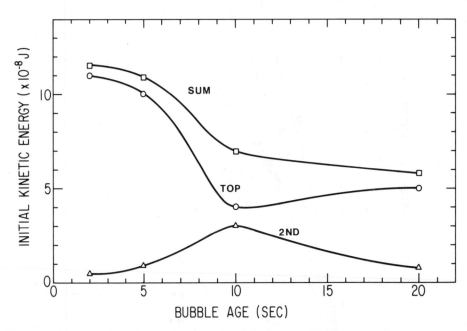

Figure 8.   Initial kinetic energy of the top and second jet
droplets as a function of bubble age derived from one set of
observations given in Figure 6.   See text for details.

The energy for the ejection of the jet set appears to come
from the surface free energy of the bubble.[11] The decrease in
combined kinetic energy with age in Figure 8 is then likely due to
the decrease in surface free energy of the rising bubble as it
adsorbs more material.   Since most of the initial jet-set kinetic
energy is in the top droplet, observations of the top droplet size
and ejection height as they change with bubble age could be used
to gauge the relative changes in total surface free energy of the
bubble.   Data on the lower members of the jet set would improve
the estimate, but is relatively more difficult to collect and the
improvement will usually be small.

In fact a rough estimate can be made by just observing the
relative changes in the top jet-droplet ejection height alone.
These usually correlate well with changes in initial kinetic energy
of the top droplet.   (Compare Figures 6 and 8.)   At the moment,
however, this is merely a hypothesis suggested by the available
data.   Total jet-set kinetic energy is only ten to fifteen percent

of the total bubble-surface free energy typically. Until the bursting of bubbles is better understood and modelled there is no proof that changes in the jet-set energies must faithfully parallel changes in the bubble-surface free energy.

SUMMARY

It has been shown that the rise speeds of air bubbles in water can be used to gain an idea of the concentrations and adsorption dynamics of surface-active species dissolved in the water. Available theory allows one to predict from a bubble rise speed the difference in surface free energy between a clean water surface and one which has a compressed surface layer of the surface-active species when the bubble has a diameter of a few tens of microns or less. For larger bubbles no general quantitative theory exists.

In addition, the time it takes a bubble of a given size (less than a millimeter or two in diameter) to reach terminal velocity can be used as a qualitative indicator of how much surface-active material is dissolved in the water. Concentrations beneath a part per billion can be detected this way, although not accurately quantified.

Observations of jet-droplet ejection heights and sizes can also be used to qualitatively infer average surface-tension changes with age over a bubble surface. In addition, jet droplets can be collected and the bubble surface layers thereby sampled. Much further work is required before truly quantitative studies of surface phenomena can be done with bursting bubbles.[26,28,29]

Finally, one additional aspect of how material adsorbs onto a rising air bubble in water and affects its bursting must be mentioned. For a bubble of a given size a terminal velocity which is observed to be that of an equivalent-size rigid sphere is reached at a certain age. However, changes in the jet-droplet ejection heights may continue to occur for greater ages. We do not yet have sufficient observations of ejection heights and jet-set droplet sizes to ascertain whether these further changes take place but the total jet-set kinetic energy remains constant, or whether the observed height changes reflect changes in the total jet-set kinetic energy and, by inference, in the bubble-surface free energy. There is, then, a possibility that a bubble may reach terminal rise speed before its surface reaches an equilibrium with respect to the surface-active species present in the water.

ACKNOWLEDGEMENTS

I would like to thank Duncan Blanchard and his co-workers and students for help in preparing this contribution. I especially appreciate the effort of Mary Haley in typing this copy. A portion of the research discussed was sponsored by the Atmospheric Research Section, National Science Foundation.

REFERENCES

1.   F. MacIntyre in "The Sea," E. D. Goldberg, Editor, Vol. 5, pp. 245-292, John Wiley and Sons, New York, 1974.
2.   D. C. Blanchard in "Applied Chemistry at Protein Interfaces," R. Baier, Editor, Advances in Chemistry Series, No. 145, pp. 360-387, American Chemical Society, 1975.
3.   R. A. Duce and E. J. Hoffman in "Annual Review of Earth and Planetary Sciences," F. A. Donath, Editor, Vol. 4, pp. 187-228, Annual Reviews, Inc., Palo Alto, California, 1976.
4.   H. Lamb, "Hydrodynamics," Dover, New York, 1945.
5.   P. Savic, "Circulation and Distortion of Liquid Drops Falling through a Viscous Medium," Mechanical Engineering Report MT-22, National Research Council of Canada, Ottawa, 1953.
6.   R. E. Davis and A. Acrivos, Chem. Eng. Sci., $\underline{21}$, 681 (1966).
7.   D. A. Saville, Chem. Eng. J., $\underline{5}$, 251 (1973).
8.   J. F. Harper in "Advances in Applied Mechanics," R. von Mises and T. van Karman, Editors, Vol. 12, pp. 59-129, Academic Press, New York, 1972.
9.   C. Bachluber and C. Sanford, J. Appl. Phys., $\underline{25}$, 2567 (1974).
10. J. F. Harper, Quart. J. Mech. Appl. Math., $\underline{23}$, 87 (1974).
11. D. C. Blanchard in "Progress in Oceanography," M. Sears, Editor, Vol. 1, pp. 71-202, MacMillan Company, New York, 1963.
12. A. Detwiler and D. C. Blanchard, Chem. Eng. Sci., $\underline{33}$, 9 (1978).
13. A. C. Best, Quart. J. Royal Meteor. Soc., $\underline{76}$, 302 (1950).
14. D. W. Moore, J. Fluid Mech., $\underline{23}$, 749 (1965).
15. R. Lemlich, Editor, "Adsorptive Bubble Separation Techniques," Academic Press, New York, 1972.
16. J. C. Scott, J. Fluid Mech., $\underline{69}$, 339 (1975).
17. F. MacIntyre, J. Phys. Chem., $\underline{72}$, 589 (1968).
18. F. MacIntyre, J. Geophys. Res., $\underline{77}$, 5211 (1972).
19. D. C. Blanchard, Pure Appl. Geophys., $\underline{116}$, 302 (1978).
20. J. V. Iribarne and B. J. Mason, Trans. Faraday Soc., $\underline{63}$, 2234 (1967)
21. R. Cheng, personal communication, Atmospheric Sciences Research Center, Albany, New York, 1978.
22. D. C. Blanchard and L. D. Syzdek, J. Geophys. Res., $\underline{77}$, 5087 (1972).
23. H. F. Bezdek and A. F. Carlucci, Limnol. Oceanogr., $\underline{17}$, 566 (1972).

24. H. F. Bezdek and A. F. Carlucci, Limnol. Oceanogr., $\underline{19}$, 126
     (1974).
25. M. Styk, M.S. Thesis, State University of New York at Albany,
     Albany, New York, 1977.
26. F. MacIntyre, Limnol. Oceanogr., $\underline{23}$, 571 (1978).
27. F. F. Abraham, Phys. Fluids, $\underline{13}$, 2194 (1970).
28. D. C. Blanchard and E. J. Hoffman, J. Geophys. Res., $\underline{83}$, 6187
     (1978).
29. D. C. Blanchard and L. D. Syzdek, Limnol. Oceanogr., $\underline{23}$, 389
     (1978).

IMPLICATIONS OF SURFACE CONTAMINATION ON MULTIUSE MILK CONTAINERS

F. W. Bodyfelt, J. D. Landsberg and M. E. Morgan

Dept. of Food Science and Technology
Oregon State University
Corvallis, Oregon 97331

Three types of multiuse milk containers were studied for retention of chemical contaminants, on inner-surfaces after treatment with 29 common household chemicals to simulate consumer abuse. Of the three container materials, glass was the most resistant to retention of chemical contaminants. The contaminant detector device failed to signal the presence of surface contaminants on nine treated polyethylene (PE) containers and on five treated polycarbonate (PC) containers. Each of these containers retained sufficient contaminants to develop off-odors in milk subsequently stored in the containers. Gas chromatographic analyses also demonstrated the retention of volatile constituents in many of the treated PE and PC containers, whereas no volatile compounds were detected in treated glass bottles. Chlordane residues of 0.007, 24, and 14.6 ppm were recovered from milk placed in glass, PE, and PC containers, which had been exposed to full strength chlordane, rinsed, washed, and refilled with milk. The absorption of 11 of 16 common chemical substances by PE containers was not detected by the required detector device. The inability of the detector to respond to significant levels of several potentially hazardous chemicals suggests possible public health problems. Milk stored in five of 11 "detector-accepted" bottles contained either pesticide residues in excess of FDA tolerance limits or had objectionable off-flavors.

INTRODUCTION

For the last 50 years, the packaging of fluid milk in the United States has traditionally been closely aligned with the relatively high levels of sanitation, public health, convenience, and efficiency practiced by the dairy industry. Generally, the processor or distributor seeks maximum product protection for the product within the limits of optimum economic parameters.

The successful development and commercial implementation of multiuse glass and plastic bottles for fluid milk products demands that the containers meet the criteria of consumer appeal and use, performance during production, and withstand storage, distribution and handling abuses. One of the more frequent and potentially serious abuses for any returnable container is consumer misuse of the bottle for storage of household chemicals, cleaning products, pesticides, herbicides, or petroleum products.

According to Roberts[1], there may be on the order of 10,000 chemical substances that may, at least potentially, be indirectly added to food through migration from packages, containers, and other food-contact surfaces. The total environment is even more complicated. Industry overall annually manufactures some 70,000 chemicals, with 1,000-2,000 additional new substances each year. This, together with the advances of science and growing societal attention has produced a plethora of warnings and concerns about contaminant hazards associated with the foods we consume. As a result of these developments, new packaging materials are subjected to intensive testing for safety. Brana[2] recently outlined the evaluation of product/bottle compatibility in the consideration of plastic resins for food packaging. Thorough analysis involves evaluation of sensory, chemical, toxicological and physical properties, with emphasis on surface migration, permeation rates, odor transfer, staining, stress cracking and/or deformation.

Specific abuses of multiuse polyethylene (PE) milk containers have been observed and reported by Wildbrett[3], Clark[4], Bodyfelt et al.[5], and Landsberg et al.[6]. The properties of PE permit the absorption of certain chemical contaminants if exposed to the container surface. Certain other plastic resins, with varied solubility characteristics, when employed as reuseable food or beverage containers, tend to exhibit similar shortcomings under misuse. Unfortunately, the food industry has no way of controlling what a consumer may place in the container before it is returned for refilling. Furthermore, the consumer has no way of knowing how a refillable container may have been misused for temporary storage of chemical substances prior to return to the processor.

LITERATURE REVIEW

In 1963, Ettling at Washington State University made initial
studies on the absorption of contaminants by refillable PE milk
containers. The WSU study concluded that absorption did occur and
that non-polar compounds (xylene, toluene, n-hexane) were absorbed
more rapidly and in greater quantity than polar compounds (ethanol,
onion juice, wine). The polar compounds were found to be difficult
or impossible to remove by washing. Polymer Service Laboratories[8]
of U.S. Industrial Chemicals Co. tested contamination levels in PE
and glass refillable milk containers in 1967. Their study confirmed
that PE did absorb contaminants and concluded that a detection
device was needed.

The Connecticut State Department of Health[9] in 1967, conducted
an evaluation of the in-line contamination detection device
(snifter). They found it capable of detecting most volatile hydrocarbon contaminants, but it did not detect phenol or formaldehyde.

Mobley[10] of the Ottawa-Carlton, Canada, Regional Health Unit
conducted the first known investigation of a contamination incident
involving public use of multiuse PE containers. The Health Unit
reported 37 complaints involving foreign substances and objects in
the containers from January to November, 1969. Supermarket surveys
indicated each store had received twice as many complaints as had
the health unit during this time period. One documented case of an
acute illness involved a 1 year old child who had consumed milk
from a returnable plastic milk container which had a strong petroleum odor. Results of a survey of Ontario province totaled 234
complaints from the 30 health units in 1971. Fifteen of the units
felt the container constituted a public health hazard and 17 of
the 30 units believed the multiuse container should be banned.
However, in 1974 the Ontario Solid Waste Task Force, Milk Packaging
Group[11] reported that returnable plastic containers did not pose a
sufficient public health hazard to ban their use. Orsage[12] reported
at the 1969 National Dairy Engineering Conference the need for an
in-line contamination detector which would be capable of detecting
and destroying contaminated PE containers.

Clark[4] reported 102 consumer complaints received by milk
regulatory officials in 1970 in Oregon, Washington and Idaho. Of
these complaints there were 14 reported cases of illness associated
with consumption of milk from multiuse plastic containers.

In Germany, Wildbrett[3], demonstrated that PE has an affinity
for lipids and repeated usage caused increased absorption of fat.
The lipids had a greater affinity for the PE surface than the
cleaning solution, making it difficult to remove deposited fat
through routine washing procedures.

The developers[12] of the linear PE milk bottle recognized that satisfactory multiple use would be dependent upon non-permanent contamination of the interior surface or absorption into the plastic by any chemical substance injurious to health or objectionable to the consumer. A series of tests conducted by the developers[13] of this packaging system indicated an uptake and retention by the PE of certain pure chemicals and commercial products or mixtures of compounds. Generally, they claimed that nonpolar substances such as gasoline, turpentine, and fuel oil were highly absorbed, while more polar substances such as wine, detergent, and onion juice were poorly absorbed. Even so, onion odor was detectable by sensory evaluation at concentration levels as low as 0.01% onion juice in the plastic. Simulated washing cycles of treated PE showed a high percentage of removal of the more polar chemical contaminants, while the hydrocarbon compounds showed a low percentage of removal. This demonstrated the need for a reliable detection system which would reject and destroy contaminated bottles, rather than attempting to clean such contaminated milk containers.

The rationale employed for selection of a detection device was that most contaminants would be expected to have some degree of volatility hence sampling the air from inside a bottle should indicate the possible absorption of such contaminants. Hydrogen flame ionization detectors (FID) are very sensitive in response to virtually all organic compounds and simultaneously insensitive to water, carbon dioxide, and the permanent gases of the atmosphere[13]; hence such a unit was selected, but modified for rapid sampling cycles. The relatively rapid rate of bottle passage on the conveyor necessitated a bottle examination approximately every 2 sec[12]. Therefore, a standard FID and its necessary adjuncts were modified to accomplish more rapid sampling and reduction of response time by reducing the physical size of tubing and other components of the unit. Essentially, a pulse of sample gas had to be withdrawn from the bottle, pass through the flame and be completely flushed from the detector within 2 sec. Likewise, the electronic portion of the detection system would have to respond and recover within 2 sec. Instrument sensitivity to some volatile organic substances was undoubtedly sacrificed to attain the required rapid sampling and cycling capacity.

Since the introduction of the multiuse PE milk container system in Oregon in 1966, the Oregon Department of Agriculture has received and recorded a significant number of consumer complaints[14] related to this method of packaging. During the 1973 energy crisis, the shortage of gasoline in Oregon resulted in apparent use by consumers of the nonbreakable, accessible PE milk container for automobile fuel storage. Subsequent consumer complaints of "petroleum-flavored" milk[14] served to point out that for one reason or another, the hydrocarbon sensing capability of the contaminant detector did not always function.

Pennwalt Corporation[15] investigated the absorption of isooctane and toluene by polycarbonate (Lexan), glass and PE multiuse milk containers. The glass and the polycarbonate (PC) containers absorbed isooctane and toluene at less than the limits of detection (0.5 ppm), while the polyethylene containers absorbed 50 ppm isooctane. PC containers were contaminated with pesticides and tritium-labeled toluene and isooctane at the WARF Institute[16]. Tritium-labeled isooctane was recovered from milk later placed in the contaminated containers and from the container material itself. The comparable retention of chemical contaminants by glass, PE, and PC multiuse milk containers was investigated in 1976 by Landsberg et al.[6]. More recently, Gasaway[17] has discussed the significance of abuse chemical contamination of returnable dairy containers. He assessed the impact of pesticide storage and detector evaluation, organoleptic and extraction studies, and hazard assessment for PC (Lexan) containers. These studies basically indicated that visible alterations in the appearance of the PC container (i.e. stress cracking, discoloration, etc.) was a more effective and reliable means of rejecting chemically abused containers than the available contaminant detector device.

Some known point sources of abuse of multiuse milk containers are: mixing of garden sprays for home use, storage of common automotive chemicals, use as refrigerator containers for fruit juices and fruit flavored drinks, mixing and storage of cleaning solutions for household purposes and use as urinals on sport boats. Some misused containers are known to be discarded into garbage cans and public dumps and are subsequently scavenged and returned for deposit and reuse. A popular newspaper column directed towards homemakers has recommended the use of PE milk containers for storage of iced tea and juice drinks. Such use can lead to contaminated containers which produce off-flavors when the container is later used for milk.

As a result of the frequency of consumer complaints, the Oregon Department of Agriculture, Foods and Dairy Division encouraged the authors to conduct an evaluation of the adequacy of the contaminant detector and the tendency for multiuse PE milk containers to absorb chemical contaminants.

The USPHS has published seven safeguards[18] which "must be met before a system utilizing multiple-use plastic milk containers could be considered as complying with the applicable provisions of the Grade "A" Pasteurized Milk Ordinance--1965 Recommendations of the USPHS." These safeguards state in part:

"4. A device shall be installed in the filling line capable of detecting in each container before it is filled, volatile organic contaminants in amounts that are of public health significance. . .
"7. The container shall not impart into the product,

pesticide residue levels or other chemical contaminants in excess of those considered acceptable by the FDA.

"If further data become available which would indicate that the use of plastic containers for fluid milk may constitute a public health hazard, such containers will no longer be considered as meeting the applicable provisions of the Grade "A" Pasteurized Milk Ordinance--1965 Recommendations of the U.S. Public Health Service"[19].

Our studies compared the retention of chemical contaminants by PE, glass, and the relatively new PC multiuse milk containers. Potential consumer abuse of the various containers was simulated to determine the effects of contamination upon the various container types and the effects upon milk subsequently placed in a contaminated container.

MATERIALS AND METHODS

Container Preparation

A supply of unused one-half gallon size PE milk containers of the multiuse type were secured from an Oregon dairy processor. The containers, constructed of high density linear PE, were subjected to a washing procedure in a commercial milk bottle washer. The milk containers were treated with 16 different commercially available beverage products and household chemicals that could conceivably serve as contaminants. The test chemical products were prepared at typical use concentrations as directed on the product label. The concentration levels for the pesticide products, toxic salts, and the detergent employed in this phase of the study are shown in Table I. The substances were retained in the treated containers for a storage period of 7 days. The bottles were next subjected to a scanning by the contaminant detector, followed by washing through a conventional bottle washer. The sensitivity and reaction of the contaminant detector to the standard test substance, methane, preceded the operation of the unit and the bottle washer. Caustic soak strength was 4.2% as NaOH (pH = 13.0) at 68.5 C for 2 min, 50 sec. Following the final rinse and sanitizing step the treated bottles were conveyed to the contaminant detector for a second scanning to detect the presence of any residual volatile contaminants. The bottle punch was disengaged to retain the integrity of the containers, but the response of the detector to each tagged bottle was noted.

New, unused PC[a], high-density PE and flint glass multiuse milk containers were treated with chemical contaminants to simulate

---

[a]Lexan, resin from General Electric Co., Pittsfield, MA.

Table I. Concentration of Certain Contaminants Used to Treat
Plastic One-half Gallon Multiuse Milk Containers[a]

Treatment Number	Contaminant[b]	% aqueous solution	Amount 1/2 gal. solution
1	DDT emulsifiable liquid (25%)	0.88	16.0 ml
2	DDT wettable powder (50%)	0.77	14.0 g
3	Malathion wettable powder (50%)	0.25	4.5 g
4	2,4-D Dimethylamine salt (50%)	0.17	3.0 ml
5	Lead arsenate, standard (14.25%)	1.82	33.0 g
6	Copper sulfate, micronized tribasic (92%)	1.65	30.0 g
11	Cleaning compound, "Pine-sol"	1.56	28.4 ml
12	Fruit punch, raspberry flavor	10.14	184.0 ml
13	Orange juice concentrate	18.74	340.0 ml

[a] One-half gallon size.
[b] All solutions retained in containers for 7 days, removed,
drained, and rinsed with cold water.

possible abuse by the consumer. The "economic poisons" selected
for treatment of containers were based on chemical hazard data and
representative pesticides and herbicides from the several compound
classifications (Table I).

Twenty-nine chemical substances (Table II) were selected to
treat the containers. Selection was based on availability to the
consumer, toxicity and/or their ability to impart off-odors or
flavors. Chemicals selected included 10 pesticides, seven herbicides,
six household chemicals, five beverages, and urine. Liquid substances both with and without petroleum solvents, emulsions, and
wettable powders were represented.

Simulation of container abuse was conducted as follows: the
containers were treated with the contaminants at full strength to
simulate storage of the contaminant as purchased and/or diluted to
the recommended usage concentration (Table II). The treatment compounds were placed in the containers and stored at room temperature
for 10 days and were then removed. The containers were rinsed until
visually clear and were then washed in a simulated bottle washing
procedure. The special cleaning compound, PBS[a], recommended for PC
containers was used. The containers were filled with a 3% solution
of PBS at 65.5 C and soaked for 8 min. The cleaning solution was
removed and the containers rinsed three times at 50 C. The con-

[a]PBS, Pennwalt Corp., Philadelphia, PA.

Table II. Compounds Selected for Treatment of Containers

Compound and manufacturer	Active ingredients (% by wt.)
Pesticides	
Isotox Insect Spray, Ortho	Carbaryl (5), Meta-systox-R (5), Kelthane (2), Aromatic petroleum derivative solvent (18)
Malathion 50 Insect Spray, Ortho	Malathion (50), Aromatic petroleum derivative solvent (33)
Fruit and Vegetable Insect Control, Ortho	Diazinon (25), Aromatic petroleum derivative solvent (57)
Liquid Sevin, Ortho	Carbaryl (27)
Ortho-Klor 74 Chlordane Spray	Technical chlordane (74), Petroleum distillate (21)
Lindane Borer and Leaf Miner Spray, Ortho	Lindane (20), Aromatic petroleum derivative solvent (59)
Vapona Insecticide Emulsible Concentrate, Shell	2,2-dichlorovinyl dimethyl phosphate (21.8), related compounds (1.6), Petroleum hydrocarbons (65.6)
Methoxychlor Mosquito Insect Spray, Black Leaf	Technical methoxychlor (25), Xylene (72)
Garden Spray, Black Leaf 40	Nicotine expressed as alkaloid (40)
Sevin Garden Spray, Ortho, WP[a]	Carbaryl (50)

[a] WP = wettable powder

Table II. Continued

Herbicides

Triox Liquid Vegetation Killer, Ortho	Pramitol (1.86), Pentachlorophenol (0.68), Other chlorinated phenols (0.08)
Contax Weed and Grass Killer, Ortho	Sodium diethylarsenate (10.40), Dimethylarsenic acid (1.77)
Amitrol-T, Amchem	3-amino-1,2,4-triazole (21.0)
Dowpon M Grass Killer, Miller's WP	Sodium salt of 2,2-dichloropropionic acid (72.5), Magnesium salt of 2,2-dichloropropionic acid (12.0)
Blackberry and Brush Killer, Miller's	2,4,5-trichlorophenoxyacetic acid, isooctyl ester (12.0), 2,4-dichlorophenoxyacetic acid, isooctyl ester (25.1)
Phaltan Rose and Garden Fungicide, Ortho, WP	Folpet (75)
Maneb Fungicide, Black Leaf, WP	Manganese ethylenebisdithiocarbamate (80)

Household chemicals

Prestone II Winter/Summer Concentrate	Ethylene glycol base (*b)
Penta, Pioneer	Pentachlorophenol (4.3), Other chlorophenols and related compounds (0.9), Mineral spirits (78.8)
Miracle Sander Paint Remover	Methanol (*), Benzol (*), Methylene chloride (*)
Pine-Sol	Pine oil (30.0), Isopropanol (10.9), Soap (10.0)
Lysol Disinfectant Deodorizing Cleaner	Soap (5.5), o-benzyl-p-chlorophenol (3.2), Isopropanol (1.1)
Crystal White Liquid Detergent, Lemon Fresh	*

Table II. Continued

Beverages

Funny Face Goofy Grape, Pillsbury	*
Funny Face Choo-Choo Cherry, Pillsbury	*
Country Time Lemonade Drink Mix, General Foods	*
Grape Juice Frozen Concentrate, Welch's	*
Orange Juice Frozen Concentrate Minute Maid	*

Other

Urine, human, pooled sample	*

_aWP = Wettable powder
_b* = Concentration of active ingredients not known.

tainers were subjected to a properly functioning, U.S.I. Polytrip
in-line contamination detection device, model A-I, serial 003,
calibrated to respond to 60 ppm methane, and the response recorded.

Immediately before filling, the containers were sanitized with
200 ppm of Cl and allowed to drain for 30 min. They were then
filled with homogenized milk and stored for 10 days at 1 C. Samples
of milk were removed and frozen for future residue analysis. Containers and the remainder of milk therein were tempered to 15 C for
evaluation of odor by the five trained judges. Samples for sensory
evaluation were coded and judged for odor differences against a
reference. Untreated coded samples were provided as internal controls. The odor difference from the reference was noted first on a
sample of milk poured into a snifter-shaped glass. A second odor
evaluation was made on the milk remaining in the coded container,
again against a reference in the same material container. Judges
knew the containers had been treated with a variety of substances;
they did not know which samples were internal controls. (Milk samples were not tasted because of the toxic nature of some of the
treatments.) Containers were then emptied, rinsed twice, and washed
using the above washing procedure to remove traces of milk before
gas chromatographic analyses were initiated.

Two separate studies were carried out in which the sensory
evaluation was conducted on the washed containers rather than on
milk samples. In these studies the containers were treated with
usage level concentrations of selected pesticides and one herbicide,
or with various concentrations of cleaner contaminant solutions.
The containers were then rinsed and washed according to the above
procedure. The empty, coded containers were presented to the judges
for evaluation of odor different from the reference. An untreated,
coded control was included.

Analysis of Container Contents

Each treated container and the untreated control containers
were manually filled with homogenized whole milk and stored for 14
days at 3.5 C. Analyses for pesticide and toxic salt residues in
the milk were done after 7 and 14 days. The methods of sample clean
up and analysis were those prescribed by AOAC[20] and the FDA[21] for
toxic salts and pesticides, respectively. Sensory evaluation of
samples was done by a panel of four experienced milk judges after
7 days of storage; except in the instance of the pesticides and toxic
salts, observations were limited to the presence or absence of a
characteristic off-odor.

Analysis of Volatile Compounds Within Containers

Volatile compounds within the containers were collected for subsequent gas chromatographic analyses essentially as described by Miller et al.[22]. The empty containers were purged with purified nitrogen (50 cc/min) for 30 min while being held in a 95 C water bath. The volatile compounds in the effluent were collected on a 102 x 6 mm o.d. stainless steel tube packed with Porapak Q. The Porapak trap was removed from the sampling device and water vapor expelled from the trap by heating with a thermostatically controlled heat gun to 55 C for 15 min under a nitrogen flow of 30 cc/min. The Porapak trap was reversed and the volatile organic compounds eluted with nitrogen (12 cc/min) at 135 C for 30 min and condensed in a 250 x 0.8 mm i.d. coiled stainless steel tube immersed in dry ice snow. This coiled trap was transferred to a modified inlet system[23] on a Varian Aerograph 1200 gas chromatograph and was heated to flash the organic compounds onto the 153 m x 0.8 mm i.d. stainless steel column coated with SF-96. The column was held at 70 C for 5 min and then programmed to 160 C at 2°/min.

Analysis for Chlordane Residue

Chlordane residue was analyzed using electron capture gas chromatography. The pesticide-containing lipid fraction was extracted from 100g milk samples[24]. Recovery of chlordane from spiked milk samples was from 98 to 101%. Hexane was used instead of pentane and an additional step to remove water from the solvent fraction was found necessary. After the residual water had been drawn off from the separatory funnel and discarded, the solvent layer was transferred to a 500 ml Erlenmeyer flask and about 20g of anhydrous sodium sulfate was added. The separatory funnel was washed with hexane (3 x 15 ml). The solvent layer was allowed to remain in contact with the sodium sulfate for about 5 min and then decanted and passed through the 4-inch glass filter prepared with a glass wool plug and a layer of sodium sulfate (about 25g). The solvent was evaporated and the residual chlordane-containing lipid fraction was retained. The extracted lipid material was stored in the freezer if the analysis was interrupted for more than overnight.

The pesticide-containing lipid fraction was purified before gas chromatographic analysis on a 25g alumina column using 5% H_2O deactivated alumina. A portion (about 0.35g) of the lipid fraction was weighed, dissolved in hexane, evaporated to about 3 ml, and placed on the column. Hexane was used to elute the pesticide from the column. The chlordane was found in the first 75 ml of eluate. This eluate was evaporated to 2 ml and then analyzed against a 0.1 µg/ml chlordane standard on a Hewlett Packard 5700A GC with a nickel 63 detector. The column was 1.22 m x 3.175 mm stainless steel

packed with 7% 4/1 OV-210/OV-17 on 100-120 Chromasorb HPW. The
injector, column, and detector temperatures were 300, 205, and 300 C,
respectively. The carrier gas flow rate was 28 cc/min, 5% argon/95%
methane.

RESULTS AND DISCUSSION

Detector Response

Nearly all observations made in this phase of the study are
summarized in Table III. The contaminant detector did not respond
to the polyethylene containers that had been treated with three of
four different pesticides (DDT, malathion, and 2,4-D) and two forms
of toxic salt solutions (lead arsenate and copper sulfate). Furthermore, the detector did not respond to washed containers treated with
four commercial products commonly available in the home or garage:
a pine-scented detergent, antifreeze, brake fluid, and motor oil.
The detector responded to an emulsified form of DDT (hexane carrier),
kerosene, gasoline, paint thinner, and a 2-cycle engine fuel mixture.
Before bottle washing, the detector signaled the presence of potential contaminants in seven of the 16 treated containers. After the
bottles were subjected to washing, the detector mechanism reacted to
only five treated containers. Presence of characteristic odors of
the given compounds was detected by the panel in 11 of the 16 plastic
bottles (Table III).

The contaminant detector response to treated containers (Table
III) indicates that only those chemical compounds classified as
volatile hydrocarbons were detected by the device. This is in complete accord with the generally recognized detection capability and
sensitivity of hydrogen flame ionization detectors[13]. Conversely,
the detector device did not respond to inorganic compounds or to non-
volatile hydrocarbon compounds.

Analysis for Chemical Residues

Results of analyses for pesticide and toxic salt residues in
the milk samples after 7 and 14 days are shown in Table III. The
emulsifiable liquid form of DDT was apparently moderately soluble in
the polyethylene plastic, as indicated by the more than three-fold
increase in the DDT residue in the milk stored an additional week
in the container treated with this material (0.27 ppm and 0.86 ppm
for 7 and 14 days, respectively). Milk from the container treated
with the wettable powder form of DDT had residues of 0.28 ppm and
0.29 ppm after storage for 7 and 14 days, respectively. Malathion
was detected at the level of 0.08 ppm for both time periods. The
2,4-D content was less than 0.01 ppm for both examinations. The

Table III. Summary of Data for Contaminant Adsorption by Multiuse Plastic Milk Containers, Contaminant Detector Device Response and Sensory Observations of Stored Product

Sample number	Container contaminant	Detector response Before washing	After washing	Characteristic odor of washed bottle	Milk flavor observations after 7 days	Residue in milk after 7 days	14 days
						(ppm on whole milk basis)	
	Glass Control	0^a	0	None	Feed flavor		
	Plastic Control	0	0	Paraffin-like	"		
1	DDT, emuls. liquid	$+^b$	+	"Chemical"	"Chemical" odor	0.27	0.86
2	DDT, wettable powder	0	0	Sl. chemical	Okay	0.28	0.29
3	Malathion, wettable powder	0	0	Typical of malathion	Okay	0.08	0.08
4	2,4-D (Dimethyamine salt)	0	0	V. sl. smell	Okay	$< 0.01^c$	$< 0.01^c$
5	Lead-arsenate	0	0	None	Not flavored	$< 0.02^c$	$< 0.02^c$
6	Copper sulfate	0	0	V. sl. odor	" "	$< 0.1^c$	$< 0.1^c$
7	Kerosene	+	+	Kerosene	Kerosene odor		
8	Gasoline	+	+	Gasoline	Gasoline odor		
9	Paint thinner	+	+	"Solvent"	"Turpentine"-like		
10	Outboard motor fuel	+	+	Gas-like	Pron. gasoline		
11	"Pine-Sol" cleaner	+	0	Def. piney	Pron. piney		

Table III Continued

12	Fruit punch	0	Mod. fruity	Pron. fruity
13	Orange juice	0	None	Feed, sl. off-flavor
14	Anti-freeze	0	None	Feed, sl. aftertaste
15	Brake fluid	+	None	Feed, sl. aftertaste
16	Motor oil	0	None	Feed flavor

^{a}0 = No detector response for a contaminant

^{b}+ = Positive detector response for a contaminant

^{c} = Residue value below detection limits of the method

lead arsenate and copper sulfate treatments showed negative values
for all analyses.

Sensory Evaluation of Samples

Following 7 days of storage, the milk in the treated containers
was evaluated organoleptically for presence or absence of off-flavors
(or off-odors) characteristic of the treatment substance. Seven of
the milk samples demonstrated definite taste and/or odor defects,
indicating that leaching of the treatment substance from the plastic
had occurred. Additionally, the milk from three containers exhibited
slight off-flavor or aftertastes, compared to that from the control
containers (untreated glass and plastic bottles). Quite obviously,
several of the test substances (fruit punch and orange juice) represent more of a reduction in flavor quality than a public health
question.

Comparison of Retention of Chemical Contaminants

The new, untreated PE containers had a hot-paraffin-like, smoky
or plastic odor. Neither the PC nor the glass containers had an
intrinsic odor.

The response of the in-line contaminant detector to the treated,
washed containers is compared with the sensory evaluation of the
milk and container in Table IV. None of the glass containers contained sufficient volatile contents to trigger a positive response
from the detector. No off-odors were found in the milk-container
combinations for the glass container with the exception of one container which had been treated with a full strength pesticide containing diazinon.

Fifteen treated PE containers did not trigger a positive
response from the detector. There were off-odors in the milk later
stored in nine of those 15 containers. A wettable powder pesticide,
an emulsified herbicide, two wettable powder herbicides, lemon
scented dishwashing detergent, three beverages, and urine had been
stored in the PE containers which were undetected by the contamination detection device, yet the combination of container and milk
contents had off-odors. These substances are mainly water based,
have low volatility, and do not contain a petroleum derived carrier
which would activate the detector. The pesticides and herbicides
have potential public health significance; the other compounds
produce an off-odor in the milk which varies from being a nuisance
to being repugnant. The detector did register a positive response
for those PE containers which had been contaminated with full
strength substances which contained a petroleum derivative solvent.

Table IV. Detector Response to Treated, Washed Containers and Sensory Evaluation of Milk-Container Combination

Treatment & (Concentration)	Container material					
	Glass		Polyethylene		Polycarbonate	
	Det. resp.	Sens. eval.	Det. resp.	Sens. eval.	Det. resp.	Sens. eval.
Isotox (full strength)	0[a]	-[b]	*[c]	+++[d]	dmgd[e]	
Malathion (full strength)	0	-	*	+++	dmgd	
Insect Control[f] (full strength)	0	+	*	+++	dmgd	
Sevin liquid (full strength)	0	-	*	+	0	-
Chlordane (full strength)	0	-	*	++	*	+
Lindane (1:2)	0	-	*	+++	dmgd	
Vapona (1:3)	0	-	*	+++	dmgd	
Methoxychlor (full strength)	0	-	*	+++	dmgd	
Black Leaf 40 (usage)	0	-	0	-	0	-
Sevin (WP)[g] (usage)	0	-	0	++	0	-
Triox (full strength)	0	-	*	++	dmgd	
Contax (full strength)	0	-	0	-	0	-
Amitrol-T (full strength)	0	-	0	+	0	+
Dowpon (usage)	0	-	0	-	0	-
Blackberry & Brush Killer (full strength)	0	-	*	+++	dmgd	
Phaltan (WP) (usage)	0	-	0	+	0	+
Maneb (WP) (usage)	0	-	0	++	0	++
Prestone (full strength)	0	-	0	-	0	-
Penta (full strength)	0	-	*	++	0	-
Miracle Sander (full strength)	0	-	*	+	dmgd	
Pine-Sol (usage)	0	-	*	+++	0	-
Lysol (usage)	0	-	0	-	0	-
Crystal White (usage)	0	-	0	+	0	-
Goofy Grape (serving)	0	-	0	++	0	++
Choo-Choo Cherry (serving)	0	-	0	-	0	-
Lemonade (serving)	0	-	*	+	0	-
Grape Juice (serving)	0	-	0	+	0	-
Orange Juice (serving)	0	-	0	+	0	-
Urine	0	-	0	+	0	+
Controls	0	-	0	-	0	-
	0	-	0	-	0	-
	0	-	0	-	0	-

[a]0 = no positive detector response for a contaminant; [b]- = no off-odor in milk-container combination; [c]* = positive detector response for a contaminant; [d]+ = slight off-odor in milk-container combination; ++ = definite off-odor in milk-container combination; +++ = pronounced off-odor in milk-container combination; [e] = container dissolved, disintegrated, cracked or otherwise damaged by treatment; [f]treatment = container passed detector yet had off-odor; [g](WP) = wettable powder

Table V. Detector Response and Sensory Evaluation of Usage-level Treated, Washed Containers

	Container material					
	Glass		Polyethylene		Polycarbonate	
Treatment	Det. resp.	Sens. eval.	Det. resp.	Sens. eval.	Det. resp.	Sens. eval.
Isotox	0[a]	-[b]	*[c]	+++[d]	*	++
Sevin, liquid[e]	0	-	0	+	0	+
Chlordane	0	-	0	++	0	+
Lindane	0	+	*	+++	*	++
Triox	0	-	*	+++	dmgd[f]	
Control	0	-	0	+[g]	0	-

[a]0 = No positive detector response for a contaminant.
[b]- = No off-odor in washed container.
[c]* = Positive detector response for a contaminant.
[d]+ = Slight off-odor in washed container.
 ++ = Definite off-odor in washed container.
 +++ = Pronounced off-odor in washed container.
[e]treatment = Container passed detector without response yet had
                off-odor.
[f]Container softened, clouded or cracked by treatment.
[g]Container had smoky, hot-paraffin-like, plastic odor.

     Nine of the PC containers were cracked, dissolved, disintegrated, or clouded by the treatment substances and further testing of these compounds was not necessary for such a container could not be reused. The chemicals which damaged the PC containers had in common a hydrocarbon solvent carrier or a chlorinated organic constituent. Five PC containers that had been treated with either an emulsifiable liquid or wettable powder herbicide, a beverage, or urine were not detected by the in-line detector, yet imparted off-odors to the milk contained therein. Again, as with PE containers, the containers which passed the detector without a positive response yet produced off-odors in milk had contained non-volatile, non-hydrocarbon water based substances.

     Five of the pesticides and herbicides tested at full strength were selected for additional testing diluted to usage concentration (Table V) with the sensory evaluation made on the containers themselves. With the compounds diluted to usage strength, all the treated, washed glass containers passed the detector without response. The panel of trained judges noted an off-odor in the glass container that had been treated with lindane. Two of the five PE containers passed the detector without a response; both of these containers, which had held liquid Sevin or chlordane diluted to the recommended concentration, were judged to have off-odors by the trained panel.

Table VI. Number of Peaks Obtained by Gas Chromatographic Head Space Analysis

Treatment	Container material		
	Glass	Polyethylene	Polycarbonate
Control 1	no significant peaks	53	5
2	no significant peaks	57	7
Chlordane			
usage	NDO[a]	+18[b]	+6
full strength	NDO	overwhelmed the method	+17
Sevin			
liquid at usage	NDO	NDO	+7
wettable powder			
at usage	NDO	+2	NDO
Fruit and Vegetable			
Insect Control,			
full strength	NDO	overwhelmed the method	NA[c]
Amitrol-T,			
full strength	NDO	+4	+4
Phaltan, usage	NDO	+6	+4
Maneb, usage	NDO	+4	NDO
Goofy Grape, serving	+2	NDO	+2
Urine	NDO	NA	NDO

[a]NDO = No difference observable from the control.
[b]+18 = 18 Peaks in addition to those present in the control.
[c]NA = Data not available.

At full strength both of these treatments were sufficient to elicit positive detector response, yet at usage strength the containers went undetected. Two PC containers passed the detector and had off-odors as determined by the panel. Again, as with PE, the chemical substances involved were Sevin and chlordane. Triox, a herbicide which contained a hydrocarbon solvent, dissolved the PC container even at usage level.

GLC Analysis of Container Volatile Constituents

The containers which were not detected by the contamination sensing device yet had off-odors in the milk later placed in the container were selected for gas-liquid chromatographic (GLC) analysis The number of peaks detected in each of the containers examined is

Table VII. Chlordane Recovered from Milk from Treated Containers

Treatment	Container Material		
	Glass	Polyethylene	Polycarbonate
		(ppm)	
Full strength	0.007	24.0	14.6
Usage strength	< 0.005*	< 0.005*	0.630
Control	< 0.005*	--	< 0.005*

* = Residue value below detection limits of the method.

shown in Table VI.

The control glass containers produced no significant peaks when analyzed by GLC. Two peaks were observed in the volatile constituents from the container which had held grape flavored punch. No peaks were detected in the volatile constituents from any of the other glass containers.

The two PE control containers yielded 53 and 57 peaks, respectively, indicating the large number of volatile compounds found within the new, unused PE container. In all but one of the treated PE containers, additional peaks were found. Chlordane at full strength and Fruit and Vegetable Insect Control containing diazinon, overwhelmed the method indicating the high absorbency of polyethylene.

The control PC containers demonstrated five and seven peaks, respectively, indicating the presence of a few volatile compounds. The PC container treated with chlordane produced the largest number of peaks, perhaps indicative of the affinity of PC for chlorinated organic compounds. Additional peaks were found in several of the other containers, indicating that PC did absorb components from non-chlorinated compounds.

Chlordane Residue Analysis of Milk

Containers holding the following pesticides and herbicides did not elicit a positive contaminant detector response, yet produced an off-odor either in the empty, washed container when treated with diluted, usage level contaminant or in the milk-container combination when treated with the full strength contaminant: Fruit and Vegetable Insect Control (diazinon) (full strength), Sevin wettable powder (WP) (usage), Sevin liquid (usage), Amitrol-T (full), Phaltan WP (usage), Maneb WP (usage), chlordane (usage), and lindane (usage). Chlordane was selected from the substances listed above for pesticide residue analysis.

Table VIII. Sensory Evaluation and Detector Response to Polyethylene Containers Treated with Cleaning Solutions

Treatment	Concentration ml/qt.	Relative concentration	Detector response	Sensory evaluation
Lemon-scented dishwashing detergent	13.2	recommended concentration	0^a	$+^b$
	16.5	1.25x	0	++
	19.8	1.5 x	0	++
	26.4	2 x	0	++
Pine-scented cleaner	0		0	$-^c$
	4.73	1/3x	$*^d$	+++
	9.74	2/3x	*	++
	14.2	recommended concentration	*	+++

^{a}0 = No positive detector response.
^{b}+ = Slight off-odor in the washed container.
 ++ = Definite off-odor in the washed container.
 +++ = Pronounced off-odor in the washed container.
^{c}- = No off-odor in the washed container.
^{d}* = Positive detector response for a contaminant.

    Chlordane residues were found present in milk (Table VII) which
had been in the PE container treated with full strength chlordane
(24 ppm) and in the PC container treated at both usage level (0.630
ppm) and with full strength chlordane (14.6 ppm). Surprisingly, no
chlordane was detected in milk from the usage level treated PE container. At full strength treatment, the milk absorbed more chlordane
from the PE container than from the PC container. The PE and PC containers treated with full strength chlordane contained sufficient
solvent to trigger a positive response from the in-line detector.
Such containers would not be refilled providing the detector was
functioning properly. As there has been no tolerance established
by the FDA for chlordane in milk, the presence of this pesticide
places it in excess of the legally established limits.

                   Detector Response to Cleaning Solutions

    Treatments with two scented household cleaning solutions
produced off-odors described as piney or lemony, respectively, in
all of the treated PE containers (Table VIII). The in-line contamination detector responded to the pine-scented cleaner at all concentrations but did not detect the lemon-scented detergent at any
level. The above results demonstrate that a refillable PE container
could conceivably become contaminated simply by soaking it in a

solution of dishwashing detergent.

CONCLUSIONS

A consideration of the basic operating principles of hydrogen flame ionization detectors indicates that it is not possible to detect the presence of numerous chemical compounds, especially those substances that are not volatile at room temperature or have low vapor pressures[13]. Many pesticide products and other toxic materials fall into this category. Many commercial pesticide products exist as wettable powders and dusts, and constitute a significant portion of the pesticides used in home gardens. Conceivably, the convenient, shatterproof plastic milk container could be temporarily used for mixing wettable powder pesticides, particularly malathion, perthane and chlordane. In this investigation residues of DDT and malathion above the permissible tolerance levels were found in milk held in polyethylene containers in which solutions of these pesticides had been stored. Milk from five of 11 containers "accepted" by the contaminant detector exhibited excessive amounts of pesticide residue or objectionable off-flavor.

The Grade "A" Pasteurized Milk Ordinance--1965 Recommendations of the USPHS, Item 7[18,19] prescribe minimum safeguards and conditions for packaging systems utilizing multiple use plastic milk containers: "the container shall not impart into the product, pesticide residue levels or other chemical contaminants in excess of those considered acceptable by the FDA. If further data become available which would indicate that the use of plastic containers for fluid milk may constitute a public health hazard, such containers will no longer be considered as meeting the applicable provisions of the Grade "A" Pasteurized Milk Ordinance--1965 Recommendations of the USPHS. The FDA has established tolerance limits of 0.05 ppm for DDT, 0.02 ppm for malathion.

The PMO[19] also states in part ". . . that all milk contact surfaces of multiuse containers shall consist of materials which are nontoxic, fat resistant, relatively nonabsorbent, and do not impart flavor or odor to the product. . ."

Both the polyethylene and polycarbonate container have been shown to absorb contaminants and to impart off-odors into the product. Pesticide residues in excess of legal tolerances established by the FDA were found.

In the opinion of the authors, the properties of the material for a multiuse milk container should match or exceed those of the conventional flint glass bottle, as related to ease of cleaning, inert characteristics, porosity, and solubility or absorption of chemical substances.

The demonstrated inability of the contaminant detector to detect objectionable amounts of several common undesirable chemical substances, raises the question as to the degree of consumer protection actually afforded by this device. In our opinion, the required contaminant detector should have the operational capability, sensitivity, and reliability to readily detect significant amounts of all or most chemical substances that are potentially injurious to human health. Additionally, it would be most desirable to detect those contaminants that tend to impart off-flavors (odor and/or taste) or otherwise affect the purity, wholesomeness, and aesthetic qualities of milk.

A properly functioning in-line contamination detector appears capable of detecting containers which have been contaminated with petroleum-derived substances. The detector is not capable of detecting those test containers which have held water based, non-hydrocarbon substances of low volatility. The detector does not appear to give adequate consumer protection for the multiuse polyethylene or polycarbonate milk container system.

Neither the polyethylene nor the polycarbonate multiuse milk containers appear to completely comply with the Grade "A" Pasteurized Milk Ordinance--1965 Recommendations of the USPHS[18].

REFERENCES

1.   H. R. Roberts, Food Technol. 32(8), 59 (1978).
2.   L. C. Brana, Package Engr. 23(8), 52 (1978).
3.   G. Wildbrett, Fette, Seifen, Anstrichm. 69, 781 (1967).
4.   F. D. Clark, J. Envir. Health 34,206 (1971).
5.   F. W. Bodyfelt, M. E. Morgan, R. A. Scanlan, and D. D. Bills, J. Milk Food Technol. 39(7),481 (1976).
6.   J. D. Landsberg, F. W. Bodyfelt, and M. E. Morgan, J. Food Protect. 40(11), 772 (1977).
7.   B. V. Ettling, (1963), unpublished data. Div. Industrial Res., Instit. Technol., Wash. State Univ., Pullman.
8.   U.S. Industrial Chemicals Co. Polymer Service Laboratories, (1967), unpublished data. Tusceola, IL.
9.   Connecticut State Dept. of Health, (1967), unpublished data. Hartford.
10. J. L. Mobley, (1972), unpublished report, Ottawa-Carlton Regional Health Unit, Ottawa, Ontario, Canada.
11. Solid Waste Task Force, in "General Report of the Solid Waste Task Force to the Ontario Minister of the Environment,"Vol. I, Part 3. Ottawa, Canada.
12. R. L. Orsage, in "Proc. 17th Annual Nat'l. Dairy Engr. Conf.," 17,52, Mich. State Univ., E. Lansing,(1969).

13.   H. M. McNair and E. J. Bonelli. "Basic Gas Chromatography."
       Consolidated Printers, Berkeley, California, 142 (1968).
14.   C. J. Wyatt, Oregon Dept. of Agriculture Consumer Protection
       Bulletin 2(5), 1 (1974).
15.   R. F. Velez, (1976), unpublished report, Pennwalt Corporation,
       Philadelphia, Pa.
16.   WARF Institute, Inc., (1976), unpublished report, Madison,
       Wisconsin.
17.   J. M. Gasaway, J. Food Protection 41(11), 863 (1978).
18.   United States Public Health Service, "Memorandum on Multi-use
       Plastic Milk Containers." Chief, Milk Sanitation Section, Milk
       and Food Branch, Washington, D.C. Sept. 3, 1965.
19.   United States Public Health Service, "Grade A Pasteurized Milk
       Ordinance--1965 Recommendations. Item 11p". Dept. of Health,
       Education and Welfare, Washington, D.C., 1965.
20.   Association of Official Analytical Chemists, "Official Methods
       of Analysis, 11th edition," Washington, D.C., 1970.
21.   Food and Drug Administration, "Pesticide Analytical Manual,
       Vol. I, Foods and Feeds," Sections 211:14, 221:12A, and 230:12.
22.   A. Miller III, R. A. Scanlan, J. S. Lee, and L. M. Libbey.
       J. Fish. Res. Bd. Canada 29, 1125 (1972).
23.   R. A. Scanlan, R. G. Arnold, and R. C. Lindsay. J. Gas Chrom.
       6,352, (1968).
24.   H. P. Burchfield and D. E. Johnson, in "Guide to the Analysis of
       Pesticide Residues," 2,160, U.S. Department of Health, Education
       and Welfare, Bureau of State Offices (Environmental Health,)
       Washington, D. C., 1965.

ABOUT THE CONTRIBUTORS

Here are included brief biographical sketches of only those authors who have contributed to this volume. Biosketches of contributors to Volume I are included in that volume.

John L. Anderson is President of the ERA Systems, Inc., Ooltewah, Tennessee. ERA Systems is the producer of the MESERAN systems used in evaporative rate analysis. He received his Ph.D. in Organic Chemistry in 1950 from the University of Illinois. He was the recipient of the Arthur K. Doolittle Award in 1970 for the development of evaporative rate analysis technique. He has published more than 25 research papers and holds 60 U.S. and foreign patents.

Ikuro Anzai is currently working for the Department of Radiological Health, Faculty of Medicine, University of Tokyo. He holds a Doctor of Technology degree, and graduated from the Department of Nuclear Engineering, University of Tokyo in 1964. He is also a part-time instructor in the Faculty of Commerical Science, University of Chuo. His major interests include theoretical aspects of radiation monitoring and dose evaluation due to internal contamination, and safety problems of medical use of radiations.

F. D. Auret is Senior Lecturer, Department of Physics, University of Port Elizabeth. He received his D.Sc. degree in Physics. His research interests include theoretical considerations of misfit in epitaxial growth. He is also doing experimental work in the field of Auger analysis.

W. L. Baun is a Research Chemist in the Mechanics and Surface Interactions Branch, Air Force Materials Laboratory, Wright-Patterson Air Force Base, Ohio. His research interests lie primarily in the areas of adhesive bonding and the characterization of solid surfaces.

P. F. A. Bijlmer is currently residing in New Zealand, and was previously employed at Fokker - VFW B.V. in Amsterdam, The Netherlands. He studied metallurgy in Delft and obtained his degree in 1964. His main interests are surface treatment of metals, and nondestructive testing and fracture toughness of metals and composites.

Floyd W. Bodyfelt is Professor of Food Science, Oregon State University, Corvallis, and Extension Dairy Processing Specialist. He received his B.S. in Dairy Manufacturing (1963) and his M.S. in Food Science in 1967. His research activity is related to flavor chemistry of cheddar cheese and cultured milk products, utilization of cheese whey for wine production, and relationship of quality and consumer acceptance of dairy products.

Ray L. Bowen was appointed Associate Director of the American Dental Association Health Foundation Research Unit in 1970. He obtained his dental degree from the University of Southern California. His work in the development of composite restorative materials and resins for use in preventive dentistry has resulted in 70 publications and 10 patents. He is the recipient of the Wilmer Souder Award by the Dental Materials Group of the International Association for Dental Research, and of the Callahan Award by the Ohio Dental Association.

Walter Brockmann is Head of the Structure and Composites Department at the Institut für angewandte Materialforschung in Bremen, Germany. He studied at the Technisché Universität Hannover and received his degree in the field of metal bonding.

J. Brous is a Senior Scientist of Alpha Metals, Inc., where he has been employed for eight years. Previously, he worked for RCA and for General Electric. He attended the City College of New York, University of Chicago, and the Polytechnic Institute of Brooklyn receiving the B.S., M.S. and Ph.D. degrees from these respectively. At Alpha Metals, he has engaged in research and development of chemical materials for applications in the Electronics Industry. He has conducted extensive research in cleaning processes as well as techniques for the measurement of cleanliness which included the invention and development of Ionograph.®

C. E. Bryson joined Surface Science Laboratories, Palo Alto, California in 1978 where he is a partner. He was with Hewlett-Packard from 1973 until 1978, and worked at Southwest Research Institue from 1964 to 1969. He received his Ph.D. in physics from the University of Missouri - Rolla in 1973.

T. W. Carr is currently manager of the Process Analysis Technology Department at IBM Corporation in Hopewell Junction, New York. He joined IBM in 1974 and was involved in the development of plasma chromatography/mass spectroscopy. He received his Ph.D. in Physical Chemistry in 1974 from Rensselaer Polytechnic Institute.

W. Chen is a graduate research assistant at the Virginia Polytechnic Institute and State University, Blacksburg, Virginia.

Robert E. Cuthrell has been employed at Sandia Labortories, Albuquerque, since 1963, and is currently studying the embrittling effects of hydrogen on inorganic solids. He received his Ph.D. degree in Inorganic Chemistry in 1964 from the University of Texas, Austin. He is a Fellow of the American Institute of Chemists and has published over 60 technical papers. He holds the rank of Lieutenant Colonel in the U.S. Air Force Reserve and serves as a research scientist in laser development at the Air Force Weapons Laboratory.

Andrew Detwiler is a graduate student at the State University of New York at Albany. He has been working on problems relating to the sea-to-air transfer of particulate material.

D. W. Dwight is Assistant Professor in the Department of Materials Engineering, VPI & SU. He obtained his Ph.D. in Colloid and Surface Chemistry from Rensselaer Polytechnic Institute. He worked for five years with du Pont's Plastics Department. He has 12 publications in the field of surface analysis and adhesion.

Albert V. Ferris-Prabhu is an Advisory Physicist at the IBM Corporation, Essex Junction, Vermont and concurrently Adjunct Professor of Solid State Physics in the Electrical Engineering Department of the University of Vermont, Burlington. Before joining IBM in 1968, he has held a number of appointments in various universities. He obtained his Ph.D. in Solid State Physics from the Catholic University of America in 1963. He has published over 70 papers in the areas of magnetism; mathematical, semi-conductor and reliability physics; memory systems performance evaluation; and the role of scientists in industry. He is a Fellow of the AAAS, Senior Member of IEEE, and is listed in Who's Who in the East and American Men of Science. He is the past Chairman of the Vermont Section of IEEE.

Howard A. Froot is a Senior Engineer at IBM Corporation, Hopewell Junction, New York, and is working on packaging diagnostics. He received his B.S. in Chemistry and Physics from the City College of New York in 1951 and M. Met. Eng. from the Polytechnic Institute of Brooklyn in 1959. He has published in the areas of solid state diffusion and optics.

Carl Garrett is employed at the Lord Corporation, Erie, Pennslyvania as a material and process specialist in the manufacture of rubber-to-metal products since mid 1977. He holds two patents for his development of process relative to telecommunications industry. He was graduated from the U.S. Naval Academy in 1964 with a B.S. in general engineering, and received his M.S. in Metallurgy and Materials Science in 1974 from Lehigh University.

Ed Good is currently a student attending Purdue University, majoring in chemical engineering. He is working at the Lord Corporation part-time under their Cooperative Education Program.

A. S. Hoffman has been Professor of Chemical Engineering and Bioengineering and Assistant Director of the Center for Bioengineering, University of Washington, Seattle, since 1972. He received his B.S., M.S., and Sc.D. degrees in Chemical Engineering from MIT between 1953 and 1957. He was a Fullbright Fellow 1957-58, spent three years in industry and was on the faculty of the Chemical Engineering Department, MIT, between 1958 and 1970. He was a Visiting Fellow at Battelle Seattle Research Center, 1970-1972.

Lloyd C. Jackson is a Consulting Engineer with the Bendix Corporation in Kansas City, Missouri. He served as Secretary and Chairman of the Committee D-14 on Adhesives of ASTM as well as on the Committee on Technical Operations. He is the author of many articles and has made numerous presentations on adhesion, adhesives, and surface cleanliness. His current activities are in the area of contamination analysis and control in clean room environments for electrical and electronic component manufacture.

James L. Jellison has been employed since 1970 at Sandia Laboratories, Albuquerque, in their Metallurgy Department and has conducted research in both solid phase welding and fusion welding. He was with the General Electric Company from 1958 to 1967 and worked on metallurgical forming and joining processes. He received his Ph.D. degree in Materials Science from Rensselaer Polytechnic Institute, Troy.

George Karady is Electrical Task Leader and Director of Engery Conversion System of the Tokamak Test Reactor (TFTR) Project (Princeton) with Ebasco Services Inc. In addition, he is a Visiting Professor at Polytechnic Institute of New York. Previously, he was Program Manager of the high voltage laboratory of IREQ. He has had a number of faculty appointments in Hungary, Baghdad, and England. He earned the Dipl. Eng. degree in 1952 and the Dr. Eng. degree in 1960 from the Polytechnic University of Budapest, Hungary. He has authored more than 55 technical papers and was elected in 1978 a Fellow of the IEEE. He is Chairman of the IEEE Working Groups: Non-ceramic and Composite

Insulators and HVDC Valve Specification, and a member of the CIGRE Working Group 33-04 on Insulator Contamination.

W. R. Kiang is a graduate Research Assistant at the Virginia Polytechnic Institute and State University, Blacksburg, Virginia.

Tohru Kikuchi is a technical official working for the Research Center for Nuclear Science and Technology, University of Tokyo. He is also a part-time teacher at the Tokyo Metropolitan Institute for Training Radiological Technologies, and graduated from the same institute. He is mainly concerned with the radiation safety problems in the medical field.

Hansgeorg Kollek has been working at the Institut für angewandte Materialforschung in Bremen, Germany since 1976. He studied analytical chemistry at the University of Goettingen. He is interested in adhesion-related problems on metallic surfaces.

Joanne D. Landsberg is Research Associate, Bend Research Inc., Bend, Oregon. Received her B.S. and M.S. degrees in Food Science from Oregon State University. Research activity includes food packaging and membrane transport systems.

Paul A. Lindfors is currently on the staff of Physical Electronics Industries, Inc., where he joined in 1976. He received his Ph.D. in Electrical Engineering from the University of Minnesota in 1975 and was a Research Associate in the Physical Electronic Laboratories for 1975-1976. Prior to pursuit of his doctorate, he worked in the areas of electronics design, power systems, control systems, and circuit theory.

W. C. McCrone is President and Director of the McCrone Research Institute, a not-for-profit laboratory devoted to fundamental research and teaching in the fields of light microscopy, crystallography and ultramicroanalysis. He has well over 200 publications with The Particle Atlas in six volumes, his best known work. He has received the Microchemical Society's Benedetti-Pichler Award in microchemistry, and the New York Microscopical Society's Ernst Abbe Award in microscopy.

Robert L. McNeely is Professor of Chemistry at the University of Tennessee, Chattanooga. He is an analytical chemist with primary research interest in the field of surfaces and instrumental analysis.

Dwarika Nath Misra is a Research Associate at The American Dental Association Health Foundation Research Unit, Washington, D.C. which he hoined in 1972. He obtained his Ph.D. degree from Howard University specializing in physical adsorption on evaporated metal films. Before joining the Research Unit, he worked at Itek Corporation. He has been working on the role of adsorption and surface-chemical reactions in the development of a true chemical bond between hydroxyapatite and dental resins, and has published 26 papers.

*Kashmiri Lal Mittal** is presently employed at the IBM Corportation in Hopewell Junction, New York. He received his B.Sc. in 1964 from Panjab University, M.Sc. (first class first) in Chemistry in 1966 from Indian Institute of Technology, New Delhi, and Ph.D. in Colloid Chemistry in 1970 from the University of Southern California. During his college days, based on his scholastic records, he was recipient of many scholarships. In the last few years, he has organized and chaired a number of very successful international symposia and in addition to this two-volume set, he has edited seven more volumes as follows: Adsorption at Interfaces, and Colloidal Dispersions and Micellar Behavior (1975); Micellization, Solubilization and Microemulsions, Volumes 1 & 2 (1977); Adhesion Measurement of Thin Films, Thick Films, and Bulk Coatings (1978); and Solution Chemistry of Surfactants, Volumes 1 & 2 (July, 1979). In addition to these volumes he has published about 40 papers in the areas of surface and colloid chemistry, adhesion, polymers, etc. He has given or is scheduled to give many invited talks on the multifarious facets of surface science, particularly adhesion, on the invitation of various societies and organizations in many countries, and is always a sought-after speaker. He is a member of many professional and honorary societies, is a Fellow of the American Institute of Chemists, is listed in American Men and Women of Science and Who's Who in the East. Recently he has been appointed a member of the editorial boards of a number of scientific and technical journals. He started the highly acclaimed short course on adhesion in the United States in 1976.

Max E. Morgan is Professor Emeritus of Food Science at Oregon State University. He received his Ph.D. in Microbiology from Iowa State in 1948. In 1969, he was awarded the Borden Award for Outstanding Dairy Foods Research. His research has been related to flavor chemistry and microbiology, and methodology and instrumentation of flavor analysis.

*As the editor of this volume, his contribution as an author is included in Volume 1.

W. E. J. Neal currently holds the posts of Senior Lecturer and Senior Tutor in Physics at Aston University, Birmingham, England. He was awarded a B.Sc. Hons. Physics degree (London 1949) and a doctorate in 1953. He is a Fellow of the Institute of Physics and a chartered engineer. His research interests include transport properties in gases, electrical superconducting and optical properties of thin films, and solar energy applications.

C. D. Needham is affiliated with IBM Corporation, Hopewell Junction, New York. He received his Ph.D. in Physical Chemistry from the University of Minnesota in 1965. He worked for two years as Staff Fellow at the National Institute for Arthritis and Metabolic Diseases before joining IBM. His research interests include polymer physics, molecular spectroscopy and plasma chromatography.

B. D. Ratner is presently a Research Associate Professor in the Department of Chemical Engineering, University of Washington, Seattle. He received his B.S. degree in Chemistry from Brooklyn College in 1967 and his Ph.D. in Polymer Chemistry from the Polytechnic Institute of Brooklyn in 1972.

J. J. Rosen is with the Chemical Engineering Department, University of Washington, Seattle. He received his Ph.D. degree in Biomaterials Science from Case Western Reserve University in 1975, and his B.S.E. and M.S. degrees from the University of Michigan. He was a Research Fellow of the NIH Heart, Lung and Blood Institute from 1975 to 1978 when he received the Young Investigators Award of the NIH to work at the University of Washington.

L. H. Scharpen has been a partner in Surface Science Laboratories, Palo Alto, California since it was founded in 1976. He was with McDonnell-Douglas as a Research Scientist and Hewlett-Packard as an Application Chemist. He received his Ph.D. in Chemistry from Stanford University in 1966.

Malcolm E. Schrader is Research Chemist and Senior Scientist at the Annapolis Laboratory of the David W. Taylor Naval Ship R & D Center. He received his Ph.D. from the Polytechnic Institute of Brooklyn in 1956, and since then he has been in various positions with the U.S. Department of Defense Laboratories. His research interests include surface science of materials, utilization of radioisotopes for lubrication-additive research and for determining the mechanism of action of compling agents in reinforced plastics, and ultrahigh vacuum techniques for studying solid-liquid interactions. He has been serving as Meeting Secretary of the Division of Colloid and Surface Chemistry of the ACS since 1977.

Steve R. Smith is presently a Research Associate at Case-Western Reserve University. He received his Ph.D. in Physical Chemistry in 1977 from Oregon State University. His research interest is the application of electron and ion spectroscopies in high temperature corrosion and semiconductor processes.

Tennyson Smith has been employed by Rockwell International for 22 years and is presently Group Leader of the Polymer Reliability Group at the Rockwell Science Center at Thousand Oaks, California. He received his B.Sc. in Chemical Engineering and his Ph.D. in Physical Chemistry from the University of Utah. He has performed fundamental applied research in adhesion, corrosion, lubrication, solid state physics, electrochemistry and related fields. He has 60 scientific publications.

L. W. Snyman is a Research Assistant in the Department of Physics, University of Port Elizabeth. He is presently working toward his M.Sc. degree.

J. S. Solomon is a Research Chemist at the University of Dayton Research Institute, Dayton, Ohio. His research includes the characterization of adhesive bonding phenomena using a number of surface characterization techniques.

G. R. Sparrow is an Application Specialist in Analytical Systems of 3M Company, St. Paul, Minnesota. He received his Ph.D. in Physical Chemistry in 1966 from Iowa State University. Before joining 3M, he spent two years in the U.S. Army, Ordnance Corps at Edgewood Arsenal/Aberdeen Proving Grounds, Maryland. He has been very active in pursuing new and extended applications of ISS/SIMS and providing technical communications of ISS/SIMS capabilities, developments, and applications on a nationwide basis.

R. K. Spears is employed by General Electric Neutron Devices Department in St. Petersburg, Florida as a development engineer. He received his B.S. and M.S. in Metallurgical Engineering from Colorado School of Mines and University of Denver, respectively, and Ph.D. in Engineering Science from the University of Florida.

T. Tamai is in the Department of Electrical and Electronic Engineering, Sophia University, Tokyo, which he joined in 1965. He graduated in 1964 from Kogakuin University. He is engaged in conducting research dealing with electrical contacts phenomena and related surface subjects.

J. S. Vermaak is Professor of Physics and Chairman of the Physics Department, University of Port Elizabeth, Port Elizabeth, South Africa. He received his D.Sc. (Physics) degree in 1964.

He has published well over 30 papers in the field of surface properties of materials and epitaxial growth, and has given many invited seminars at various research organizations and international conferences.   In 1967 and again in 1974 he was, on invitation, visiting scientist in the Department of Materials Science, University of Virginia, and at Philips Research Laboratories, Briarcliff Manor, New York, respectively.

Judith A. Wagner is currently working in the Chemistry of Organic Materials Division at Sandia Laboratories, Albuquerque. She received her B.S. degree in 1967 from Colorado State University, and her M.S. in Organic Chemistry in 1973 from the University of New Mexico. In addition to R & D activity with polymers, she interfaces with the engineering staff in many areas including contamination control, and compatability studies of weapons systems.

Bernard Wargotz received his B.S. and M.S. degrees from the University of Michigan and his Ph.D. from the University of Rochester. After directing the Materials Science and Engineering Department at General Cable Corporation, he returned to Bell Telephone Laboratories, Whippany, New Jersey in 1973 where he is presently concerned with materials technology related to inter-communications. He has received eleven patents, awards and authored over twenty publications.

J. P. Wightman is Professor of Chemistry at VPI & SU.   He received his Ph.D. in 1960 from Lehigh University, and was Visiting Professor at the University of Bristol (1975-1976). His research interests are principally in surface chemistry in which he has 38 publications.

P. L. Zajicek joined Surface Science Laboratories, Palo Alto, California in 1977, and before this time she worked for Syntex and Alza. She received her B.S. in Chemistry from San Jose State University in 1975.

SUBJECT INDEX

Pages 1-524 appear in Volume 1
Pages 525-1042 appear in Volume 2

_____

†Ionograph is a trademark of
Alpha Metals, Inc.

---

†Meseran is a trademark of ERA
Systems, Inc.

---

†Omega Meter is a trademark of
Kenco Chemical Company